ESOPHAGEAL CANCER RESEARCH DEVELOPMENTS

ESOPHAGEAL CANCER RESEARCH DEVELOPMENTS

HENRY K. KINNER
EDITOR

Nova Biomedical Books
New York

For permission to use material from this book please contact us:
Telephone 631-231-7269; Fax 631-231-8175
Web Site: http://www.novapublishers.com

Library of Congress Cataloging-in-Publication Data

New Research on esophageal cancer / Henry K. Kinner, (editor).
p. ; cm.
Includes bibliographical references and index.
ISBN-13: 978-60021-309-0
ISBN-10: 1-60021-309-X
1. Esopaghus—Cancer—Research. I. Kinner, Henry K.
[DNLM: 1.Esophageal Neoplasms—genetics. 2. Esophageal Neoplasms—epidemiology. 3. Gene Expression. 4. Gene Therapy. WI 250 N532 2006]
RC280.E8N492 2006
616.99'4320072-dc22 2006019486

Published by Nova Science Publishers, Inc. ✢ New York

Contents

Preface

Esophageal cancer is a malignant tumor of the esophagus, the muscular tube that transports food from the mouth to the stomach. Esophageal cancer occurs most often in men over 50 years old. There are two main types of esophageal cancer, squamous cell carcinoma and adenocarcinoma. These two types are distinguished by the way they look under the microscope. Squamous cell cancer is associated with smoking and alcohol consumption. The incidence of this disease in the United States has remained relatively constant, while the incidence of adenocarcinoma of the esophagus has risen dramatically. Barrett's esophagus, a complication of gastroesophageal reflux disease (GERD), is a risk factor for the development of adenocarcinoma of the esophagus. Risk factors for adenocarcinoma of the esophagus include male gender, obesity, western diet, and smoking. This new book presents new advances in this field of cancer research.

Chapter I - Esophageal cancer (EC) is one of the most common malignancies worldwide with half of them happened in China, esophageal squamous cell carcinoma (ESCC) is the predominant histological subtype of esophageal cancer and characterized by high mortality rate and regional variation in incidence in China. We have previously shown that a Mendelian autosomal recessive major gene may contribute to genetic susceptibility to ESCC in a moderate high-incidence area of Northern China. To identify genes that are differentially expressed in human ESCC, we have developed a cDNA microarray to analyze gene expression profiles in ESCC. The results showed that genes involved in squamous cell differentiation were coordinately down-regulated, including annexin I, small proline-rich proteins (SPRRs), calcium-binding S100 proteins (S100A8, S100A9, S100A14), transglutaminase (TGM3), cytokeratins (KRT4, KRT13), gut-enriched Krupple-like factor (GKLF) and Stefin A etc. Interestingly, most of the down-regulated genes encoded Ca^{2+}-binding or Ca^{2+}-modulating proteins that constitute the cell envelope.

These differentiation-associated genes were further characterized as c-Jun/AP-1 target genes essential for epithelial differentiation program. All of the genes contain AP-1 DNA binding sites in the promoter regions. Coincident with the down-regulation of these genes, c-Jun expression was obviously decreased in cancerous tissues, whereas c-Fos was equally expressed between normal and esophageal cancerous tissues. Our results suggested that c-Jun/AP-1 activated the expression of differentiation-associated genes via MAPK pathways.

In addition, our studies revealed that GKLF was also able to regulate the expression of differentiation-associated genes SPRR1A, SPRR2A and KRT4 in human ESCC.

The down-regulation of the genes could be caused by deletion, mutation and methylation, since most of the down-regulated genes were located on chromosome 1q21, we firstly investigated whether loss of heterozygosity (LOH) on chromosome 1q21 occurred in ESCC, and if LOH occurred, the correlation of chromosome 1q21 LOH with clinical and histopathological parameters. We analyzed LOH in 61 ESCC samples using 18 microsatellite markers on chromosome 1q, 46 of 61 (75.4%) tumors presented LOH at one or more loci. A significant association was found between chromosome 1q LOH and histopathological grade. These results suggest this region harbor putative tumor suppressor gene(s) contributing to tumorigenesis and differentiation in ESCC.

Functional studies on the differentially expressed genes revealed that Stefin A plays an important role in the growth, angiogenesis, invasion and metastasis of human ESCC; S100A8, S100A9 and S100A14 might partially reverse malignant behavior of ESCC cells by inducing cell differentiation, suppressing cell proliferation and inducing apoptosis; Furthermore, by analyzing the cDNA array data, we found a new candidate oncogene, Stomatin-like Protein 2 (SLP-2), which is over-expressed in cancers and involved in regulating cell growth and cell adhesion in human ESCC.

These studies provide new insights into the role of cell differentiation in ESCC initiation and progression.

Chapter II - Esophageal carcinoma is characterized by sharp differences in incidence among different countries and , even, in short distances. Tobacco smoking and alcohol drinking are important risk factors in most areas in the World. Nevertheless, countries at high risk are characterized by lack of these effects and diet is suspected as playing an important role in esophageal carcinogenesis. More precisely, diet poor in fresh vegetables and citrus fruits appears to be a major determinant in the etiology of esophageal cancer. Perhaps nitrosamines are the major initiators in the process of esophageal carcinogenesis. Since the eighties adenocarcinoma of the esophagus have experimented a sharp increase in incidence. Obesity, body mass index and esophagogastric reflux have play a an importan role in the development of this histologic type.

Chapter III - Esophageal cancer is a male-predominant disease which is globally among the ten most common malignant tumours. The two histological main types of esophageal cancer are squamous-cell and adenocarcinoma. Between individual countries and areas, there is great variation in the incidence of esophageal cancer. This phenomenon is explained by different dietary, environmental, and cultural risk factors. The highest incidences of squamous-cell carcinoma have been reported from developing countries, whereas adenocarcinoma incidence has increased rapidly in some western countries. The risk factor profiles for these two histological cancer types are different. Low intake of B vitamins (especially riboflavin and niacin), magnesium, zinc, selenium as well as eating hard and scratchy, spicy or highly salted food, and smoked meat or fish are risk factors for squamous-cell carcinoma. Heavy alcohol use together with smoking is a strong risk factor for squamous-cell carcinoma. Risk factors for esophageal adenocarcinoma are male gender, gastroesophageal reflux disease, Barrett's esophagus, white race, obesity, tobacco smoking, family history, and lower esophageal sphincter pressure lowering medication. Barrett's esophagus is the strongest risk factor for adenocarcinoma. Endoscopic surveillance of Barrett's esophagus to reduce mortality from esophageal adenocarcinoma is recommended by many experts, although definite scientific evidence for this strategy is not available. Effective tobacco and alcohol control would probably be the most efficient way to reduce the global burden of esophageal cancer.

Chapter IV - Although more than 70 cell lines have been established from esophageal cancer, the growth properties of these cell lines have not compared extensively with their normal counterpart. Comparative culture of normal and malignant cells is important for understanding the growth properties of esophageal cancer. We succeeded in establishing a serum-free culture system for normal human esophageal epithelial cells (HEE cells). We also established 15 cell lines from esophageal cancer (TE-cell lines) using serum-containing medium. Among 15 TE-cell lines, 14 cell lines grew well in serum-free medium. The growth properties of HEE cells and 14 TE-cell lines were characterized upon stimulation with various growth-related factors. Among these factors, acidic fibroblast growth factor (aFGF) was the most effective stimulant for both the HEE cells and all the TE-cell lines. Transforming growth factor-β1 (TGF-β1) inhibited the growth of HEE cells and 7 TE-cell lines; however, the other 7 TE-cell lines were resistant to the inhibitory effect of TGF-β1. Interestingly, epidermal growth factor (EGF) had a much greater stimulatory effect on the TGF-β1-resistant cells than the TGF-β1-sensitive cells. Although ethanolamine enhanced the growth-promoting ability of EGF or aFGF in the TGF-β1-sensitive cells, it had no effect on the TGF-β1-resistant cells. These findings suggested a possible cross talk between TGF-β1 and EGF signaling, and an important role of ethanolamine in the signaling pathways of growth factors. We also analyzed the expression of cyclin-dependent kinase inhibitor 2B (CDKN2B) mRNA, which was reported to be related to TGF-β signaling. Induction of CDKN2B mRNA by TGF-β1 treatment was observed in HEE cells using the reverse transcription polymerase chain reaction (RT-PCR). In contrast, 12 TE-cell lines didn't express this mRNA. Although remaining 2 TE-cell lines barely expressed this mRNA, these cells were resistant to the inhibitory effect of TGF-β1. Basic fibroblast growth factor (bFGF) had a much lower growth-promoting activity than did aFGF for HEE cells and all the TE-cell lines with the exception of one TE-cell line (TE11). In TE11 cells, however, the growth promoting activity of bFGF was higher than that of aFGF. Most of TE-cell lines required higher concentration of calcium for their optimal growth, especially two cell lines (TE2 and TE10) required calcium essentially for their growth over 10 holds. Conditioned medium of TE3, TE9 and TE14 stimulated the growth of all TE-cell lines; however had inhibitory effect on HEE cells.

Chapter V - Atypical epithelial regeneration (AER) is one of the most common changes in surgical pathology. Among locations where not infrequently can AER be seen, are gastric mucosa, pancreas, urinary bladder, prostate gland, gallbladder, mesothelial lining of pleura and peritoneum, skin, squamous epithelium of the cervix and eventually the esophagus [1-7] (table 1). Only slight or moderate cellular atypia is usually evident in the majority of cases with epithelial regeneration, and therefore the condition can then be easily recognized as benign. However, in some other cases atypical regeneration may closely resemble carcinoma or mesothelioma in both cytological samples and biopsies.

In the present chapter we deal with atypical regeneration from several of the locations cited above and in particular squamous epithelium of the esophagus. We also briefly comment on the difficulties of the interpretation of atypical reglands with generative changes in biopsies of the gastro-esophageal junction, which can result in false positive results for dysplasia or malignant neoplasia.

Chapter VI - Esophageal cancer is the major common cancer of the digestive tract, and recent evidence shows that its incidence is increasing. Esophageal cancers are highly lethal neoplasms, with an overall 5-year survival rate of less than 10%. Recent studies have revealed that alterations of specific chromosome loci, leading to gains of oncogenes, or losses of tumor suppressor genes, including p53 and p16, as well as epigenetic alterations, are early events in the carcinogenesis. Those alterations are associated with loss of cell cycle checkpoints, and evade apoptosis by down- or up-regulation of pro-apoptotic or anti-apoptotic proteins. Ongoing genomic instability leads to cumulative genetic errors and thereby the generation of multiple transformed clones, which are generally refractory to conventional multidisciplinary interventions. In this article, we have noted recent studies of the molecular biology of esophageal cancer, which may characterize its genomic instability in the carcinogenesis scenario.

Chapter VII - Despite improvement of surgical treatment and application of multi-modality therapies to advanced esophageal cancer, the prognosis is extremely poor in patients with un-resectable tumors. Based on the genetic background of esophageal cancer, we have developed various gene therapy strategies against human esophageal cancer. In this article, we reviewed molecular events of esophageal cancer and gene therapy approaches for its treatment. First, we analyzed p53 genetic alterations and angiogenesis in esophageal cancer. Second, we tested a p53 recombinant adenoviral vector (Ad5CMV-p53). Significant growth suppression was observed following infection with Ad5CMV-p53 in human esophageal squamous cell carcinoma cell lines. This observation suggests that Ad5CMV-p53 may be a potentially effective therapeutic agent for locally advanced esophageal cancer. Promising avenues for investigation include double gene therapy and adjuvant use of gene therapy with radiation therapy. Third, based on recent reports of clinical trials of p53 gene therapy for lung cancer and head and neck cancer, we developed a clinical protocol for p53 gene therapy for un-resectable advanced esophageal cancer. This clinical trial was planned to evaluate vector tolerability and efficacy. A total of 10 patients were enrolled into this phase I/II trial. The overall responses were stable in six and progressive in four cases. Using PCR analyses, gene transfer and *p53* specific transgene expression were detected in tumor biopsy tissue from all patients. mRNA levels of *p53*, *p21* and *MDM2* increased in all but one case. Three patients showed absence of disease upon repeat biopsies. No serious adverse events related to Ad5CMV-p53 have occurred in these patients, and the trial has been safely conducted. Substantial improvement in swallowing was observed in one patient with stenotic lesions. Intra-tumoral injection of Ad5CMV-*p53* is safe, feasible, and biologically active when administered in multiple doses to patients with esophageal cancer. Observations from this study indicate that this treatment results in local antitumor effects in chemo-radiation-resistant esophageal squamous cell carcinoma.

Chapter VIII - Oesophageal tumours generally bear a poor prognosis. Radical surgery is generally the only curative method available but is not feasible in the majority of patients;

palliative therapy with stent placement is generally performed. It has been demonstrated that high intensity ultrasound (HIU) can induce rapid, complete and well-defined coagulation necrosis. Thus, for the treatment of this cancer, we designed an ultrasound applicator that uses an intraluminal approach to fill up this therapeutic gap. Thermal ablation is performed with water-cooled ultrasound transducers operating at a frequency of 10 MHz. Elementary lesions extend from the transducer surface up to 10 mm in depth when applying an intensity of 16 W/cm2 for 10s. A lumen inside the therapy applicator provides path for an endoscopic ultrasound imaging probe. The rotation of the applicator around its axis enables treatment of sectorial or cylindrical volumes. This method is thus particularly suitable for oesophageal tumours that may develop only on a portion of the oesophageal circumference. Feasibility experiments were conducted from bench to *in vivo* studies on pig oesophagi. Four patients with oesophageal tumours included in a pilot study have been treated using this ultrasound applicator. The treatment of oesophageal tumours is done under fluoroscopic guidance and ultrasound imaging. Tumour response was observed in all cases. This clinical work demonstrated the efficacy of interstitial high intensity ultrasound therapy for local tumour destruction in the oesophagus.

In: Esophageal Cancer Research Developments
Editor: Henry K. Kinner, pp. 1-97
ISBN: 1-60021-309-X

Chapter I

The Identification, Regulation and Functional Study of the Differentially Expressed Genes in Human Esophageal Squamous Cell Carcinoma

Zhihua Liu
State Key Lab of Molecular Oncology, Cancer Institute, Chinese Academy of Medical Sciences and Peking Union Medical College, Beijing, P.R.China

Abstract

Esophageal cancer (EC) is one of the most common malignancies worldwide with half of them happened in China, esophageal squamous cell carcinoma (ESCC) is the predominant histological subtype of esophageal cancer and characterized by high mortality rate and regional variation in incidence in China. We have previously shown that a Mendelian autosomal recessive major gene may contribute to genetic susceptibility to ESCC in a moderate high-incidence area of Northern China. To identify genes that are differentially expressed in human ESCC, we have developed a cDNA microarray to analyze gene expression profiles in ESCC. The results showed that genes involved in squamous cell differentiation were coordinately down-regulated, including annexin I, small proline-rich proteins (SPRRs), calcium-binding S100 proteins (S100A8, S100A9, S100A14), transglutaminase (TGM3), cytokeratins (KRT4, KRT13), gut-enriched Krupple-like factor (GKLF) and Stefin A etc. Interestingly, most of the down-regulated genes encoded Ca^{2+}-binding or Ca^{2+}-modulating proteins that constitute the cell envelope.

These differentiation-associated genes were further characterized as c-Jun/AP-1 target genes essential for epithelial differentiation program. All of the genes contain AP-1 DNA binding sites in the promoter regions. Coincident with the down-regulation of these genes, c-Jun expression was obviously decreased in cancerous tissues, whereas c-

Fos was equally expressed between normal and esophageal cancerous tissues. Our results suggested that c-Jun/AP-1 activated the expression of differentiation-associated genes via MAPK pathways. In addition, our studies revealed that GKLF was also able to regulate the expression of differentiation-associated genes SPRR1A, SPRR2A and KRT4 in human ESCC.

The down-regulation of the genes could be caused by deletion, mutation and methylation, since most of the down-regulated genes were located on chromosome 1q21, we firstly investigated whether loss of heterozygosity (LOH) on chromosome 1q21 occurred in ESCC, and if LOH occurred, the correlation of chromosome 1q21 LOH with clinical and histopathological parameters. We analyzed LOH in 61 ESCC samples using 18 microsatellite markers on chromosome 1q, 46 of 61 (75.4%) tumors presented LOH at one or more loci. A significant association was found between chromosome 1q LOH and histopathological grade. These results suggest this region harbor putative tumor suppressor gene(s) contributing to tumorigenesis and differentiation in ESCC.

Functional studies on the differentially expressed genes revealed that Stefin A plays an important role in the growth, angiogenesis, invasion and metastasis of human ESCC; S100A8, S100A9 and S100A14 might partially reverse malignant behavior of ESCC cells by inducing cell differentiation, suppressing cell proliferation and inducing apoptosis; Furthermore, by analyzing the cDNA array data, we found a new candidate oncogene, Stomatin-like Protein 2 (SLP-2), which is over-expressed in cancers and involved in regulating cell growth and cell adhesion in human ESCC.

These studies provide new insights into the role of cell differentiation in ESCC initiation and progression.

Introduction

Esophageal cancer (EC) is a common disease in many areas of China, especially in north of China, it is the fourth most frequent cause of death from malignant tumors in China. Almost half of the esophageal cancer worldwide occurred in China. Remarkable variations in incidence and sex ratio are seen both between countries and by graphical and ethnic divisions within countries. In China, rates of EC vary more widely than those of any other common cancers, with some areas of remarkably high incidence [Liu and Li. 1984]. High incidence areas in northern China are mainly in the areas where borders of three provinces (Henan, Hebei and Shanxi) meet on the south side of Taihang Mountains. The mortality rate can be >100/100,000, such as in Linzhou, Henan Province. The highest country mortality rates are 254.77/100,000 for males and 161.11/100,000 for females in Linzhou, which are 1075-fold and 671-fold higher than the lowest sex-specific mortality in the country, respectively. The mortality rate in males is higher than that in females, most cases occur between the ages of 60 and 64 years. Since the prognosis of EC is very poor, prevention of this disease is very important [Zhang et al. 2000].

Esophageal squamous cell carcinoma (ESCC) is the predominant histological subtype of esophageal cancer and characterized by high mortality rate and regional variation in incidence in China [Parkin et al. 1993], more than 95% EC in China is ESCC. All the studies followed are focused on ESCC only.

Genetic or Environmental Factors, Which is More Important in ESCC Etiology?

Epidemiological studies showed that certain environmental risk factos are associated with EC, though the odds ratios (ORs) vary considerably from different studies. The etiology of EC is related to a variety of smoking and alcohol-exposure history. Other factors include nutritional deficiencies, hot food, pickled vegetable intaking [Dayne et al. 1982; Li 1982; Day et al. 1994, Zhang et al. 2000].

In the last two decades, increasing numbers of studies have shown that genetic susceptibility may influence the risk of EC, familiar aggregation of EC was found in high-incidence areas of northern China [Hu et al. 1992; Chang-Claude et al. 1997]. Genetic susceptibility to EC was considered one of the most important causes for the high prevalence and familial aggregation in northern China [Wu et al. 1989]. The previously published segregation analysis of EC suggested an autosomal recessive Medndelian inheritance of EC in Linzhou, the county with the highest mortality rate in China [Carter et al. 1992]. In order to explore whether genetic factors play the same role in the etiology of EC in a moderate high-incidence area of northern China, we had conducted a pedigree survey on 225 patients affected by EC in Yangquan, Shanxi Province, in 1990s. The crude mortality from EC in this city is 40.12/100,000, which is in a middle level among the high-risk areas in northern China. EC accounts for >40% of cancer deaths occurring in this city. Previous study showed a positive familiar aggregation of EC in this city and the results favor genetic etiology, at least as important as known environmental factors [Li et al. 1998]. To further explore the model of inheritance of EC in this area and compare the previous study in Linzhou, the family data were collected and subjected to complex segregation analysis.

A total of 228 unrelated EC patients (probands) who were newly diagnosed with EC between 1989 and 1994 were identified in the survey of 132,039 people in Yangquan city. Segregation analysis was performed using the REGTL program of S.A.G.E. The results showed that Mendelian autosomal recessive inheritance of a major gene that influences susceptibility to esophageal cancer provided the best fit to the data. In the best-fitting recessive model, the frequency of the disease allele was .2039. There was a significant sex effect on susceptibility to the disease. The maximum cumulative probability of esophageal cancer among males with the AA genotype was 100%, but, among females, it was 63.5%. The mean age at onset for both men and women was 62 years. The age-dependent penetrances for males with the AA genotype by the ages of 60 and 80 years were 41.6% and 95.2%, respectively, whereas, for females, they were 26.4% and 60.5%, respectively. Incorporating environmental risk factors—such as cigarette smoking, pipe smoking, alcohol drinking, eating hot food, and eating pickled vegetables—into the models did not provide significant improvement of the fit of the models to these data. The results suggest a major locus underlying susceptibility to esophageal cancer with sex-specific penetrance [Zhang et al. 2000].

Many studies showed that most of the common cancers-such as colon, breast, prostate, lung and ovarian cancer-have one or more rare autosomal dominant gene(s) that increase the risk of these cancers, with susceptibility-allele frequencies in the range 0.002-0.006, penetrance >60%, and the proportion of patients affected because of these genes decresing with the increasing age at onset [Claus 1995].

Although the EC motality rates could be hundreds of times higher or lower in two different regions in China, the cause of the difference is not clear. There are some environmental factors (e.g., local economic status, geographical factors, nitrosamines, micronutrients, fungi and even virus) that might differ in different areas. A great deal of evidence showed that environmental factors are not the only important causes of EC: (1) EC mortality rates in high-incidence areas are very stable, even though some environmental risk factors for EC were identified many years ago. For example, the nitrosamine concentration was found higher in high-incidence areas such as Linzhou. During the past 30 years, the change of water sources, and the reduction of pickled-vegetable eating have reduced nitrosamine exposure but have not change the EC mortality in those areas significantly. (2) Incidence of EC among immigrants from high-incidence areas to lower incidence areas remained much higher than that of local residents [Wu et al. 1989]. The differences of EC incidences between ethnic groups are at least as high as the differences between different areas. (3) With the same environmental exposures, EC occurrence differs a great deal from family to family, typified by the existence of highly aggregated "cancer families". Familiar aggregation may be more significant in high-incidence areas than in low-incidence areas [Wang et al. 1992]. It is hard to say which factor (environmental or genetic) is more importamt. Both may play important roles in the etiology of EC. The mutation of EC gene(s) may act as a trigger for EC onset, or people with the gene(s) may tend to be more susceptible to EC under the effects of certain environmental risk factors [Zhang et al. 2000].

Some cytogenetic studies also revealved that chromosomal instability, as measured by increased chromosomal aberration rates, fragile sites, and/or sister chromatid exchanges, is much more common in ESCC cases and their healthy relatives than in healthy persons from families without a history of ESCC [Wu et al. 1989]. In order to search for genes involved in the development and/or progression of ESCC in patients from Shanxi Province, Hu et al in Dr. Taylor's lab, have initiated a search for a major ESCC gene by conducting a genome-wide scan in ESCC tumors. Chromosome 13 showed loss of heterozygosity (LOH) in 95% of microsatellite markers, the highest frequency of LOH on any chromosome. They established a high-resolution deletion map using 107 markers on 13q and compared LOH frequency by family history of upper gastrointestinal cancer. Overall allelic loss was significantly higher in those with a positive (versus negative) family history, suggesting the presence of an inherited tumor suppressor gene on 13q in ESCC [Hu et al. 2003]. This is further confirmed with the subsequent fine mapping of LOH on this chromosome [Li et al. 2001]. These data also showed two common deletion regions (including one on 13q12.3–q13.1, where BRCA2 is located) and a significantly higher frequency of LOH in patients with a positive family history [FH (+)] of upper gastrointestinal (UGI) cancer compared to patients with no family history [FH (-)] [Hu et al. 2004].

Hu et al further observed a significant difference in the rate of germline mutations in BRCA2 between familial and sporadic ESCC patients. These data support a role for germline mutations in BRCA2 in the genetic susceptibility of familial ESCC in this high-risk area of China, and suggest a possible link with the high rate of genetic instability observed on chromosome 13q [Hu et al. 2004]. Germline mutations in BRCA2 in ESCC patients from this high-risk area of China are more frequent in FH (+) than FH (-) cases, suggesting that BRCA2 may play a role in genetic susceptibility to familial ESCC.

By now, whether BRCA2 or any other gene(s) on chromosome 13 or even other chromosomes is the inherited esophageal cancer susceptibility gene still remains unclear and needs further investigation. On the other hand, environmental factors have been implicated by epidemiologic studies as being very important in EC development. Epidemiological studies revealed that, different from the adenocarcinoma (ADC) which was developed from Barrett's esophagus, ESCC is thought to be more related with dietary habits, tobacco smoking and alcohol drinking. However, the nature of the relationship between genetic and other risk factors remains unclear. We do believe that esophageal cancer, like many other cancers, is also a complex disease with both genetic and environmental factors involved. The aim of our study is to find the ESCC susceptibility gene(s), the genes with the altered expression in esophageal tumorigenesis, along with the function and the regulation network of the genes. We have found that many of the calcium-relevant and differentiation associated genes were down-regulated in ESCC by cDNA microarray. We are now studying the function and regulation network of the genes.

2. Discovery of Ca^{2+}-relevant and Differentiation Associated Genes Down-regulated in Esophageal Squamous Cell Carcinoma using cDNA Microarray

It is known that complicated genetic alterations occur frequently in ESCC, including allelic loss on chromosomes (17p, 13q, 9p21-22) and mutations of certain known genes (p53, p16) [Ko et al. 2001; Lam et al. 1997]. In addition, ESCC is related to nutritional deficiencies in calcium, vitamin D_3 and other minerals [Li et al. 1998]. However, the nature of the relationship between genetic and other risk factors remains unclear.

Multiple genes are involved in ESCC, including cyclin D1, EGFR, annexin I, cystatin B, stefin A, int-2, CASK, keratin 4, SPRR, Transglutaminase and SLP-2 [Nagasawa et al. 2001; Xia et al. 2002; Li et al. 2005; Suzuki et al. 1997, Wang et al. 2001, 2002; Chen et al. 2000, 2002; Zhang et al. 2006]. Several studies on differential gene expression in ESCC using commercially available microarray membranes have been reported [Xu et al. 2002; Hu et al. 2001, Lu et al. 2001; Zhi et al. 2003; Tamoto et al. 2004; Gomes et al. 2005]. In this study, we analyzed the expression profiles of 34,176 clones in ESCC using a high-density cDNA microarray derived from our own cDNA libraries. We identified genes differentially expressed in ESCC, and confirmed the altered gene expression in selected genes by Northern blot and immunohistochemistry analysis. We found that a number of genes involved in squamous cell differentiation were down-regulated in ESCC, and we also observed significant down-regulation of genes involved in transcription, translation, protein folding and protease inhibitors. In addition, several up-regulated genes are involved in ESCC invasion or proliferation. Although further study on the regulation of differentially expressed genes in ESCC is necessary, these results will help us better understand the process of esophageal tumorigenesis and provide additional markers for the diagnosis and treatment of this disease.

Materials and Methods

Patients and Samples

For Northern blot and cDNA microarray analysis, esophageal cancer and matched adjacent normal mucosa tissue samples were taken from patients who had not been treated with radiotherapy or chemotherapy before surgery. Fresh samples were dissected manually to remove mixed connective tissues and were immediately stored in liquid nitrogen. The samples used for immunohistochemistry analysis were fixed in 4% polyformaldehyde and completely embedded in paraffin.

cDNA Library Construction

Equal quantities of high-purity undegraded 5μg of poly$(A)^+$ RNA were extracted using the mRNA purification Kit (Amersham Pharmacia, Inc, Piscataway, NJ, USA) from both cancerous and normal esophagus tissues. The cDNA libraries, ESPN (esophageal normal) and ESPC (esophageal cancer), were constructed by using the λ-*ZAP* express vector system (Stratagene Inc. Kirkland, WA, USA) according to the previously described method [Liew et al.1997]. Briefly, the first strand cDNA was synthesized using a restriction endonuclease (RE1)-adapted oligo-dT primer. After synthesizing the second strand, RE2 adaptors were then ligated onto the cDNA using T4 DNA ligase. Following adaptor ligation, cDNA was digested, a RE1 site at the 3'-end and a RE2 site at the 5'-end were subsequently generated. To increase the fraction of cDNA clones and remove excess adaptor, the cDNA was size fractionated by gel filtration chromatography. After the libraries were constructed, 1584 clones were randomly picked for sequence analysis.

cDNA Array Construction

The λ-*ZAP* express vector has the advantage of *in vivo* excision of the insert and flanking sequences into a phagemid. Generally, the libraries had a set of highly abundant genes which were not complex enough to be exploited further by cDNA microarray analysis. Approximately 6.57% of the clones from ESPC and ESPN libraries contained cDNA inserts representing KRT4 and KRT13 according to the ESTs sequencing, thus we subtracted these two genes, and then randomly picked 34,176 cDNA inserts from ESPN and ESPC libraries by PCR amplification with T3 (forward) and T7 (reverse) primers. The amplified cDNA inserts were visually inspected on 1% agarose gels to ensure quality and quantity of the DNA prior to depositing onto two 8X12 Hybond-N^+ nylon membranes (Amersham Pharmacia Inc, Piscataway, NJ, USA) using a specialized arrayer (Bio Robotics Inc. Haslingfield, Cambridge, UK). Each cDNA fragment was placed in two different spots (i.e., double – offset). Lambda phage and pUC18 vector DNA were spotted as negative controls. Eight housekeeping genes were evenly distributed (12 times each) onto each membrane as an intramembrane control, as previously described by Xu et al [Xu et al., 2001].

Hybridization and Image Processing

Total RNA was extracted from frozen tissues by using the TRIzol reagent (Invitrogen, Carlsbad, CA, USA), poly(A)$^+$ mRNA was then isolated from total RNA using a poly(dT) resin. Approximately 1μg of mRNA was labeled in a reverse transcription reaction in the 200μCi [α-^{33}P]dATP using Superscript^{II} reverse transcriptase (Invitrogen, Carlsbad, CA, USA). The hybridization and subsequent imaging was performed using a previously described method [Xu et al. 2001]. Genes with intensities greater than 10 were considered as positive signals to ensure that they were distinguished from background with statistical significance >99.9%. Normalization among arrays was performed by the intensities of the housekeeping genes on the membranes.

Sequencing Analysis of the Differentially Expressed Genes

The clones were chosen for sequencing analysis when the ratio of the signals between the same spot on different membranes was greater than 3. DNA was sequenced using the ABI377 automated sequencer (Perkin Elmer/Applied Bioasia, Shanghai, China); homology comparison was performed using the BLAST software. The sequences were considered to be part of the known genes if they were at least 95% homologous. ESTs without significant homology were considered as novel genes.

Northern Blot and RT-PCR

Genes identified by cDNA microarray were further analyzed using Northern blot to confirm their expression profiles. For Northern blot analysis, total RNA (20μg) was electrophoresed through 1% agarose formaldehyde gel and transferred to Hybond-N$^+$ nylon membrane. The membrane was hybridized with the [α-^{32}P]dCTP-labeled, random-primed cDNA fragments and visualized by autoradiography. Moreover, five micrograms of total RNA was converted to cDNA using Superscript^{II} reverse transcriptase (Invitrogen, Carlsbad, CA, USA). The relative expression of genes was determined using RT-PCR with ß-actin or GAPDH as internal control.

Cell Culture and Transient Transfection

Human esophageal squamous cell carcinoma cell line *KYSE-70* was cultured in RPM1640 medium supplemented with 10% fetal bovine serum. The open reading frame of the GKLF cDNA was subcloned into the mammalian expression vector *pcDNA3.1* (Invitrogen, Carlsbad, CA, USA) to generate *pcDNA3.1-GKLF*. *pcDNA3.1-GKLF* plasmid DNA was transiently transfected into *KYSE-70* cells using lipofectamine™ 2000 as recommended by the manufacturer (Invitrogen, Carlsbad, CA, USA). *pcDNA3.1* was used as a control for the transfection.

Results

Overview of Library Quality and EST Expression Pattern in ESCC

Our in-house library consists of normalized and size-filtered cDNA derived from esophageal cancer (ESPC) and esophageal normal (ESPN) tissue, the average insert size is approximately 1.2-1.5kb (data not shown), in line with the mean size of human mRNAs. After sequencing the transcripts and eliminating repetitive elements and mitochondrial DNA, a total of 1,093 genes were analyzed. Using a public database search, these were classified as known genes (617, 56.45%), known ESTs (221, 20.22%), and previously uncharacterized ESTs (255, 23.33%). We sorted the 617 known genes into seven categories (or functional groups) according to their biological roles as defined by the GENECARDS database. We used EST data from both the ESPC and ESPN libraries to construct our gene expression profiles. These profiles clearly reflect the functional characteristics of esophagus in that the majority of the 10 genes were differentially expressed according to their EST copy numbers (Table 2-1).

Table 2-1. The known genes highly expressed in ESPC and ESPN, from ESTs sequencing

Gene Name	ESPN Copies (%*)	ESPC Copies (%*)
Hs.78225(Annexin I)	6(0.38)	1(0.06)
Hs.242057(C1orf10)	5(0.32)	2(0.13)
Hs.181165(EEF1A1)	25(1.58)	14(0.88)
Hs.111334(Ferritin L chain)	0(0.00)	4(0.25)
Hs.90093(Heat shock protein 70)	0(0.00)	3(0.19)
Hs.3235(Cytokeratin 4)	23(1.45)	4(0.25)
Hs.334309(Cytokeratin 6A)	0(0.00)	6(0.38)
Hs.74070(Cytokeratin 13)	73(4.61)	0(0.00)
Hs.139332(SPRR3)	7(0.44)	1(0.06)
Hs.279682(RHCG)	4(0.25)	1(0.06)

* Percentage of given EST in all ESTs obtained from cDNA libraries. ESPC, esophageal cancer cDNA library; ESPN, normal esophagus cDNA library. C1orf10, chromosome 1 open reading frame 10; EEF1A1, eukaryotic translation elongation factor 1 alpha 1; SPRR3, small proline-rich protein 3; RHCG, Rh type C glycoprotein.

cDNA Microarray Analysis

cDNA microarray technology was used to analyze the gene expression profiles of five esophageal cancer tissue samples and their matched adjacent normal mucosal tissue samples. Differential expression was considered to be significant when the ratio of the signals from tumorous and normal tissues was greater than 2. The cDNA clones with an expression ratio of 3 or greater in four or more paired samples were sequenced for further analysis. Sequence analysis showed 240 clones which formed 92 unique genes. Of these, 60 were previously known genes and the remaining 32 were novel genes. In addition, 77 genes (including 31 novel genes) were down-regulated, whereas 15 genes (including 1 novel gene) were up-regulated in five cancer tissue samples. Table 2-2 summarizes of the database search results of the genes with three-fold differential expression in ESCC.

Table 2-2. Summary of the genes with three fold differential expression in ESCC

Recovery Frequency	Ratio*	GenBank GenBank Identity	Accession Number
26	0.12	small proline-rich protein 3	NM_005416
16	0.18	annexin I	NM_000700
15	0.22	S100 calcium-binding protein A8	NM_002965
14	0.23	small proline rich protein II	M20030
11	0.13	chromosome 1 open reading frame 10	NM_01619
10	0.33	S100 calcium-binding protein A9	NM_002965
9	0.16	transglutaminase 3	NM_003245
8	0.16	Rh type C glycoprotein	NM_016321
8	4.34	keratin 17	NM_000422
6	0.23	epithelial membrane protein 1	NM_001423
6	0.27	polycystic kidney disease 1	NM_000296
5	0.24	cystatin B	NM_000100
4	3.98	1-phosphatidylinositol-4-phosphate 5-kinase	BC018034
3	0.15	cystatin A	NM_005213
3	6.89	IgM heavy chain constant domain	X67301
3	0.13	T-cell differentiation protein	M15800
3	3.71	ferritin, light polypeptide	NM_000146
3	3.74	M2-type Pyruvate kinase M23725	
3	4.64	ribosomal protein L23a	NM_000984
3	3.56	secreted protein, acidic, cysteine-rich	NM_003118
2	0.32	D component of complement	NM_001928
2	0.13	eukaryotic translation elongation factor 1 alpha 1	NM-001402
2	0.24	dolichyl-phosphate beta-glucosyltransferase	AF102850
2	3.18	immunoglobulin lambda locus	AF058075
2	0.28	inhibitor of DNA binding 2, dominant negative	NM_002166
2	3.22	protein phosphatase 4, catalytic subunit	AF097996
2	0.33	Ribosomal protein L17	NM_000985
2	0.26	serine (or cysteine) proteinase inhibitor	NM_030666
2	0.24	small proline-rich protein I	M19888
1	0.22	small proline-rich protein 2B	NM_006945
1	0.15	keratin 13	NM_002274
1	0.11	keratin 4	NM_002272
1	0.25	keratin 15	NM_002275
1	4.21	keratin 16	AF061812
1	0.22	adenylate kinase 3 alpha like	NM-016282
1	11.99	Immunoglobulin heavy constant gamma 3	AJ294732
1	0.21	calpastatin	AF327443
1	0.10	cAMP responsive element binding protein 3	NM_006368
1	3.00	Cathepsin B	NM_001908
1	0.19	cathepsin L	NM_001912
1	0.15	calponin 1	NM_001299
1	0.30	creatine kinase B	NM_001823
1	0.14	CSRP2 binding protein	AK023634
1	0.13	DNA (cytosine-5-)-methyltransferase 1	NM_001379
1	0.31	emopamil-binding protein	NM_006579
1	4.45	fibronectin	NM_002026

Table 2-2. (Continued)

Recovery Frequency	Ratio*	GenBank Identity	GenBank Accession Number
1	0.10	interferon regulatory factor 3	NM_001571
1	0.29	interleukin 1 receptor antagonist	AJ005835
1	0.27	microfibrillar-associated protein 1	NM_005926
1	3.91	N-myc downstream regulated gene 1	AF004162
1	0.17	nuclear receptor coactivator 1	NM_003743
1	0.18	peptidylprolyl isomerase A	NM_021130
1	0.12	protein tyrosine phosphatase type IVA2	NM_003479
1	0.18	SET binding factor 1	BC009268
1	0.22	solute carrier family 1	NM_006671
1	0.23	sublingual gland mucin	BC033831
1	0.31	SUMO-1 activating enzyme subunit 1	AF090385
1	0.12	thioredoxin-like 2	AJ010841
1	0.20	ubiquinol-cytochrome c reductase hinge protein	M36647
1	0.25	gut-enriched kruppel-like factor	NM_004235
1	0.11	EST	NM-004454
1	0.12	EST	AK000507
1	0.12	EST	XM202199
1	0.13	EST	BC000936
1	0.14	EST	BC041179
1	0.15	EST	AF322067
1	0.15	EST	NM-024632
1	0.15	EST	XM-166549
1	0.15	EST	AK021468
1	0.16	EST	AC104506
1	0.16	EST	NM_014040
1	0.17	EST	NM000970
1	0.17	EST	BC017972
1	0.18	EST	AK022451
1	0.18	EST	AL353948
1	0.18	EST	NM_017584
1	0.18	EST	AB014585
1	0.19	EST	AK022517
1	0.20	EST	NM_017634
1	0.20	EST	AB059410
1	0.21	EST	AK025716
1	0.22	EST	AK000920
1	0.22	EST	AK097417
1	0.24	EST	NM_020195
1	0.24	EST	AF102850
1	0.25	EST	NM_003413
1	0.29	EST	AL109698
1	0.30	EST	BC003560
1	0.31	EST	AK092686
1	0.33	EST	NM_014837
1	0.33	EST	XM-048825
1	8.71	EST	AB014589

* Ratio represents the fold of differential expression of genes in esophageal cancer tissue samples compared to normal tissue samples. For genes recovered multiple times, the ratio listed is the average of all the hits.

Hierarchical Clustering Analysis

Differential gene expression analysis revealed that 168 genes (including 95 novel genes) were down-regulated whereas 21 genes (including 5 novel genes) were up-regulated in five cancer tissue samples. Hierarchical clustering reveled a gene cluster of squamous cell differentiation marker genes in chromosome 1q21, including cystatin A and several individual members of keratin, small proline-rich protein (SPRRs) and the S100 family, are known to be the components of the cell envelope, cross-linked by transglutaminase.

To confirm the altered expression level, five genes were randomly selected and verified by Northern blot. As shown in Figure 2-1, Northern blot results were consistent with cDNA microarray results in terms of differential gene expression. Moreover, cDNA microarray and randomly sequenced ESTs consistently showed altered expression in several genes, including KRT4, KRT13, annexin I, eukaryotic translation elongation factor 1alpha-1 (EEF1A1) and SPRR3.

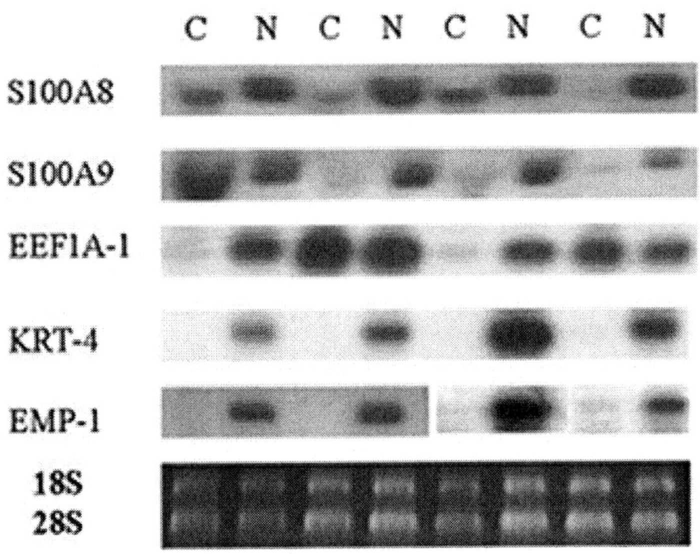

Figure 2-1. Northern blot analysis of the differentially expressed genes. S100A8, S100A9, EEF1A1, KRT4 and EMP1 were down-regulated in esophageal cancer tissues (C) compared to the matched almost normal tissues (N).

Alterations of Gene Expression in ESCC

A group of differentiation related genes are coordinately down-regulated in ESCC tissue samples, including SPRRI, SPRRII, SPRR3, EEF1A1, chromosome 1 open reading frame 10 (C1orf10), Mal T-cell differentiation protein, polycystic kidney disease 1 (PKD1), epithelial membrane protein 1 (EMP1), annexin I, S100A8, TGM3, GKLF, and cystatin A. All of these genes are highly expressed in normal esophagus. Interestingly, most of these genes encode Ca^{2+}-binding or -modulating proteins, which constitute the cell envelope (CE). Moreover, SPRRs and S100 calcium-binding proteins belong to the epidermal differentiation complex (EDC), which is located on chromosome 1q21 and is involved in the terminal differentiation of epithelial cells [Marenholz et al., 1996].

Cytokeratin expression is a reliable biochemical indicator of epithelial differentiation process [van Dorst et al., 1998]. Three different cytokeratins, KRT4, KRT13 and KRT15 were found to be down-regulated, whereas KRT16 and KRT17 were up-regulated in ESCC. It was revealed that expression of cytokeratins (KRT4 and KRT13) and the cross-linked envelope were aberrant during the different stages of terminal differentiation in ESCC [Banks-Schlegel et al., 1986]. The most striking feature of KRT4 -/- mice is that esophageal epithelial cells do not terminally differentiate but have abnormal proliferation [Ness et al., 1998]. Our data suggested that the altered expression of several cytokeratins might be associated with the abnormal differentiation and proliferation of ESCC.

Our data also showed that several genes involved in transcription and translation, were down-regulated in ESCC. Genes associated with transcription were down-regulated, including GKLF and inhibitor of DNA binding 2. EEF1A1 was down-regulated, this gene modulates the dynamic behavior of microtubules assembled from tubulin sources and regulates of cell cycle [Moore et al., 1998]. In parallel with the down-regulation of EEF1A1, the chaperone, heat shock protein 70 was also down-regulated.

The metastatic spread of cancer is a complex process that involves the combination of different cellular actions including cell adhesion to the extracellular matrix (ECM) and breakdown of the ECM by specific matrix-degrading proteinases. The protease of cathepsin B was up-regulated, whereas cathepsin L was down-regulated in ESCC. On the other hand, several protease inhibitors, which were important negative regulators of protease action *in vivo*, were down-regulated in ESCC, including cystatin A, cystatin B and serine (or cysteine) proteinase inhibitor. In contrast, fibronectin and SPARC were up-regulated in ESCC, These genes are involved in a variety of diverse biological processes including matrix synthesis/turnover, differentiation, cell adhesion and tumor cell migration and invasion. Our data suggest that down-regulation of protease inhibitors might be associated with esophageal cancer invasion and metastasis.

Expression of GKLF in KYSE-70 stimulates SPRRs and KRT4 Transcription

GKLF, an epithelial cell-enriched zinc finger-containing transcription factor, transcriptionally activates two esophageal epithelial promoters (Epstein-Barr virus ED-L2 and keratin 4), which play an important role at the transition toward differentiation [Jenkins et al., 1998]. Since our data show that GKLF, SPRR1A, SPRR2A and KRT4 are all down-regulated in ESCC (Figure 2-2A), we speculate that GKLF expression might trigger the expression of its target genes in ESCC. The results show that the expression of SPRR1A, SPRR2A and KRT4 are increased in response to the up-regulation of GKLF in *KYSE-70* cells when transiently transfected with the *pcDNA3.1-GKLF* vector (Figure 2-2B and C).

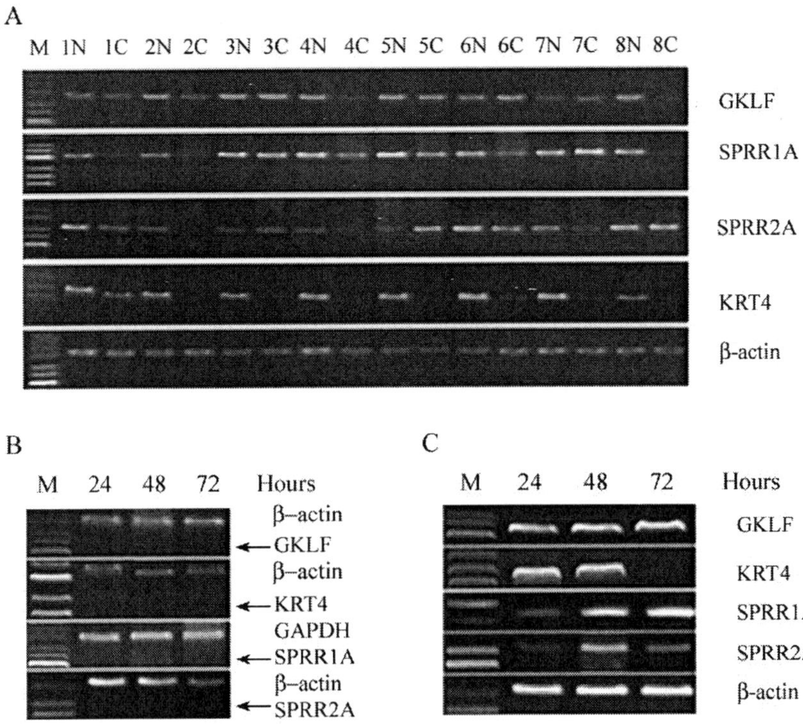

Figure 2-2. RT-PCR analysis showed the differential expression of GKLF, SPRR1A, SPRR2A and KRT4 in 8 pairs of esophageal cancer tissue samples and the matched normal tissue samples (A), and the transient transfection of GKLF in *KYSE-70* cell activates the expression of SPRR1A, SPRR2A and KRT4 (B and C). Transient transfection with 2μg of *pcDNA3.1* (control) or *pcDNA3.1-GKLF* vector in *KYSE-70* cell, then total RNA was isolated at the indicated time points and converted to cDNA using Superscript^{II} reverse transcriptase. Expression of SPRR1A, SPRR2A and KRT4 was increased in response to the up-regulation of GKLF (C), whereas no changes were observed in *KYSE-70* cell transfected with *pcDNA3*.1 vector control (B). ß-actin or GAPDH was used as internal control.

Discussion

Esophageal mucosa is lined by a stratified squamous epithelium which functions in epithelial proliferation, differentiation, turnover and serves as a barrier [Geboes et al., 1978]. Cellular proliferation is frequent in the epibasal layers, and loss of contact with the basement membrane eventually triggers differentiation [Seery et al., 2000] and the sequential expression of differentiation markers [Jankowaski et al., 1993]. Squamous cell differentiation requires the coordinated activation and repression of genes specific to the differentiation process, and disruption of this program accompanies neoplasia [Jones et al., 1997]. In this study, we used cDNA microarray analysis to profile the changes in gene expression that accompany these processes.

Our genetic profile aims to find correlations between genetic changes and esophageal squamous cell differentiation. Squamous cell differentiation associated genes were down-regulated in ESCC including TGM3 and its substrates SPRRs, as well as calcium binding

S100 protein, annexin I, EMP1, and cystatin A, all of which are incorporated into the cell envelope (CE) [Matsuki et al., 1998; Marenholz et al 1996; Robinson et al., 1997; Steven et al., 1994]. The cell envelope performs a vital barrier function, and it is constructed through the sequential expression, processing and deposition of several distinct proteins that are cross-linked by transglutaminase. Recently, knockouts of the major cell envelope constituent loricrin and GKLF have both failed to significantly disrupt the barrier function of the skin, and die shortly after birth due to dehydration [Koch et al., 2000; Matsuki et al; 1998]. Our data suggest the down-regulation of squamous cell differentiation associated genes might lead to the disability of the formation of the cell envelope and sustenance of toxic material.

The reduction of cell envelope related genes might be relevant to metastasis. To a significant extent, these genes are controlled at the transcriptional level, thus they may have several regulators that can be identified. One possible regulator is calcium, since cells require calcium to induce cell envelope formation and express a number of differentiation-related proteins such as membrane-bound transglutaminase and cytokeratins [Bikle et al., 2001]. The switch between proliferation and differentiation is modulated by calcium [Dotto et al., 1999, review]. Intracellular calcium concentration is transduced by the "activation" of Ca^{2+}-binding proteins and calcium- and phospholipid-binding proteins of the annexin family [Kerkhoff et al., 1998, Review]. Our data show that several members of the annexin and S100 family have significantly decreased expression levels in ESCC. Annexin I and annexin II form Ca^{2+} channels in the plasma membrane, thereby regulating Ca^{2+} homeostasis [Frey et al., 1999]. Annexin I-S100A11 or annexin II-S100A10 form heterotetramer [Thiel et al., 1992] and modulate intracellular calcium concentration. In addition, the COOH-terminal cytoplasmic tail of PKD-1 increases the activity of a Ca^{2+}-permeable cation channel [Hanaoka et al., 2000].

Epidemiological data show that calcium levels are lower in wheat and maize grown in an area of China whose population is at high risk for ESCC (Lianzhou) compared to the level in wheat and maize from a lower risk area (Yuzhou) in China [Chang et al., 1999]. Our results indicate that calcium might be involved in the progress of ESCC and in esophageal epithelium proliferation and differentiation.

The aberrant expression of squamous cell differentiation associated genes prompted us to ask whether the regulators of cell envelope formation are defective. Our data show that the expression of GKLF, SPRRs and cytokeratins is altered in ESCC. It is known that GKLF plays an important role in epithelial cell proliferation and/or differentiation [Katz et al., 2002], previous studies showed that SPRR1A and SPRR2A are down-regulated in response to the down-regulation of GKLF in transgenic mice over-expressing claudin 6, and GKLF -/- mice [Segre et al., 1999; Turksen et al., 2002]. SPRR1A and SPRR2A, the precursor proteins of the cell envelope, are more abundant in the cell envelope of oral and esophageal epithelium [Hohl et al., 1995]. Thus, GKLF may be responsible for the cascade of changes that lead to abnormal expression of differentiation associated genes and disruption of the cell envelope cross-linking process. Moreover, we showed the first time that GKLF can physically bind to keratin 13 promoter and regulate the expression of keratin 13 directly (data not shown), as it is well known that keratin 13 and keratin 4 are specific differentiation markers of esophagus, thus we believe that GKLF is involved in the regulation of the differentiation associated genes in ESCC.

Furthermore, a common feature of the promoter region of differentiation associated genes including SPRR1, SPRR3, cystatin A and KRT4, is the presence of AP-1 binding sites [Sark et al., 1998; Fischer et al., 1999; Yamazaki et al., 1997; Ma et al., 1997]. Further evidence shows that the expression of S100A8, S100A9, cystatin A and SPARC is indirectly regulated by AP-1 [Gebhardt et al., 2002; Arnould et al., 1998; Briggs et al., 2002], thus we speculate that AP-1 may participate in the process of altering expression of squamous cell differentiation-associated genes in ESCC. We have more detailed study in this chapter.

Very interestingly, another report from University of Chicago about Barrett's esophagus and esophageal adenocarcinoma revealved that transformation of Barrett's esophagus to adenocarcinoma was also associated with suppression of the genes involved in epidermal differentiation, including genes in *1q21* loci and corresponding to the epidermal differentiation complex which is very similar to our results on ESCC eventhough esophageal adenocarcinoma and squamous cell carcinoma are quite different [Kimchi et al., 2005]. Barrett's esophagus is a specialized intestinal metaplasia of normal squamous to columnar epithelium, it is thought to be a premalignant transformation and has been identified in 80% to 100% of esophageal adenocarcinoma of the distal esophagus [Stein et al. 1993; Kimchi et al., 2005]. Chronic gastroesophageal reflux is considered as a major risk factor [Lagergren et al., 1999]. The presence of Barrett's esophagus increases the risk of developing adenocarcinoma 40- to 125-fold [Shaheen et al., 2002]. Kimchi et al examined differential gene expression patterns in resected esophageal specimens composed of normal esophageal epithelium, Barrett's esophagus, and adenocarcinoma obtained from the same individual patients [Kimchi et al., 2005]. They found that transition from normal esophageal epithelium to Barrett's esophagus and esophageal adenocarcinoma is associated with the suppression of epidermal differentiation programs, including genes localized in the epidermal differentiation complex (EDC), including S100 proteins, small proline-rich proteins, involucrin, periplacin, desmocollins, and desmoglein 3. Many of these genes are in the specific chromosomal loci called the *EDC* [Williams et al., 2002; Marenholz et al., 2001]. The *EDC* is the chromosomal loci containing >40 genes involved in epidermal differentiation and localized in *1q21* [Williams et al., 2002; Marenholz et al., 2001; Elder et al., 2002]. Other two genes are also involved in epidermal differentiation: *transglutaminase-3* [Hitomi et al., 2003] and *NICE-1*, with the latter also localized in *1q21* [Marenholz et al., 2001]. They suggest that such dedifferentiation is important for tumor clones to release from tight junctions with the surrounding cells through desmosomes and cornified envelope. This is a potential mechanism for invasive growth and formation of metastasis [Kimchi et al., 2005].

S100 Gene Family in Human Esophageal Squamous Cell Carcinoma

In the above cDNA array analysis, we found that several S100 family members were down-regulated in ESCC, among them, S100A8, S100A9 and S100A14 are significantly down-regulated in ESCC tissue samples. Recent global transcriptional profile analyses by others also showed that S100A9 is one of the genes most discriminant between the normal and carcinomatous esophagus [Su et al., 2003]. Therefore, genetic alterations of these genes may contribute to esophageal tumorigenesis.

The first S100 calcium-binding protein was isolated from bovine brain in 1965 by Moore [Moore, 1965]. Subsequent studies identified many members of this family based on their homology of amino acid sequences and their feature of calcium-binding properties. S100 family become major interest because of their deregulated expression in human diseases especially in cancer; the cluster organization of S100 genes on human chromosome 1q21 which is a region frequently rearranged in tumors [Watson et al., 1998]. Up to now, 21 genes encoding S100 calcium-binding proteins have been identified, and a cluster of 16 S100 genes is mapped to human chromosome 1q21 region [Schafer et al. 1995; Weterman et al. 1996; Pietas et al. 2002]. Early studies showed that S100A2, S100A4, S100A12, and S100P proteins were deregulated in esophageal cancer tissues and cell lines and could influence cell differentiation and cell cycle [Kyriazanos et al., 2002; Ninomiya et al., 2001; Hitomi et al., 1998; Sato et al., 2002]. Moreover, it has been reported that many S100 genes were expressed at high level in stratified epithelia of the upper digestive tract. All these indicate that S100 gene family may be correlated with ESCC. In addition, recent studies demonstrated that many EF-hand proteins correlate with the progression of certain tumors [Ikura et al., 2000]. As one major subclass, S100 proteins have attracted substantial interest due to their tissue- and cell type-specific expression patterns and their involvement in tumorigenesis and development of various tumors [Donato et al., 2001; Heizmann et al., 2002; Ohuchida et al., 2005; Garrett et al., 2006]. The challenge is to identify some specific members of S100 family involved in the development of ESCC.

S100A8 and S100A9, namely MRP8 and MRP14, are expressed during myeloid differentiation, abundant in granulocytes and monocytes, and form a heterodimeric complex calprotectin in a Ca^{2+}-dependent manner. Typically, S100A8 and S100A9 are differentially expressed at sites of acute and chronic inflammation [Kerkhoff et al., 1998; Hunter et al., 1998; Seeliger et al., 2003]. Subsequently, it has been shown that the deregulation of S100A8 and/or S100A9 is associated with several common human malignancies [Gebhardt et al., 2002; El-Rifai et al., 2002; Stulik et al., 1999]. However few detailed data are available about the altered expression of S100A8 and S100A9 during esophageal carcinogenesis. S100A14 is a new member of S100 family [Pietas et al. 2002], by now, less information was available about its relationship with tumorigenesis. We have shown that S100A14 is significantly down-regulated in human ESCC, its down-regulation is closely related with cell differentiation, and overexpression of the gene would cause ESCC cells go to apoptosis process.

In this study, we investigated the differential expression of 16 S100 genes in 62 pairs ESCC samples, and the differential expression and intracellular localization of S100A8 and S100A9 in human esophageal dysplasia, cancer tissues and cancer cell lines. Moreover, we have characterized the function of S100A8, S100A9 and S100A14 in ESCC.

Materials and Methods

Patients and Samples

Esophageal cancer and matched adjacent normal mucosa were taken from surgical specimens of 62 patients who had not been treated with radiotherapy or chemotherapy before surgery. Cancer tissue was taken from the tumor tissue where there was no hemorrhage or putrescence, whereas the matched normal mucosa was taken from the surgical cutting edge where was approximate 3-5cm away from the cancerous lesion. Fresh samples were dissected manually to remove mixed connective tissues and stored in liquid nitrogen immediately. The clinical diagnosis of all 62 patients had been approved by histological diagnosis after surgery.

Primer Design

It was very important to choose the most specific primers for each S100 gene because of their high homologies each other. The sequences of the PCR primer pairs of 16 S100 members listed in table 3-1 were designed as follows: (1) selection and design of primer by Software Primer 3 according to standard criteria (http://www-genome.wi.mit.edu/cgi-bin/primer/primer3_www.cgi); (2) test of the primer specificity through nucleotide BLAST comparison analysis with all known human sequences; (3) test of whether both primer-sequences of each paired primer were from 2 different exons in order to distinguish the cDNA and residual genomic DNA in total RNA.

Semi-quantitative PCR and Image Analysis

Every PCR reaction was performed in 25μl final volume, containing 1×PCR buffer, 1.5mM $MgCl_2$, 0.2mM each dNTPs, 1.25 unit Taq DNA Polymerase (TakaRa, Dalian, China), 1μl RT product, 0.3-0.5μM each primer. The housekeeping gene, glyceraladehyde-3-phosphate dehydrogenase (*GAPDH*), was used as an internal control. The conditions of PCR reaction were as follows: initial denaturalization for 4 minutes at 94°C, amplification for 26-32 cycles (30 seconds at 94°C, 30 seconds at 55°C-65°C, 30 seconds at 72°C) which were varying in different genes and listed in table 3-1, and final extension for 4 minutes at 72°C. The RT-PCR reaction products were analyzed by electrophoresis on 1.5% agarose gel. Then the electrophoresis images were scanned with Fluor-S MultiImager (Bio-Rad, California) and the original intensities of specific bands were quantified by Multi-Analyst software (Bio-Rad, California). The final data were obtained after normalized by the intensity of *GAPDH*, which included the expression amount of every gene in each case of ESCC (R_C) and the corresponding normal esophageal mucosa (R_N), and the ratio of R_C to R_N ($R_{C/N}$).

Statistical Analysis

All data were analyzed with SAS 8.1 software for Windows 98. With the Wilcoxon paired signed-rank test (non-parametric variance for pair comparisons), the difference between R_N and R_C or the differential expression of 16 S100 genes in ESCC versus normal tissues was examined. After every $R_{C/N}$ was obtained, the correlation of each S100 gene with the histological differentiation of ESCC was analyzed with the Kruskal-Wallis test. When $R_{C/N}$ of two different S100 genes in the same case was used as variable, the possible correlations between every two S100 genes were carried out with the Pearson test. A p-value of <0.05 was considered as significant.

Immunohistochemistry

Paraffin sections (thickness 5μm) from ESCC patients were treated as described previously [Hse et al., 1981] followed by incubation with primary antibodies (S100A8 and S100A9, Santa Cruz Biotechnology, Inc. Santa Cruz, CA, USA) overnight at 4°C in a humid chamber, at the dilutions of 1:200 (S100A8) and 1:200 (S100A9) respectively. For negative controls, primary antibody was replaced by non-immune serum. After immunostaining, the sections were imaged at magnification (200X) by a single investigator who was not informed of the clinical characteristics; Three high-power fields were examined for each section. The value of integral intensity in these fields was measured by quantification software developed at the Center of Imagine Analysis, Beijing University of Aeronautics and Astronautics. Specimens were classified as "positive" if >10% of cells displayed a staining intensity. The statistical significance of the relationship between the gene expression pattern and the grade of differentiation was analyzed using SPSS 10.0 software.

Results

Primer Design

With Software Primer 3 and BLAST analysis, 16 specific primer pairs were obtained (Table 3-1) which could discriminate the different members of S100 gene family. At the same time, both primer-sequences of each S100 gene were from two different exons respectively so that each primer pair also was able to discriminate the cDNA and genomic DNA. Thus, residual genomic DNA in total RNA could not influence the RT-PCR results.

Semi-quantitative RT-PCR

With semi-quantitative RT-PCR, the differential expression of 16 S100 genes was examined in 62 cases of ESCC. The results evaluated with the Wilcoxon paired sign-rank test showed that 12 S100 genes were significantly deregulated in ESCC versus the normal counterparts, which were *S100A1* ($p=0.0007$), *S100A2* ($p<0.0001$), *S100A4* ($p=0.0003$), *S100A7* ($p=0.0200$), *S100A8* ($p<0.0001$), *S100A9* ($p<0.0001$), *S100A10* ($p<0.0001$), *S100A11* ($p<0.0001$), *S100A12* ($p<0.0001$), *S100A14* ($p<0.0001$), *S100B* ($p=0.0006$) and *S100P* ($p<0.0001$). However, the expression of other genes such as *S100A3*, *S100A6*, *S100A13* and *S100Z* was not significantly different in ESCC and the counterparts (Table 3-2). Partial electrophoresis images of semi-quantitative RT-PCR were shown in Figure 3-1. At the same time, the criteria of a gene over-expression or under-expression in ESCC were decided by three independent researchers. The criteria were that a gene was considered as up-regulated in one case of ESCC when a $R_{C/N}$ was higher than 4/3, as down-regulated when a $R_{C/N}$ was lower than 3/4, and the alteration of the gene was not considered as significant when $R_{C/N}$ was between 3/4 and 4/3 (including 3/4 and 4/3). With this criteria, the number of under-expression cases was found in significant down-regulated genes such as *S100A8* (51 cases, 82.3%), *S100A14* (49 cases, 79.0%), *S100A9* (48 cases, 77.4%), *S100A12* (39 cases, 62.9%),

S100A10 (37 cases, 59.7%), *S100P* (36 cases, 58.1%), *S100A2* (34 cases, 54.8%), *S100A4* (33 cases, 53.2%), *S100A1* (31 cases, 50.0%), *S100A11* (31 cases, 50.0%) and *S100B* (31 cases, 50.0%), and that of over-expression cases was found in significant up-regulated gene, *S100A7* gene (35 cases, 56.5%) in ESCC (Table 3-2, Figure 3-2).

Table 3-1. Primers used for RT-PCR and Conditions of PCR reaction

S100 genes	Dir*	Primer sequences (5'-3')	Start point	Product Size	Ref*	Anne (°C)*	PCR Cycles
S100A1	F	AGGAGCTGAAAGAGCTGCTG	208	279	NM_006271	65.5	29
	R	AGGGATAAGTGGGGTGAGGT	486				
S100A2	F	CACTACCTTCCACAAGTACT	79	229	NM_005978	56	27
	R	GAAGTCATTGCACATGAC	307				
S100A3	F	GCAGGCGGTAGCTGCCATC	101	255	NM_002960	59	29
	R	TTGAAGTACTCGTGGCAGTAG	355				
S100A4	F	CAAGTACTCGGGCAAAGAGG	120	250	NM_002961	58	27
	R	CTTCCTGGGCTGCTTATCTG	369				
S100A6	F	GGACCGCTATAAGGCCAGTC	2	285	NM_014624	60	27
	R	GGTCCAAGTCTTCCATCAGC	286				
S100A7	F	TGAGCAACACTCAAGCTGAG	73	336	NM_002963	63	27
	R	GGGTCTCTGGAGGCCCATTG	408				
S100A8	F	TGTCAGCCGTCTTTCAGAAG	9	286	BC005928	58	27
	R	ACGCCCATCTTTATCACCAG	294				
S100A9	F	GGGAATTCAAAGAGCTGGTG	149	267	NM_002965	58	26
	R	CACTGTGATCTTGGCCACTG	415				
S100A10	F	CATCTCAAATGGAACACGCC	116	321	NM_002966	56	27
	R	TATCAGGGAGGAGCGAACT	436				
S100A11	F	ATCGAGTCCCTGATTGCTGT	160	331	NM_005620	59	27
	R	AGAAAGGCTGGAAGGAAAGG	490				
S100A12	F	TTGAAGAGCATCTGGAGGG	79	269	NM_005621	56	27
	R	CTACTCTTTGTGGGTGTGG	347				
S100A13	F	ACCACCTTCTTCACCTTTGC	125	254	NM_005979	60	28
	R	AGGCGGCTTTACTTCTTCCT	378				
S100A14	F	CATGAGCCATCAGCTCCTCT	23	276	NM_020672	59	27
	R	TTCTCTTCCAGGCCACAGTT	298				
S100B	F	ATGTCTGAGCTGGAGAAGG	73	338	NM_006272	58	28
	R	CTGTCTGCTTTCTTGCATG	410				
S100P	F	AGGTGCTGATGGAGAAGGAG	122	223	NM_005980	60	28
	R	ATGGCTCTGCAGGAATCTGT	344				
S100Z	F	TGGAGTGGTCAGTTCTGCTG	307	230	NM_130772	62	32
	R	CGTTGTCCTTATTGGCATCC	536				

F: forward; R: reverse; Dir: direction; Ref: accession number from GenBank; Anne: annealing temperature

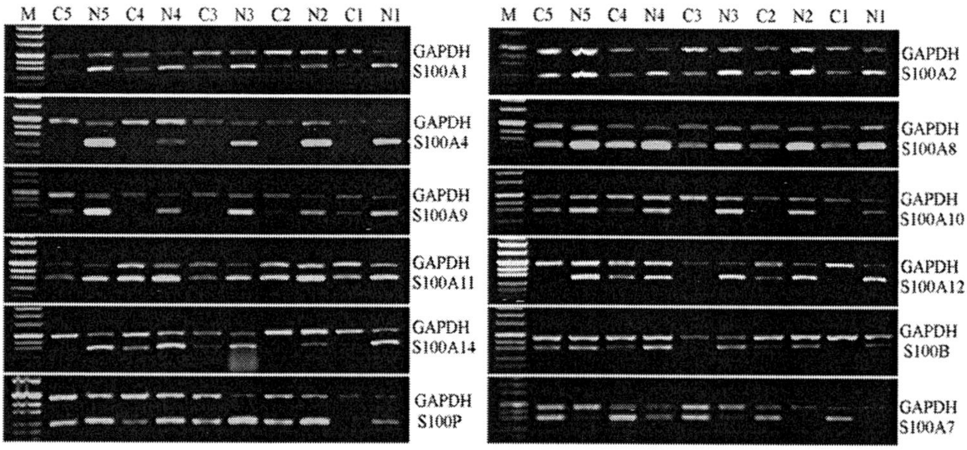

Figure 3-1. Partial electrophoresis images of semi-quantitative RT-PCR results. (M: marker; N: normal; C: cancer) The expression of *GAPDH* was used as internal control and the size of the amplified fragment was 452bp. The pUC Mix Marker 8 (MBI, Lithuania) was used as molecular weight marker.

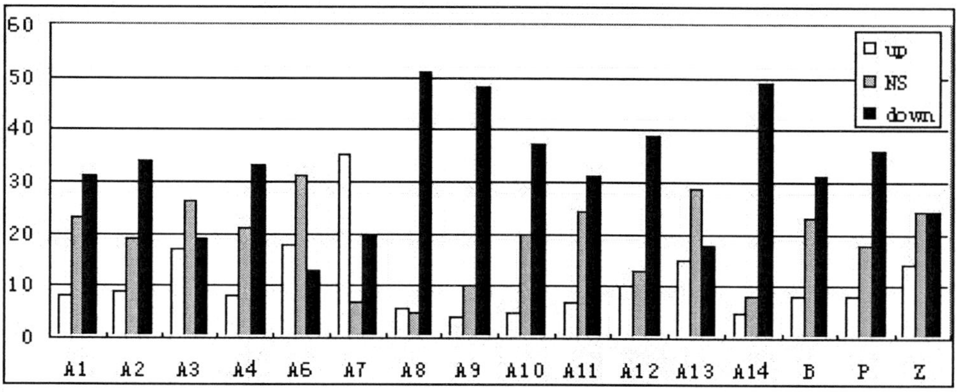

Figure 3-2. Differential expression of 16 S100 genes in 62 cases of ESCC examined by RT-PCR (up: up-regulation, white bars; NS: not significance, grey bars; down: down-regulation, black bars). Every column expressed the number of up-regulated or down-regulated or insignificantly deregulated cases.

Correlation Analysis of 16 S100 Genes' Deregulation

Table 3-3 showed that the deregulation of many S100 genes in ESCC was significantly correlated ($p<0.05$). Among *S100A2*, *S100A8*, *S100A14* and *S100P* genes, the deregulation of each other was markedly correlated ($p<0.0001$), and their correlation coefficients (r) were 0.6829 (*S100A2/S100A8*), 0.6818 (*S100A2/S100A14*), 0.6549 (*S100A2/S100P*), 0.6744 (*S100A8/S100A14*), 0.6422 (*S100A8/S100P*) and 0.5801 (*S100A14/S100P*) respectively. In addition, *S100A10/S100A11* (r=0.7559), *S100A9/S100A10* (r=0.5928), *S100A7/S100A10* (r=0.5900), *S100A9/S100A14* (r=0.5427) and *S100A8/S100A9* (r=0.5374) etc, these genes' deregulation was also significantly correlated ($p<0.0001$) (Table 3-3).

Frequent Loss of S100A8 and S100A9 mRNA and Protein Expression in ESCC Tissues

Besides RT-PCR analysis for RNA expression, we further examined the expression of S100A8 and S100A9 using Western blotting and immunohistochemistry analysis. All samples examined showed a specific band at 8.0 KDa corresponding to S100A8 monomer and at 14.0 KDa to S100A9 monomer, respectively. S100A8 (19/22) and S100A9 (16/22) protein expression was decreased in the esophageal cancer tissues compared to normal (Figure 3-3). The results also indicated that S100A8 and S100A9 did not form covalently crosslinked high molecular weight species and two antibodies did not crossreact each other. The down-regulation of S100A8 and S100A9 mRNA expression was consistent with the decreased expression of the corresponding proteins.

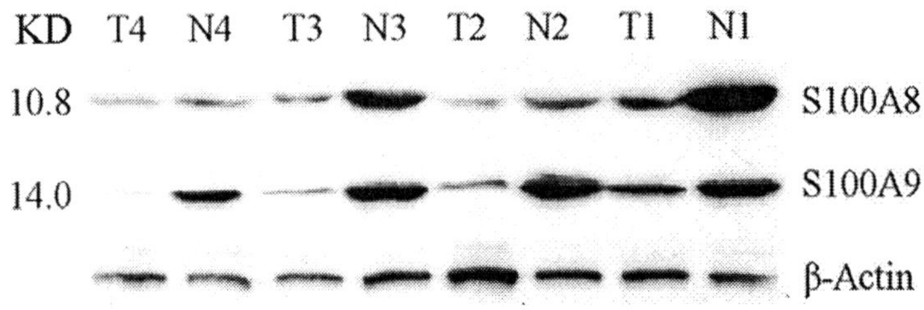

Figure 3-3. Western blot analysis showing the expression of S100A8 and S100A9 in human ESCC tissues and matched normal epithelium. β-actin was used as loading control. The low expression of S100A8 and S100A9 was found in ESCC samples compared with matched normal epithelium. T: ESCC tissues, N: matched normal esophageal epithelium. KD: molecular weight (Mt) in thousands.

Table 3-2. The differential expression of S100 genes

S100	Differential expression				
Genes	Up-cases*	NS-cases*	Down-cases*	S	p-value
S100A1	8(12.90%)	23(37.10%)	31(50.00%)	467.5	0.0007
S100A2	9(14.52%)	19(30.65%)	34(54.84%)	529.5	<0.0001
S100A3	17(27.42%)	26(41.94%)	19(30.65%)	-23.5	NS*
S100A4	8(12.90%)	21(33.87%)	33(53.23%)	498.5	0.0003
S100A6	18(29.03%)	31(50.00%)	13(20.97%)	-193.0	NS
S100A7	35(54.45%)	7(11.29%)	20(32.26%)	-328.5	0.0200
S100A8	6(9.68%)	5(8.06%)	51(82.26%)	830.5	<0.0001
S100A9	4(6.45%)	10(16.13%)	48(77.42%)	871.5	<0.0001
S100A10	5(8.06%)	20(32.26%)	37(59.68%)	768.5	<0.0001
S100A11	7(11.29%)	24(38.71%)	31(50.00%)	610.5	<0.0001
S100A12	10(16.13%)	13(20.97%)	39(62.90%)	544.5	<0.0001
S100A13	15(24.19%)	29(46.77%)	18(29.03%)	-24.5	NS
S100A14	5(8.06%)	8(12.90%)	49(79.03%)	864.5	<0.0001
S100B	8(12.90%)	23(37.10%)	31(50.00%)	472	0.0006
S100P	8(12.90%)	18(29.03%)	36(58.06%)	660.5	<0.0001
S100Z	14(22.58%)	24(38.71%)	24(38.71%)	230.5	NS

up-cases: the number of upregulated cases; NS-cases: the number of no significant deregulated cases; down-cases: the number of downregualted cases; S: statistic of the Wilcoxon paired signed-rank test; NS: not significant;

Table 3-3. Correlation analysis of 16 S100 genes differential expression

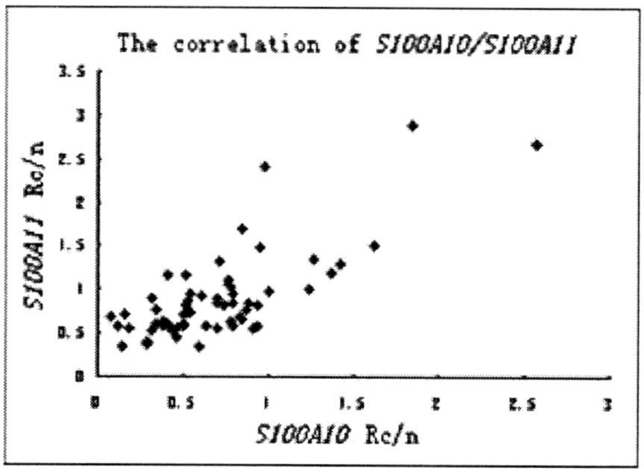

		A1	A2	A3	A4	A6	A7	A8	A9	A10	A11	A12	A13	A14	B	P
A1	r p	1														
A2	r p	NS	1													
A3	r p	NS	NS	1												
A4	r P	NS	NS	NS	1											
A6	r P	NS	NS	NS	NS	1										
A7	r P	NS	NS	NS	NS	NS	1									
A8	r p	NS	0.6829 <0.0001	NS	NS	NS	NS	1								
A9	r p	NS	0.3587 0.0042	NS	NS	NS	0.4639 0.0001	0.5394 <0.0001	1							
A10	r p	NS	0.2754 0.0303	NS	NS	NS	0.5900 <0.0001	0.4204 0.0007	0.5982 <0.0001	1						
A11	r p	NS	NS	NS	NS	NS	0.4841 <0.0001	0.4919 <0.0001	0.4645 0.0001	0.7559 <0.0001	1					
A12	r p	NS	NS	NS	NS	NS	0.2771 0.0292	0.3886 0.0018	0.3473 0.0057	0.3984 0.0013	0.4616 0.0002	1				
A13	r p	NS	NS	NS	NS	NS	NS	NS	0.3178 0.0118	0.3019 0.0171	0.2885 0.0229	NS	1			
A14	r p	NS	0.6818 <0.0001	NS	NS	NS	NS	0.6744 <0.0001	0.5427 <0.0001	0.4545 0.0002	0.3452 0.006	0.3136 0.0131	NS	1		
B	r P	0.2576 0.0433	NS	NS	NS	NS	NS	NS	NS	NS	NS	NS	0.3225 0.0106	NS	1	
P	r P	NS	0.6549 <0.0001	NS	NS	NS	NS	0.6422 <0.0001	0.3727 0.0028	0.4249 0.0006	0.3804 0.0023	0.2972 0.0190	NS	0.5801 <0.0001	NS	1
Z	r p	NS	NS	NS	NS	NS	NS	NS	NS	NS	NS	NS	0.3209 0.0110	NS	0.3389 0.0070	NS

r: correlation coefficient; p: p-value; NS: not significant; The figure of the top right corner was the scanning spot of the correlation of *S100A10* and *S100A11* (r=0.7559, p<0.0001). In this figure each spot expressed the $R_{C/N}$ of *S100A10* and *S100A11* genes in each case of ESCC.

S100A8 and S100A9 Expression in Normal Esophageal Epithelium, Premalignant Lesions and Carcinomas

The finding of abnormal S100A8 and S100A9 expression in a majority of primary ESCCs raised the question as to the timing of the alteration in their expression during the development of esophageal cancer. To address this issue, we examined S100A8 and S100A9 expression by immunohistochemistry in a series of clinical (pre)malignant specimens. Significant differences were found between early stages of malignancics and matched benign tissues (S100A8: $P \leq 0.0001$, S100A9: $P \leq 0.0001$), between advanced cancers and adjacent benign tissues (S100A8: $P = 0.0027$; S100A9: $P \leq 0.0001$) (Table 3-4, 3-5), respectively. All normal esophageal epithelium showed strong immunostaining for S100A8 and S100A9 from the deepest suprabasal cells to the apical, more differentiated, epithelial cells, but no staining was observed in columnar epithelial cells in the stratum basale (Figure 3-4A, a and e). Prominent S100A8 and S100A9 expression was also observed in the monocytes and granulocytes. However, there was a wide variation in staining intensity for S100A8 and S100A9 among dysplasia and early stages of ESCCs examined. In the mild and moderate dysplasia cases, the staining for S100A8 or S100A9 was observed in the suprabasal layers above the one third or half of the epithelium. With increasing degree of lesions, both S100A8 and S100A9 immunoreactivity shifted to more superficial layers, where differentiated cells were still present (Figure 3-4A, b and f). All the 22 CIS cases showed focal and weak positive immunostaining in certain well differentiated areas, and undetectable in other less differentiated sections within the altered epithelium (Figure 3-4A, c and g). For the 19 intraepithelial and submucosal carcinomas, S100A8 and S100A9 immunostaining was markedly decreased compared to adjacent CIS regions (Figure 3-4A, d and h). A significant difference in the expression levels for S100A8 and S100A9 was found among the dysplasia, CIS and early invasive carcinomas (S100A8: P = 0.0131; S100A9: P = 0.000171) (Table 3-4).

Table 3-4. Expression of S100A8 and S100A9 in early stages of esophageal malignancies and normal epithelium

		S100A8			S100A9		
Group	***N***	**Median**	**Mean±SD**	**P-value**	**Median**	**Mean±SD**	**P-value**
Normal*	71	0.750	0.824±0.518	≤ 0.0001	1.600	0.824±0.414	≤ 0.0001
Early stages	71	0.250	0.319±0.470		1.000	0.938±0.735	
Dysplasia	34	0.250	0.346±0.354	0.0131	1.000	1.075±0.576	0.000171
CIS	18	0.250	0.431±0.599		1.000	1.333±0.838	
Early ISCC	19	0.000	0.184±0.513		0.000	0.389±0.620	

*Mann-Whitney rank sum test between early stages of malignancies and normal controls.

†Kruskal-Wallis one-way analysis analysis followed by Dunn's multiple comparison test among dysplasia, CIS and early invasive SCC.

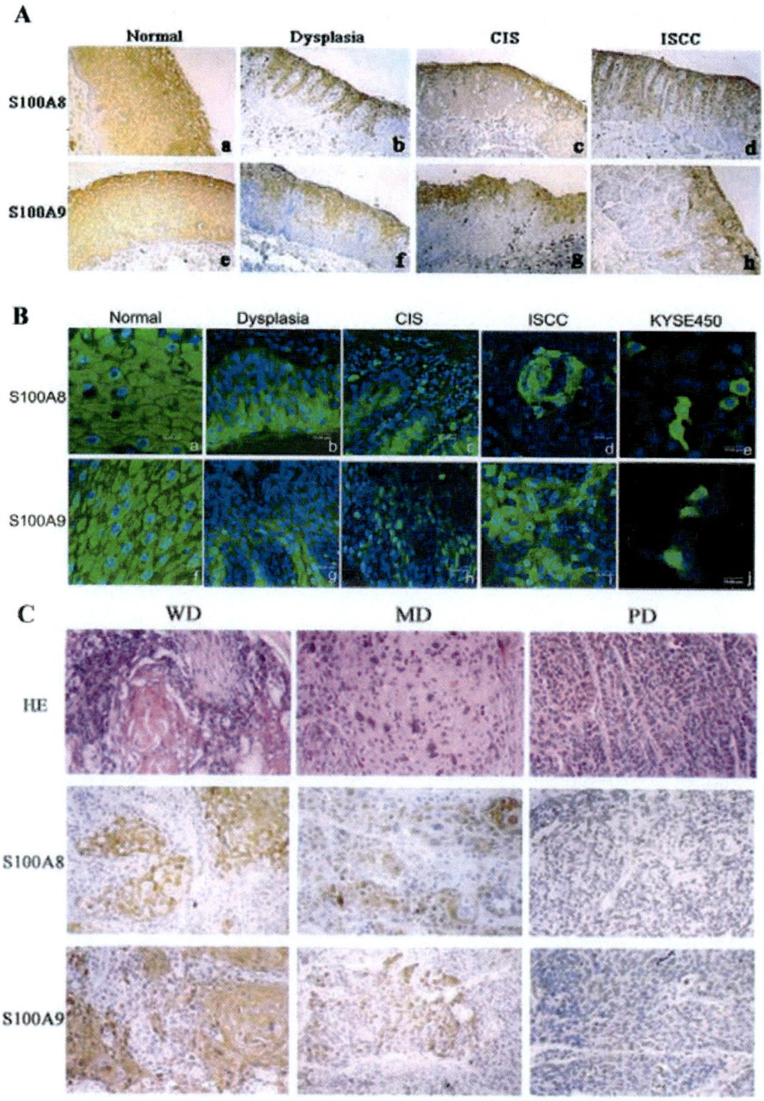

Figure 3-4. Immunohistochemical staining and localization of S100A8 and S100A9 in esophageal cancer tissues, premalignant lesions and matched normal specimen. A: Immunohistochemical expression of S100A8 and S100A9 in normal esophageal epithelium (a,e), dysplasia (b,f), CIS (c,g) and early ISCC (d,h); **B:** Localization of S100A8 and S100A9 immunoreactivity in normal esophageal epithelium, various stages of esophageal malignancies and cancer cell lines by immunofluorescence. Normal esophageal epithelium showed intense immunostaining for S100A8 (a) and S100A9 (f) mainly in the cytoplasm and plasma membrane of the superficial layers. The majority of positive cells in dysplasia, CIS and ISCC showed combined nuclear and cytoplasmic staining for S100A8 (b-d) or S100A9 (g-i), while cytoplasmic S100A8 or S100A9 expression without nuclear expression was less frequently seen. For the cell line KYSE450, most of cells were apparently devoid immunostaining, but only minimal amount of reaction product in the nuclei and cytoplasm could be seen in some sporadic positive cells, as shown in e and j. The two-color merged images showed S100A8 or S100A9 in green with a nuclear counter stain in blue. **C:** Well differentiated carcinoma showed positive S100A8 and S100A9 expression at the center of tumor foci. In moderately differentiated carcinoma staining for S100A8 and S100A9 showed weak and sporadic distribution; whereas poorly differentiated carcinoma showed absent staining for S100A8 and S100A9 in the majority of tumor areas. Consecutive sections for immunostaining respectively showed S100A8 differential expression in the majority of esophageal areas consistent with that of S100A9. Granulocytes and monocytes stain positive in all the sections.

Intracellular Localization of S100A8 and S100A9 in Esophageal Tissues and Cancer Cell Lines

By immunofluoresence analysis, we further examined the subcellular localization of S100A8 and S100A9 in a subset of clinical specimens and cancer cell lines. As shown in Figure 3-4B, in normal epithelial cells the staining for S100A8 and S100A9 was present mainly in the cytoplasm and plasma membrane. However, majority of positive cells in ESCC and its precursor lesions showed combined nuclear and cytoplasmic staining for S100A8 or S100A9, while cytoplasmic S100A8 or S100A9 expression without nuclear expression was less frequently observed. For the cell line KYSE450, only minimal amount of reaction product in the nuclei and cytoplasm could be seen in some sporadic positive cells. These findings might indicate that both nuclei and cytoplasm of malignant cells contain S100A8 and S100A9 protein.

S100A8 and S100A9 Expression Correlates with the Cell Differentiation

Overall, 81 primary advanced tumors showed significantly less reactivity of S100A8 and S100A9 than adjacent normal tissues. As shown in Figure 3-4C, in well differentiated carcinoma, the staining for S100A8 and S100A9 was positive in keratinized areas at the center of tumor foci, but weak or undetectable in the marginal areas. However, moderately and poorly differentiated carcinomas showed rarely weak and sporadic positive cells belonging to the well or moderately differentiated group, whereas other areas were completely negative. The immunostaining for S100A8 and S100A9 was frequently heterogeneous even within one specimen, both in terms of percentage of positive cells and staining intensity. A significant difference in S100A8 and S100A9 expression was found among well, moderately and poorly differentiated carcinomas (S100A8: $P = 0.0001$; S100A9: $P \leq 0.0001$) (Table 3-5). Although both S100A8 and S100A9 were down-regulated in various stages of carcionogenesis, there was no significant difference about S100A8 and S100A9 expression between early stages and late stages of malignancies ($r = 0.207$, $P > 0.05$).

Table 3-5. Expression of S100A8 and S100A9 in relation to cellular differentiation

Group	*N*	S100A8			S100A9		
		Median	**Mean±SD**	**P-value**	**Median**	**Mean±SD**	**P-value**
Normal*	81	0.900	1.078±0494	0.0027	1.800	1.740±0.407	≤ 0.0001
Advanced SCC	81	0.750	0.778±0.616		0.925	1.000±0.771	
WD†	37	1.000	1.047±0.549	<0.0001	1.600	1.505±0.597	≤ 0.0001
MD	25	0.500	0.656±0.585		0.050	0.880±0.683	
PD	19	0.100	0.416±0.562		0.050	0.283±0.470	

*Mann-Whitney rank sum test between advanced carcinomas and normal controls.

†Kruskal-Wallis one-way analysis followed by Dunn's multiple comparison test among well, moderately and poorly differentiated carcinomas.

Discussion

S100 gene family is a family encoding calcium-binding proteins, which contains one canonical EF-hand in the C-terminal half and a specific motif in the N-terminal half. Early studies have shown that some S100 members are associated with ESCC such as S100A2, S100A4, S100A12 and S100P proteins. In this work, the differential expression of 16 S100 genes was examined in ESCC at mRNA level by RT-PCR. The results showed that 11 genes were significantly down-regulated ($p<0.05$) and only S100A7 gene was markedly up-regulated ($p<0.05$) in ESCC. These indicate that S100 gene family is closely correlated with ESCC and some deregulated S100 genes may be important ESCC-associated genes.

S100A2 protein was proposed as a valuable prognostic marker in tumors including ESCC [Kyriazanos et al. 2002]. Due to the under-expression in some tumors and the stable expression in normal epithelia of *S100A2* gene, it was considered as a candidate tumor suppressor gene [Nagy et al. 2001; Liu et al. 2000; Hitomi et al. 1996]. S100A2 interacts with p53 and modulates its transcriptional activity [Mueller et al., 2005]. Our results showed that *S100A2* gene was significantly down-regulated (54.8%, $p<0.0001$) in ESCC as compared with normal mucosa so that it also might be a candidate tumor suppressor gene in ESCC. It has been shown that S100A4 protein was up-regulated in ESCC versus the normal counterparts, which was contrary with our result that *S100A4* gene was down-regulated [Ninomiya et al. 2001]. With regard to *S100A4* primer specificity, plentiful cases and proper statistical analysis, our result was convincible. The reason of S100A4 inconsistent expression at protein and gene level has not been reported yet. The inconsistency might indicate the modulation of translation from mRNA to protein was important and complex in the process of tumorigenesis. S100A4 was widely known to be a metastasis-associated protein and could induce angiogenesis through interaction with Annexin II and accelerated plasmin formation [Semov et al., 2005], the C-terminal region of S100A4 is important for its metastasis-inducing properties [Zhang et al., 2005]. Tumor suppressor p53 protein is a target for S100A4 protein [Grigorian et al., 2001]. S100P and S100A12 have been found to be down-regulated in ESCC at protein level and produced by esophageal epithelial cells in the process of cell differentiation. They were considered to be associated with the differentiation of ESCC [Hitomi et al. 1998; Sato et al. 2002]. S100P protein positive cells were located mainly in the second to fourth layers just above the basal cells [Sato et al. 2002], and S100A12 protein was located in the superbasal squamous epithelial cells undergoing differentiation, but not in the cells in the proliferating basal layer [Hitomi et al. 1996]. All these suggest that the deregulation of S100P and S100A12 protein may be an earlier event in the terminal differentiation of human ESCC.

S100B gene was first detected significantly down-regulated (50.0%, $p=0.0006$) in ESCC. Its under-expression may be accordant with its possible tumor-suppressor activity that S100B could protect p53 from thermal denaturation and aggregation and cooperate with p53 to cause cell growth arrest and apoptosis [Baudier et al. 1992; Scotto et al. 1998]. Moreover, *S100B* gene was significantly correlated with the histological differentiation ($p=0.0247$). This just approved the hypothesis of S100B possible tumor-promoting activity that S100B could actually block p53 based on structural and functional analysis of the S100B-p53 interaction [Rustandi et al. 2000; Lin et al. 2001]. According to these authors the decrease of *S100B* gene expression in ESCC would result in the increase of p53 function, and lead to a decreased cell proliferation rate.

S100A1, S100A7, S100A8, S100A9, S100A10, S100A11 and *S100A14* genes were also first detected significantly deregulated in ESCC. The precise mechanism of their deregulated expression in ESCC was still unknown, the deregulation may influence tumorigenesis by affecting on the cell cycle or proliferation, cell differentiation and apoptosis. It has been reported that nuclear S100A11 inhibited cell growth. In normal cells, S100A11 protein was phosphorylated and moved to nuclei from cytoplasm to inhibit DNA synthesis. Whereas in immortalized cells, S100A11 was neither phosphorylated nor imported to the nuclei, but remained in the cytoplasm so that it could not inhibit DNA synthesis [Sakaguchi et al. 2000]. In B16 melanoma cells, S100A4 protein has been shown to sequester p53 to form a complex of S100A4 with p53. The complex could abrogate cell G1-S checkpoint control and result in the increase of the size of the S-phase fraction [Parker et al. 1994]. Moreover, the cells induced by S100A4 protein to enter S-phase could successfully negotiate the G2-M checkpoint control to transit into mitosis [Cajone et al. 1999]. All these suggest that in the initiation and development of ESCC it may be an important mechanism that S100 family influences on the cell cycle.

Our data showed that there was significant correlation in many genes such as *S100A2/S100A8*, *S100A2/S100A14* and *S100A2/S100P* genes etc. The correlation coefficients of some genes were more than 0.5349, which was the correlation coefficient of *S100A8* and *S100A9* genes (Table 3-3). This meant that these genes' interaction might be closer than that of *S100A8* and *S100A9* genes. All these correlation indicate that some S100 genes may be modulated by some common factors or modulate each other in ESCC. Although the precise modes of interaction remained unknown, the correlation of some S100 genes provided an important clue for illustrating S100 members' functions and the roles they played in the mechanisms of ESCC initiation and progression. Considering the close association of the chromosome 1q21 region and S100 gene family with ESCC, and the possible interaction of S100 members, it would be worthwhile to further analyze the S100 genes and proteins in ESCC.

We report here that the S100 calcium-binding protein S100A8 and S100A9 are down-regulated in human esophageal carcinoma compared to the normal esophageal epithelium. To extend these results and to determine whether the abnormalities of S100A8 and S100A9 occurred during esophageal carcinogenesis, we analyzed their expression and intracellular localization in human esophageal mucosa, dysplasias and ESCC by immunohistochemistry and immunofluoresence.

In normal esophageal epithelium, S100A8 and S100A9 were preferentially expressed and localized in almost all esophageal epithelial cells (EECs) from parabasal to superficial layers. Both S100A8 and S100A9 have been identified to be expressed in a tissue/cell-specific and differentiation-dependent manner, restricted to a specific stage of myeloid differentiation and some mucosal epithelium [Ross KF, 2001]. Warner-Bartnicki *et al* reported that the expression of S100A8 and S100A9 was upregulated during mycophenolic acid and 1α, 25-dihydroxyvitamin D_3-mediated differentiation of HL-60 leukemia cells [Warner-Bartnicki AL, 1993]. Consistent with these findings, our observation of constituent expression in esophageal epithelium also supports that S100A8 and S100A9 are closely associated with the early events in the terminal differentiation of human normal EECs and the epithelial expression of these proteins is thought to require a certain state of proliferation and/or differentiation.

In epithelial carcinogenesis, progressive accumulation of molecular lesions is in parallel with a morphological journal from normality via progressively more severe abnormalities (dysplasia) to full malignancy. Altered expression of S100A8 and S100A9 throughout skin carcinogenesis is suggestive of their potential role in tumorigenesis [Gebhardt C, 2002]. To our knowledge, this is the first study about S100A8 and S100A9 expression in esophageal premalignant lesions. S100A8 and S100A9 expression was reduced or lost before the development of ESCC, which seemed to be an early event of carcinogenesis. EECs may lose the ability to enter terminal differentiation with increasing severity of dysplasia, yet whether the early neoplastic changes in esophageal epithelia lead to reduction of both genes or vice versa is not clear. In addition, frequent loss of S100A8 and S100A9 also occurred in various stages of ESCC as tumors advance in stage, suggesting that loss of S100A8 and S100A9 might be an important event for ESCC. One explanation for the molecular mechanism is that the consistent reduction in S100A8/S100A9 mRNA and protein expression implicates an inactivation at the transcriptional level. Of special interest is the fact that transcriptional regulation orchestrates the expression of another differentiated-associated gene keratin 4 (K4) during esophageal epithelial cell differentiation [Opitz et al., 1998]. Presumably, the transcriptional regulatory mechanisms of S100A8 and S100A9 may be important for the tumor cells to abandon the differentiated states in normal esophageal tissues. In the current study, we found that the expression pattern of S100A8 was in general similar to S100A9 with the exception of the expression intensity, but without significant correlation between them. Although S100A8 and S100A9 share extensive sequence similarity and appear to have very similar biological effects, the individual family members display distinct patterns of expression and regulation [Heizmann et al., 2002]. Therefore, the characterization of the molecular mechanisms about S100A8 and S100A9 deregulation will require further exploration.

Experimental data suggest that there is a significant relationship between S100A8 and S100A9 expression and clinicopathological grade. Poorly differentiated carcinoma tissues showed significantly reduced expression of S100A8 and S100A9 than did well and moderately differentiated carcinomas. Moreover, sporadic positive staining cells were mainly present in those cell lines established from well and moderately differentiated ESCC. It has been established that the defect in the pathway of terminal differentiation is one of the most important abnormalities in esophageal carcinogenesis [Kadowaki et al., 1999]. Thus, we propose that loss of S100A8 and S100A9 may have a role in the dedifferentiation during the development of ESCC. In contrast, S100A9 protein level has a positive correlation with the differentiation degree of some tumors, such as hepatocellular carcinoma, cholangiocellular carcinoma and pulmonary adenocarcnioma [Kazumori et al., 2001; Arai et al., 2000]. These differences may reflect the involvement of different pathways for S100A8 and S100A9 as well as the relative mechanisms in contribution to tumorigenesis of different cell origin.

One of the most interesting findings of the present study is the intracellular localization of S100A8 and S100A9. Squamous epithelial cells constitutively express both proteins in the cytoplasm [Kerkhoff C, 1998]. Consistent with those reports, we observed that the staining for S100A8 and S100A9 was mainly localized in cytoplasm and cell membrane in normal EECs. S100A8 and S100A9 may serve as transglutaminase substrates in normal tissues and become incorporated as cornified envelope components to facilitate calcium- and differentiation-dependent, membrane-associated responses, e.g., innate host defense, gene regulation and cell differentiation [Mischke et al., 1996]. However, in involved ESCC and precursor lesions, we noticed an evidently increased nuclear immunostaining rather than cytoplasma staining. Translocation of S100 proteins may play an important role in assembly of signaling complexes to activate specific transduction pathways. Some pathways such as tubulin-filament dependent and calcium-dependent translocation have been reported in secretion and translocation of S100A8 and S100A9 in myelomonocytes and in epidermal epithelial cells [Broome et al., 2003; Roth et al., 1993]. The translocation of S100A8 and S100A9 observed in esophageal cancer cells may also represent a differentiation-associated, calcium-dependent movement. These findings raise a question of biological function of nucleic S100A8 and S100A9. Therefore, further studies will be necessary to elucidate the mechanism and function of nuclear S100A8 and S100A9 in esophageal tumorigenesis and progression.

Meanwhile, we observed a very similar expression pattern of S100A14, and a close correlation between the cell differentiation and S100A14 expression, S100A14 is highly expressed in normal esophageal epithelium cells as well as the well differentiated ESCC cells, whereas the expression was decreased in moderately differentiated cells, and became undetectable in poorly differentiated cells. Furthermore, calcium could induce the S100A14 accumulation, and evenmore, S100A14 was able to induce apoptosis in ESCC cells (data not shown here), this part of work is still ongoing.

Transcription Factor c-Jun/AP-1 Regulates the Expression of Differentiation-associated Genes via MAPK Pathway

Transcriptional activation of eukaryotic genes depends on the precise and ordered recruitment of activators, chromatin modifiers, and general transcription factors to the promoters of target genes. In the above cDNA array analysis, we identified a group of differentiation-associated genes were coordinately down-regulated in human ESCC including SPRRs (small proline-rich proteins), keratins, cystatin A and involucrin, all of which contain AP-1 binding sites in the promoter regions. However, the mechanisms of down-regulation of these genes were poorly understood. We investigated the endogenous expression pattern of AP-1 in paired ESCC tissue specimens. Coincident with the down-regulation of these genes, c-Jun expression was obviously decreased in cancer tissues, whereas c-Fos was equally expressed between normal and cancerous tissues. The underexpression of c-Jun in esophageal cancer tissues was further confirmed by tissue microarray analysis. Using KYSE450 cells as a useful model system, we determined whether transcription factor c-Jun/AP-1 regulated the expression of these genes. It was shown that the expression of these genes increased in response to TPA-mediated induction of c-Jun/AP-1 expression. Activation of MAPK pathways was followed by acquisition of AP-1 DNA binding and transactivation capacities. Inhibition of JNK and PKC mainly prevented expression and phosphorylation of c-Jun, while inhibition of ERK pathway had a weak effect. Chromatin immunoprecipitation (ChIP) assay indicated that c-Jun and c-Fos were mainly involved in regulation of the differentiation-associated genes *in vivo*. c-Jun activated the transcription of these genes by enhancing the binding of c-Jun to the respective promoters. Taken together, these results suggested that c-Jun/AP-1 activated the expression of differentiation-associated genes via MAPK pathways.

Introduction

Esophageal squamous epithelium can be divided into two zones: basal zone and differentiated zone [Seery et al., 2002]. Cell proliferation is frequent in the epibasal layers, and migration from this area eventually triggers differentiation [Jankowski et al., 1993]. The above cDNA array study indicated that a group of differentiation-associated genes were coordinately down-regulated in human ESCC. Further analysis on the promoter region illustrated that a common feature of the promoter regions of these genes, including SPRRs (small proline-rich proteins), keratins, cystatin A and involucrin, is the presence of AP-1 DNA binding sites [Sark et al., 1998; Gibbs et al., 1990; Fischer et al., 1999; Takahashi et al., 1998; Lin et al., 2002; Welter et al., 1995].

Transcription factor AP-1 is mainly composed of Jun (c-Jun, JunB, and JunD) and Fos (c-Fos, FosB, Fra-1 and Fra-2) [Angel et al., 1987; Lee et al., 1987; Angel et al., 1991]. c-Jun is the critical component of transcription factor AP-1 that composed of homo- or heterodimers of basic region-leucine zipper (bZIP) proteins that belong to the Jun, Fos, ATF and Maf subfamilies [van Dam et al., 2001; Dorsey et al., 1995; Kouzarides et al., 1988; Smeal et al., 1989]. The AP-1 transcription complex is associated with diverse cellular processes such as differentiation, proliferation, transformation and apoptosis. Dimers recognize the 12-o-tetradecanoylophorbol-13-acetate response elements (TRE, 5'-TGAG/CTCA-3') in the gene promoter based on their ability to mediate transcriptional induction in response to phorbol esters (TPA). c-Jun/AP-1 activity is regulated by posttranslational modifications, such as phosphorylation, acetylation and ubiquitination [Smeal et al., 1992; Pulverer et al., 1991; Deng et al., 1994; Karin et al., 1997; Yamaguchi et al., 2005; Nateri et al., 2004; Wertz et al., 2004]. The mechanism of posttranslational control is most extensively documented in the case of mitogen- and cellular stress-induced hyperphosphorylation, in particular, the activation of c-Jun through the c-Jun N-terminal kinase (JNK).

Mitogen-activated protein kinase (MAPK) pathways consist of extracellular signal-regulated kinase (ERK1/2), JNK and p38 [Chang et al., 2001]. JNKs comprise three isoforms: JNK1, JNK2 and JNK3, and are activated by various extracellular stimuli including growth factor, heat shock, phorbol esters and UV light [Davis et al., 2000; Raingeaud et al., 1995; Derijard et al., 1994; Zhou et al., 2000]. c-Jun becomes transcriptionally active upon phosphorylation at Ser63 and Ser73 within its N-terminal transactivation domain by JNK, thereby enhanced its transactivation potential [Pulverer et al., 1991; Papavassiliou et al., 1995]. Upon activation, c-Jun/AP-1 regulated the expression of a large number of AP-1 target genes in both a positive and negative manner and had a role in a variety of cellular processes, including growth, differentiation, apoptosis and survival [Minden et al., 1994; Ameyar et al., 2003].

In this study, we investigated whether the induction of AP-1 expression and activity by TPA might contribute to the expression of differentiation-associated genes in ESCC cell line KYSE450 cells. TPA, a potent activator of PKC, can activate expression of differentiation-associated genes. Activation of MAPK pathways was followed by acquisition of c-Jun/AP-1 DNA binding and transactivation capacities. Based on this finding, we further evaluated how the mitogen-activated protein kinase (MAPK) signaling pathways regulate AP-1 activity. Our strategy was based on the use of TPA and pharmacological reagents to induce or block c-jun expression and activation. The results demonstrate that JNK is essential for the induction of c-Jun expression and for c-Jun phosphorylation in response to TPA. Overall, these data provide important insights into the mechanisms that ultimately determine the function of c-Jun as a regulator of cell differentiation in ESCC cells.

Materials and Methods

Cell Culture and Materials

The human esophageal squamous cell carcinoma cell line KYSE450 was generous gift from Dr. Shimada Y (Kyoto University, Japan). 12-o-tetradecanoylophorbol-13-acetate (TPA) was purchased from Sigma (Sigma-Aldrich, Inc. St. Louis, USA). PKC inhibitor (GF109203X), MEK inhibitor (PD98059) and JNK inhibitor (SP600125) were obtained from Calbiochem (EMD Biosciences, Inc. San Diego, CA, US).

Preparation of Protein Extracts and Immunoblotting

Cell lysates were prepared using the lysis buffer (50mM Tris pH8, 150 mM NaCl, 1% NP-40, 0.1% SDS, 100μg/ml phenylmethylsulfonyl fluoride (PMSF), 2μg/ml each of aprotinin, leupeptin and pepstatin A (Sigma-Aldrich, Inc. St. Louis, USA). Protein concentration was determined using BCA Protein Assay Kit (Pierce Biotechnology, Inc. Rockford, USA). For immunoblotting analysis, equal amounts of proteins were loaded onto SDS-PAGE gels for electrophoresis and then transferred to PVDF membranes (Amersham Biosciences, Inc. Buckinghamshire, UK). The membranes were incubated at 4° C with primary antibodies against JNK (sc-7345), p-JNK (sc-6254), c-Jun (sc-1694), p-c-Jun(sc-822), p-ERK (sc-7383) (Santa Cruz Biotechnology, Santa Cruz, CA), keratin 4 (ab11215) (Abcam Ltd, Cambridge, UK) involucrin (MS-126) (Lab Vision Corporation, Fremont, CA, USA) and β-actin (A5441) (Sigma-Aldrich, Inc. St. Louis, USA) respectively. The blots were probed with the ECL Western blot detection system according to the manufacturer's instructions.

Sample Collecting and Immunohistochemistry

Fresh surgical resection specimens of esophagus including tumor mass and adjacent mucosa of the tumor mass were taken from 110 patients with esophageal carcinomas who had not received chemotherapy or radiotherapy before the operation. All specimens were collected from Anhui Medical University, China. The specimens were fixed in 10% formalin, paraffin embedded, sectioned to 4 μm thickness, and HE stained. The diagnosis of esophageal squamous cancer was made by two pathologists. The expression of c-Jun proteins was analyzed using c-Jun (sc-1694) rabbit polyclonal antibody (1:100) (Santa Cruz Biot Co, USA) and DAB visualization methods according to the manufacturer's instructions (Zhongshan Company, China). Evaluation of the staining was achieved by multiplication of the extent of positivity and intensity [Hao et al., 2001]. The total number of specimens for analysis is 98 out of 110 because of samples stripping.

Plasmid Constructions and Luciferase Assays

pAP-1-luc was a generous gift from Dr Zhang L (Beijing Institute of Radiation Medicine, China). 10^4 cells were seeded in 96-well plate and transfected with promoter constructs using lipofectamine™2000 (Invitrogen, Inc, Carlsbad, CA, USA) according to the manufacturer's recommendation. The activities of both firefly and Renilla luciferases were determined 48 h after transfection with the dual luciferase reporter assay system (Promega, Inc. San Luis Obispo CA, US). The luciferase activities were normalized to the Renilla luciferase activity of the internal control. The data were representative of three independent experiments done in triplicate (mean values±standard error, $*p<0.05$).

Electrophoretic Mobility Shift Assays (EMSA)

Nuclear extracts were obtained from KYSE450 cells using the NE-PER Nuclear and Cytoplasmic Extraction Reagents (Pierce Biotechnology, Inc. Rockford, USA) according to the protocol. After determining protein concentrations using BCA Protein Assay Kit (Pierce Biotechnology, Inc. Rockford, USA), EMSA was performed according to the protocol (Pierce Biotechnology, Inc. Rockford, USA). Briefly, 5μg of equal amounts of nuclear extracts was incubated with 20μl reaction buffer (50ng/μl poly-[dI-dC], 5mM $MgCl_2$, 2.5% glycerol, 0.05% NP-40, 1X binding buffer) at room temperature for 20min. 3' end biotin labeled double-stranded oligonucleotide 5'-CGCTTGATGAGTCAGCCGGAA-3' was synthesized (Takara Biotechnology, Inc. Dalian, China). For specific competition assay, a 100-fold excess of unlabeled oligonucleotide was added to the binding reaction mixture. The samples were electrophoresed on a 6% nondenaturing polyacrylamide gel. The membranes were subjected to chemiluminescent nucleic acid detection module.

Northern blot

Total RNA was extracted with Trizol reagent (Invitrogen, Inc, Carlsbad, CA, USA) according to the manufacturer's recommendation. 15μg of total cellular RNA each lane was separated electrophoretically on a 1% denaturing formaldehyde-agarose gel, transferred to Hybond N^+ nylon membrane (Amersham Biosciences, Inc. Buckinghamshire, UK). The membranes were hybridized with biotin-labeled cDNA probe synthesized using Biotin DecaLabel™DNA Labeling kit (Fermentas, Inc. Hanover, MD, USA) and subjected to chemiluminescent nucleic acid detection module.

Chromatin Immunoprecipitation (ChIP)

This was performed with the ChIP assay kit (Upstate Biotechnology, Inc, Lake Placid, NY) according to the manufacturer's instructions. KYSE450 cells were cross-linked in a 1% formaldehyde solution for 10 min at 37°C. Cells were incubated in SDS lysis buffer and sonicated to generate about 500bp DNA fragments. After centrifugation, the cleared supernatant was diluted 10-fold with ChIP buffer and incubated with 3μg antibodies against c-Jun, c-Fos and Fra-1 respectively at 4°C overnight. Immune complexes were precipitated, washed and eluted as recommended. The complexes were reversed at 65°C for 4 h. The DNA fragments were purified and dissolved. Each sample was analyzed by PCR using the oligonucleotide primers: SPRR1A, 5'-TCTAGCTTCCACCTTCACC-3' and 5'-TCTCGGGCACTCTGTTTTC-3'; SPRR2A, 5'-AGGGAAGGTGTAGACAGG-3' and 5'-CATTTCAGGGCACAAAGAAC-3'; SPRR3, 5'-GCCTGGCTCGTTCTAGTC-3' and 5'-AGGTGTTCCTAATGCGGAG-3'; cystatin A, 5'-GGGGCTTCCTCCATATCTG-3' and 5'-AGGATGAACAAGTGGGCAAG-3'; keratin 13, 5'-AGAGCCTGCACAGTGAACTC-3' and 5'-GCAGATAGAAGCTTGCTTGG-3'; involucrin , 5'-AGCAGCAGAGGACTCCTAG-3' and 5'-GTGATGGAC AGGTTTCACC-3'.

JNK kinase assay

This was performed with the JNK activity assay kit (EMD Biosciences, Inc. San Diego, CA, US) according to the manufacturer's instructions. In brief, KYSE450 cells were treated with 100ng/ml TPA for 2h to 24h and lysed with 200μl of ice-cold JNK extraction buffer on ice for 5 min. 200μg of cellular extracts were incubated with JNK antibody at 4°C.The immunoprecipitates (IP) were then subjected to SDS-PAGE and probed with antibody against p-c-Jun.

Statistical analysis

Statistical analyses were performed using paired-samples *t* test and ANOVA test. *P* value less than 0.05 was considered to be significant.

Results

The Endogenous Expression and Phosphorylation Levels of the Components of AP-1 Proteins

The above cDNA array analysis showed that differentiation-associated genes including SPRRs, keratins, cystatin A are coordinately down-regulated in ESCC. All the promoter sequences of these genes contain AP-1 binding sites, thus we speculate that transcription factor AP-1 may participate in the process of altering expression of differentiation-associated genes in ESCC. Initially, we investigated the expression of endogenous transcription factor AP-1 in ESCC and the matched normal esophageal epithelium tissue specimens. The expression level of c-Jun was significantly down-regulated in ESCC, whereas Fra-1 was obviously over-expressed in ESCC tissue specimens compared with their normal counterparts. However, c-Fos was equally present between normal and cancer tissues (Figure 4-1A). Previous studies showed that Fra-1, a member of the Fos family of transcription factors, was overexpressed in a variety of human tumors, and contributes to tumor invasion, angiogenesis and cell proliferation [Belguise et al., 2005; Kustikova et al., 1998]. Transcription factor AP-1 mediated expression of the keratinocyte terminal differentiation markers such as SPRRs, keratins and cystatin A, and predominantly involved in the regulation of the differentiation program rather than in the proliferation control. We performed tissue microarray analysis of c-Jun in paired esophageal cancer tissue specimens and found that c-Jun was localized in cytoplasm and nucleus of epibasal, intermediate and superficial layers of normal esophageal epithelium, whereas it was predominantly located in the nucleus in the tumor tissue with more diffuse location (Figure 4-1B, Table 4-1). Consistent with the expression of c-Jun in epidermis, the upper layer location and overexpression of c-Jun in normal esophageal epithelium suggest that it may play a critical role in regulating the equilibrium of the process between differentiation and proliferation. In addition, we observed that c-Jun and cystatin A were coordinately expressed in the differentiation zone including the spinous and granular cell layers in normal esophageal epithelium (data not shown). Thus, we speculated the decreased expression of c-Jun might be associated with the down-regulation of differentiation-associated genes in ESCC. Next, a number of esophageal carcinoma cell lines were tested for expression of c-Jun and the differentiation-associated genes. c-Jun was weakly expressed in KYSE180, KYSE410, KYSE450, KYSE510 and Colo680N, whereas it was high in all the other esophageal carcinoma cell lines (Figure 4-1C). Based on the expression of c-Jun and differentiation-associated genes in response to TPA stimulation (data not shown), we selected KYSE450 cell line as a useful model system to study whether c-Jun/AP-1 resulted in regulation of the differentiation-associated genes.

c-Jun/AP-1 regulates Transcription and Expression of the Differentiation-associated Genes

It is known that transcription factor AP-1 translocates to the nucleus following exposure of cells to a variety of extracellular stimuli such as TPA, leading to activation and transcription of its target genes. To determine the effective concentration of TPA on the activation of c-Jun/AP-1, we analyzed the expression of p-ERK, p-JNK and p-c-Jun in KYSE450 cells treated with 10, 50, 100, 200ng/ml TPA for 4h. As it was shown in Figure 4-2, AP-1 activity was induced in a dose-dependent manner and reached maximal levels with 100ng/ml TPA addition.

To further elucidate whether c-Jun/AP-1 as transcriptional activator regulates expression of the differentiation-associated genes, we treated KYSE450 cells with 100ng/ml TPA for different time (2h to 24h) and determined expression of these genes by Northern blot and Western blot. The increase in the transcription of c-Jun was sustained for a long period of time (2h to 12h), peaking between 4h to 12h and returning to basal levels 24h after TPA treatment. Consistent with c-Jun/AP-1, transcription of differentiation-associated genes including SPRR3, keratin 4 and involucrin, was induced rapidly in response to TPA stimulation. The transcription of keratin 4 was transient and returning to basal levels 24h after TPA addition, whereas the transcription of involucrin and SPRR3 was sustained and reached maximal levels 12h to 24h after TPA addition (Figure 4-3A). The kinetics of expression of c-Jun protein was analyzed by Western blot. c-Jun protein expression was induced 2h to 12h after TPA stimulation. In agreement with p-c-Jun, involucrin and keratin 4 proteins expression peaked at 8h (Figure 4-3B). Taken together, c-Jun/AP-1 might be involved in the regulation of differentiation-associated genes in KYSE450 cells.

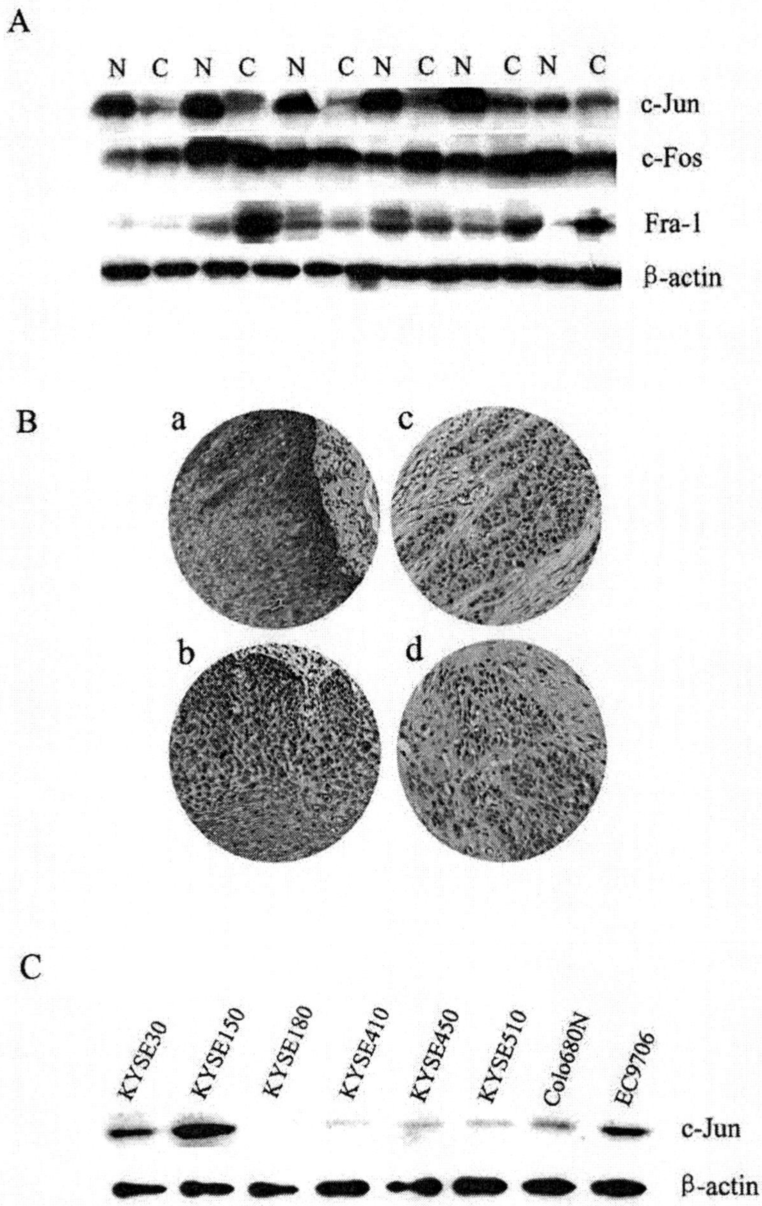

Figure 4-1. A, The expression pattern of transcription factor AP-1 in paired ESCC tissue specimens and various ESCC cell lines. Immunoblots showing the expression of c-Jun, c-Fos and Fra-1 in esophageal carcinoma (ESCC) and the matched normal esophageal epithelium tissue specimens. B, Tissue microarray analysis of the expression of c-Jun in esophageal cancer and normal esophageal epithelium (a,b: normal esophageal epithelium; c,d: ESCC tissues). C, Immunoblots of c-Jun expression in ESCC cell lines including KYSE30, KYSE150, KYSE180, KYSE410, KYSE450, KYSE510, Colo680N, EC9706. As a control, expression of β-actin is shown.

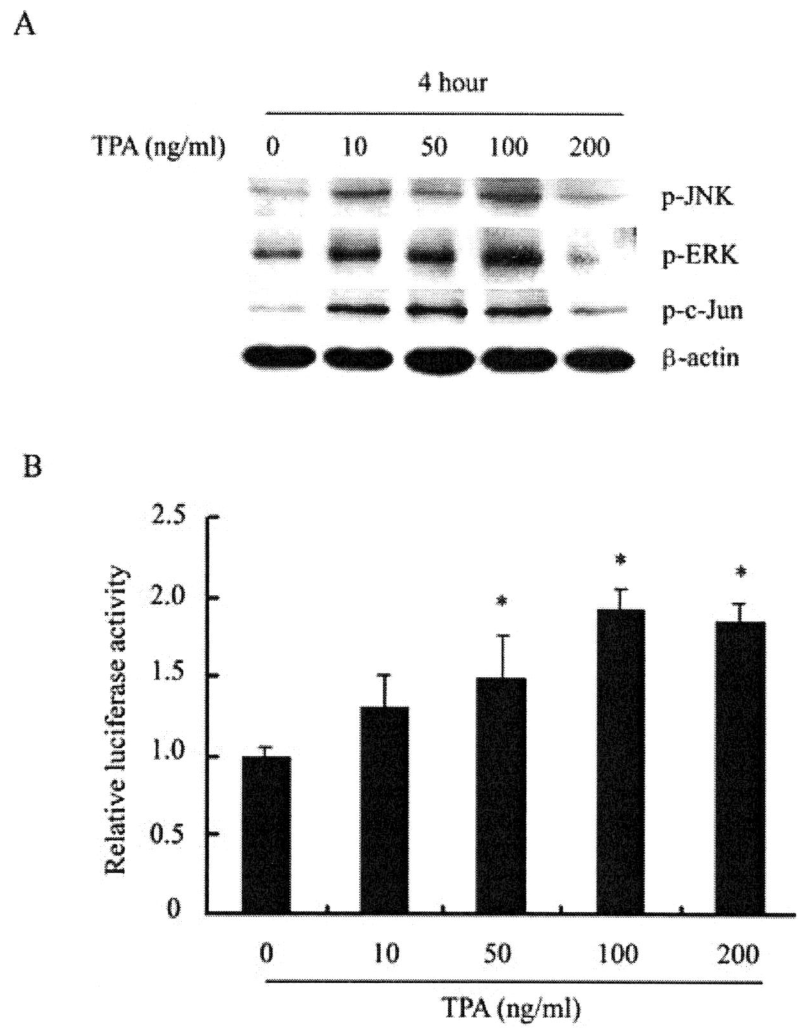

Figure 4-2. TPA induces phosphorylation and activation of c-Jun/AP-1, ERK and JNK. A, Lysates from KYSE450 cells treated with TPA (10, 50, 100, 200ng/ml) for 4h were probed with antibodies against p-JNK, p-ERK and p-c-Jun respectively. B, AP-1 transactivation activity was analyzed by dual luciferase reporter assay system. KYSE450 cells were transiently co-transfected with 200ng pAP-1-luc and 20ng pRL-CMV. After 24h, cells were treated with different concentrations of TPA (10, 50, 100, 200ng/ml) for 4h. CMV-driven Renilla luciferase vector was used as an internal control. AP-1 activity was quantified in terms of the firefly luciferase activity. The data are representative of three independent experiments done in triplicate (mean values±standard error, *p<0.05).

Table 4-1. Statistical analysis of the expression of c-Jun in esophageal cancer and normal esophageal epithelium (paired-samples *t* test)

Group	Number	Mean±SD	*P* value
Normal esophageal epithelium	98	4.844±1.949	<0.0001
Esophageal cancer	98	2.944±1.379	

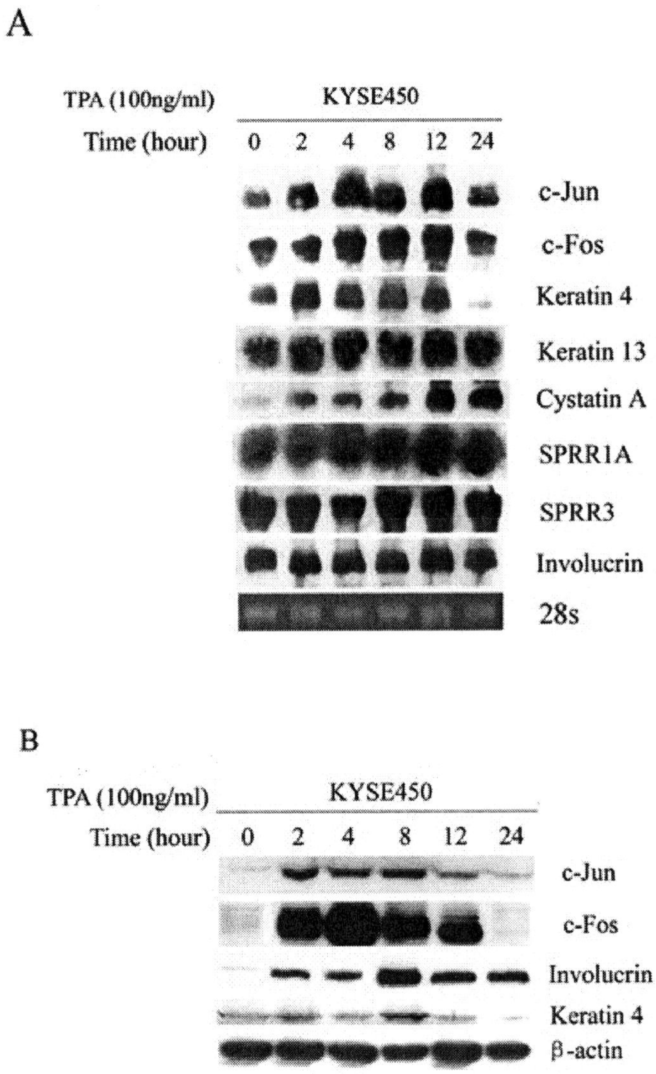

Figure 4-3. c-Jun/AP-1 regulates transcription and expression of differentiation-associated genes. A, Northern blot analyzed the time-dependent induction of c-Jun/AP-1 and differentiation-associated genes in response to TPA in KYSE450 cells. 28s rRNA was used as a loading control. B, Cytoplasmic and nuclear proteins were extracted from KYSE450 cells treated with 100ng/ml TPA at the indicated time and probed with antibodies against c-Jun, c-Fos, involucrin and keratin 4. β-actin served as a loading control.

c-Jun/AP-1 DNA Binding Activity and Transactivation Activity are Mediated by TPA

To investigate whether increased expression of AP-1 is followed by acquisition of AP-1 DNA binding activity, we performed EMSA. AP-1 DNA binding activity was prominently induced after TPA addition. Activation was observed within 2h to 8h of TPA treatment (Figure 4-4A). Next, to test whether AP-1 DNA binding activity had functional consequences, we performed transient transfections with pAP-1-luc into KYSE450 cells to analyze AP-1 transactivation activity. Treatment with TPA caused a significant increase in

AP-1 transactivation (Figure 4-4B). Thus, the results suggested TPA induced AP-1 activity in a time-dependent manner.

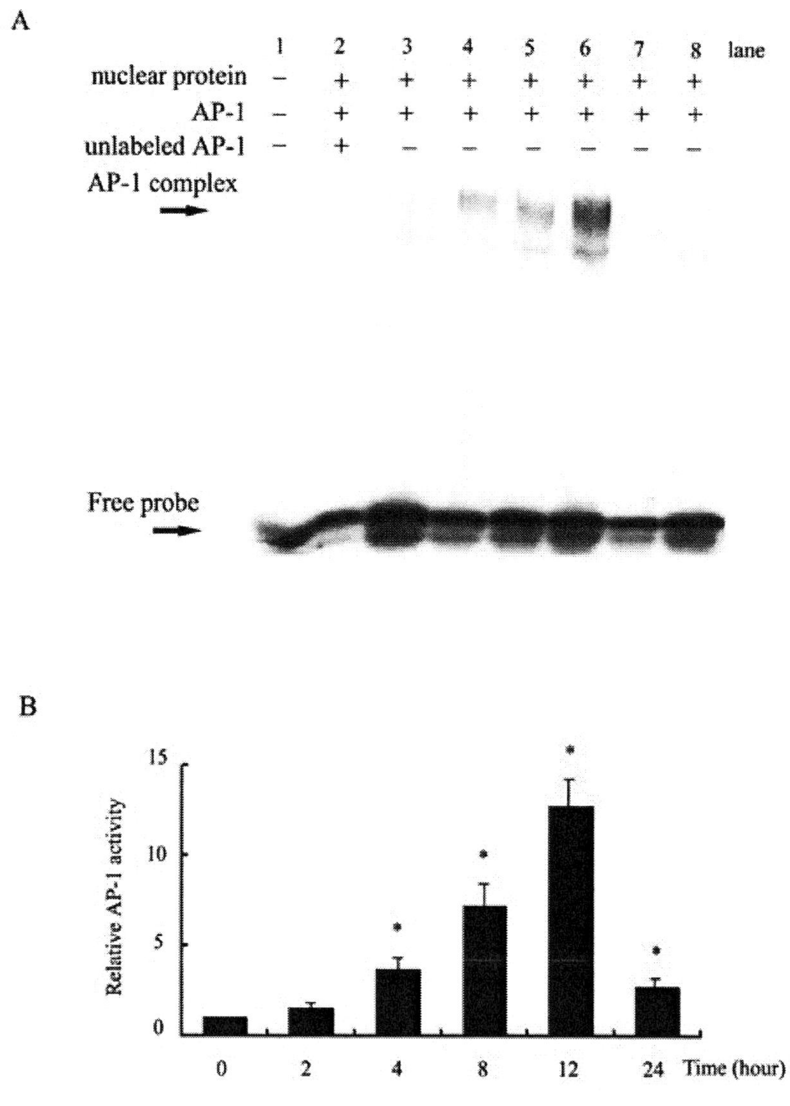

Figure 4-4. c-Jun/AP-1 DNA binding activity and transactivation activity are induced by TPA. A, 5μg nuclear extracts prepared from TPA-treated KYSE450 cells were used for electrophoretic mobility shift assays, as indicated. Lane 1, the negative control; lane 2, nuclear extracts from cells treated with TPA for 8 hours were incubated with a biotin-labeled AP-1 oligonucleotide and a 100-fold excess of unlabeled AP-1 oligonucleotide; lane 3, nuclear extracts from cells treated with vehicle were incubated with a biotin-labeled AP-1 oligonucleotide; lanes 4-8, nuclear extracts from cells treated with TPA for different time (2h, 4h, 8h, 12h, 24h) were incubated with a biotin-labeled AP-1 oligonucleotide. B, AP-1 transactivation activity was analyzed by dual luciferase reporter assay system. KYSE450 cells were transiently co-transfected with 200ng pAP-1-luc and 20ng pRL-CMV. After 24 hours, cells were treated with 100ng/ml TPA for different time (2h to 24h).

TPA-triggered Activation of c-Jun/AP-1 via MAPK Pathways

MAPKs are critical for the transmission of stimuli to the nucleus, resulting in modification of cell growth-related transcription factors such as c-Jun and c-Fos. To gain insight into the relationship between MAPK and c-Jun activation in KYSE450 cells, we investigated the phosphorylation state over a time course following TPA addition. The phosphorylation of endogenous JNK, ERK and c-Jun proteins was coordinately increased and reached maximal levels from 4h to 8h after TPA treatment of KYSE450 cells (Figure 4-1A; Figure 4-4A). Consistently, JNK activity was elevated upon exposure of cells to TPA using protein kinase assay (Figure 4-5B). Although the kinetics and amplitude of JNK and ERK activity were compatible with the phosphorylation of c-Jun upon TPA treatment, to seek the direct evidence of MAPK pathways involved in the TPA-stimulated expression of c-Jun/AP-1, we utilized PKC inhibitor (GF109203X), MEK inhibitor (PD98059) and JNK inhibitor (SP600125) respectively to prevent MAPKs in response to TPA stimulation. SP600125 dose dependently inhibited phosphorylation of c-Jun. The different doses of GF109203X similarly inhibited phosphorylation of c-Jun, whereas partial inhibition of phosphorylation of c-Jun was observed at 10, 20μM, at a concentration of 50μM, PD98059 weakly blocked phosphorylation of c-Jun. The inhibition of keratin 4 was consistent with the change of p-c-Jun (Figure 4-6A). We observed that 20μM PD98059, SP600125 and GF109203X blocked the activity of c-Jun/AP-1 (~ 50 %) (Figure 4-6B).

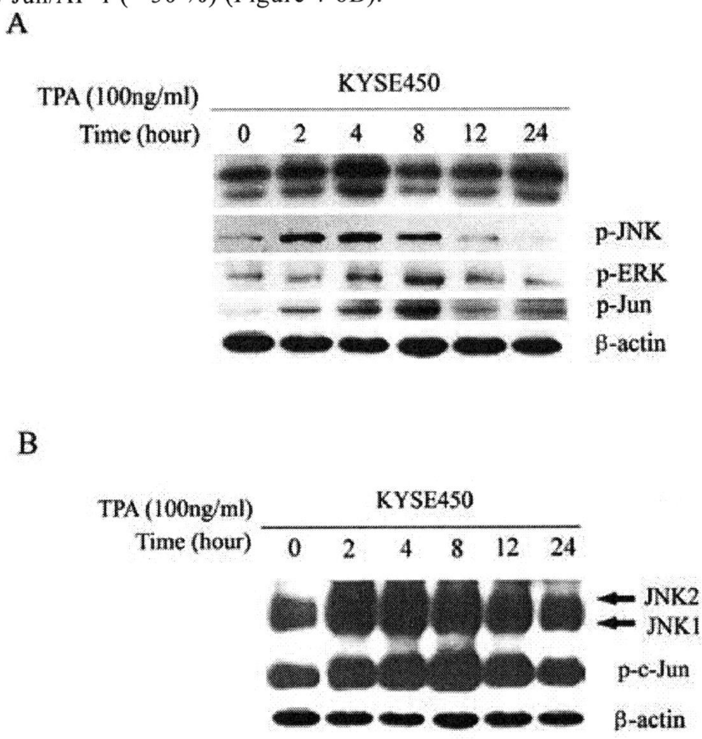

Figure 4-5. TPA-triggered activation of c-Jun/AP-1 via MAPK pathways. A, Lysates from KYSE450 cells treated with 100ng/ml TPA for different time (2h to 24 h) were probed with antibodies against JNK, p-JNK, p-ERK, p-c-Jun and β-actin respectively. B, JNK activity was measured by protein kinase assay. Cellular lystates were immunoprecipitated with antibody against JNK, as indicated. The immunoprecipitates (IP) were then subjected to SDS-PAGE analysis and probed with antibody against p-c-Jun.

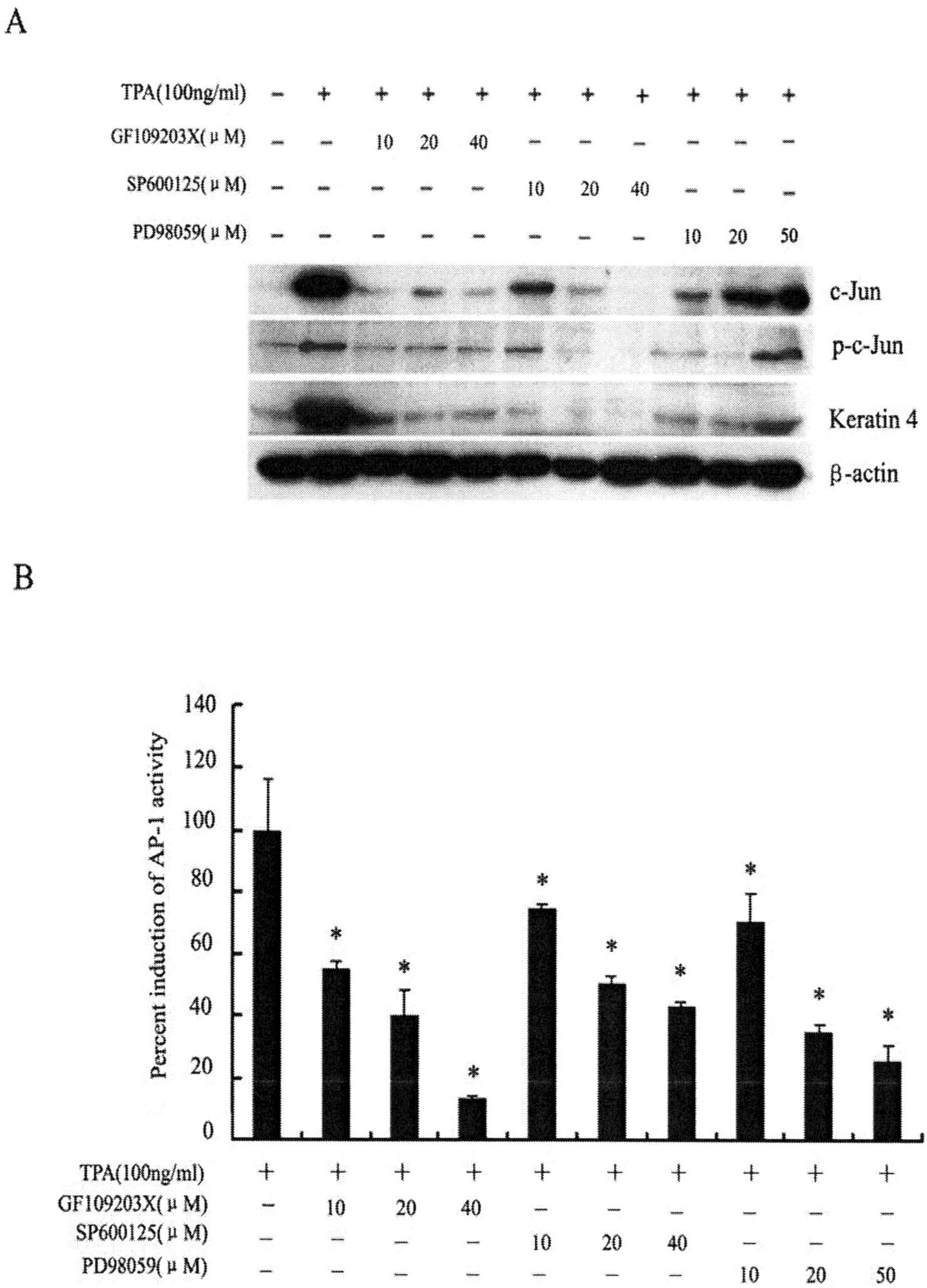

Figure 4-6. MAPK inhibitors suppressed TPA-triggered activation of c-Jun/AP-1. A, KYSE 450 cells were pretreated with PKC inhibitor GF109203X (10, 20, 40μM), JNK inhibitor SP600125 (10, 20, 40μM) and MEK inhibitor PD98059 (10, 20, 50μM) respectively for 1h and then treated with TPA for 4h. Cellular lysate protein was extracted and probed with antibodies against c-Jun, p-c-Jun, keratin 4 and β-actin. B, AP-1-driven luciferase reporter activity was assayed in KYSE450 cells transiently co-transfected with pAP-1-luc and pRL-CMV, After 24h, cells were pretreated with various concentrations of inhibitors as indicated for 1h and then treated with 100ng/ml TPA for 4h.

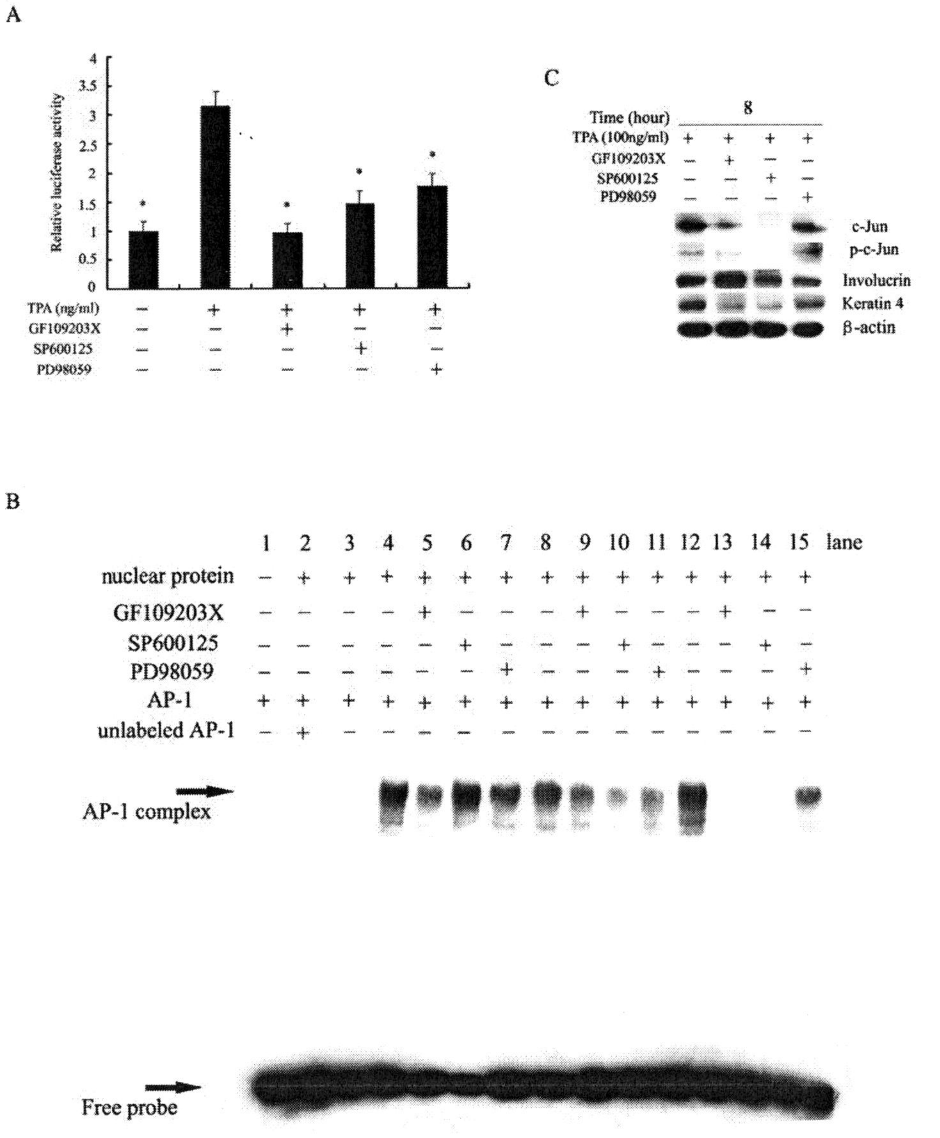

Figure 4-7. MAPK inhibitors block TPA-mediated induction of differentiation-associated genes expression. A, Transient co-transfection was made with pAP-1-luc and pRL-CMV in KYSE450 cells. After 24h, cells were pretreated with GF109203X (20μM), SP600125 (20μM) and PD98059 (20μM) respectively for 1h and then treated with 100ng/ml TPA for 8h. The AP-1 activity was detected using the dual luciferase reporter assay system. B, Nuclear extracts (5μg) were used for electrophoretic mobility shift assays. lane 1, the negative control. lane 2, nuclear extracts from cells treated with TPA for 8h were incubated with a biotin-labeled AP-1 oligonucleotide and a 100-fold excess of unlabeled AP-1 oligonucleotide. lane 3, nuclear extracts from cells treated with vehicle were incubated with a biotin-labeled AP-1 oligonucleotide. lanes 4-7, lanes 8-11, lanes 12-15, nuclear extracts from cells pretreated with PKC inhibitor GF109203X (20μM), JNK inhibitor SP600125 (20μM) and MEK inhibitor PD98059 (20μM) for 1h before treated with 100ng/ml TPA for 2h, 4h and 8h respectively were incubated with a biotin-labeled AP-1 oligonucleotide. C, KYSE450 cells were pretreated with PKC, JNK and MEK inhibitors respectively for 1 hour and then treated with 100ng/ml TPA. Nuclear or cytoplasmic extracts were subjected to immunoblotting for detection of the expression of c-Jun, p-c-Jun and target genes such as involucrin and keratin 4.

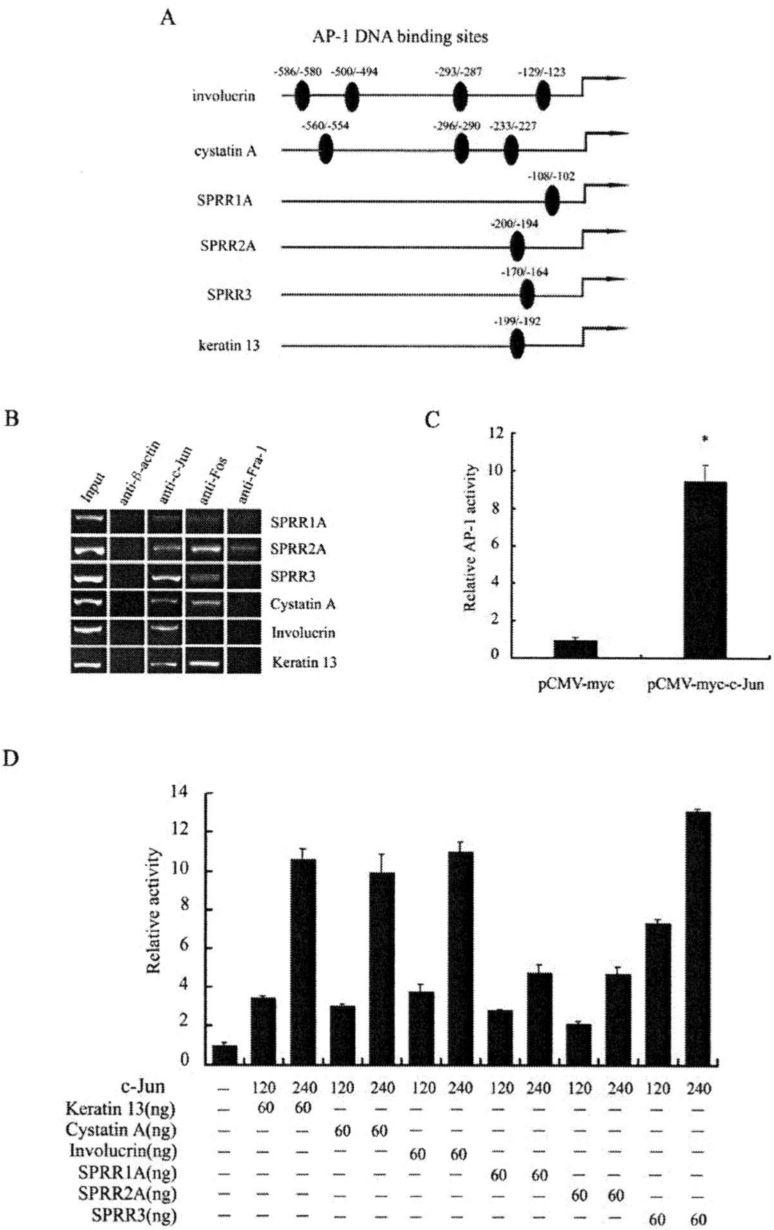

Figure 4-8. c-Jun/AP-1 transactivates the differentiation-associated genes by enhancing the binding of c-Jun to promoters. A, The promoter regions of the SPRR1A, SPRR2A, SPRR3, Keratin 13, involucrin and cystatin A contain AP-1 DNA binding sites, as indicated. B, Chromatin immunoprecipitation (ChIP) was performed to identify the target genes of AP-1. Chromatin fragments from KYSE450 cells were immunoprecipitated with antibodies against c-Jun, c-Fos and Fra-1. The SPRR1A, SPRR2A, SPRR3, Keratin 13, involucrin and cystatin A promoter containing AP-1 DNA binding sites was amplified by PCR. C, AP-1-driven luciferase reporter activity was assayed in KYSE450 cells co-transfected with pAP-1-luc and pCMV-myc-c-Jun or empty expression vector pCMV-myc. D, KYSE450 cells were co-transfected with pCMV-myc-c-Jun and a series of human SPRR1A, SPRR2A, SPRR3, keratin 13, involucrin and cystatin A promoter constructs. The AP-1 target genes transactivation was detected using the dual luciferase reporter assay system.

To further investigate whether PKC and MAPK inhibitors can block expression of differentiation-associated genes, we pretreated KYSE450 cells with 20μM PD98059, SP600125 and GF109203X respectively. The inhibitors resulted in a strong though not complete inhibition of TPA-stimulated induction of c-Jun/AP-1 at 4h after TPA addition (Figure 4-7A). c-Jun/AP-1 DNA binding activity was obviously blocked by GF109203X and SP600125, whereas weakly inhibited by PD98059 at 4h to 8h after TPA addition. Moreover, AP-1 DNA binding activity was weakly inhibited by GF109203X at 2h after TPA addition (Figure 4-7B). Inhibition of JNK and PKC mainly prevented expression and phosphorylation of c-Jun, while inhibition of ERK pathway had a weak effect. Indeed, JNK inhibitor SP600125 suppressed expression of keratin 4 and involucrin. GF109203X can inhibit expression of keratin 4, whereas PD98059 can inhibit the expression of neither keratin 4 nor involucrin (Figure 4-7C).

Taken together, the results strongly suggested that activation of JNK pathway was mainly required for the expression of c-Jun/AP-1 and differentiation-associated genes.

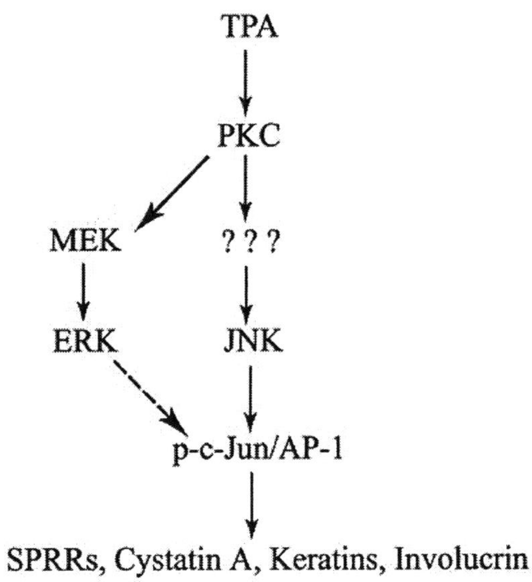

Figure 4-9. Model of c-Jun/AP-1 regulating the expression of differentiation-associated genes via MAPK pathways. TPA activates ERK and JNK, leading to the subsequent increase of c-Jun/AP-1 expression and activity. Whereas only JNK pathway triggers c-Jun phosphorylation, ERK pathway shows a weak effect. Phosphorylation of c-Jun/AP-1 can bind to AP-1 DNA binding sites on the promoter regions of differentiation-associated genes. The basal transcription machinery leads to an increase in the expression of SPRRs, keratins and involucrin in KYSE450 cells.

c-Jun/AP-1 Transactivates the Differentiation-associated Genes by Enhancing the Binding of c-Jun to Promoters of the Genes

Although it appears that TPA-induced c-Jun/AP-1 plays a role in the regulation of differentiation-associated genes via MAPK-dependent pathways, the manner of c-Jun in the regulation of the differentiation-related target genes was elusive. A potential limitation of EMSA is that it can't truly reveal the interaction of transcription factor and its endogenous target DNA sequence. The promoter regions of the SPRR1A, SPRR2A, SPRR3, Keratin 13, involucrin and cystatin A contain AP-1 DNA binding sites, as indicated (Figure 4-8A). To verify that differentiation-associated genes are the target genes of c-Jun/AP-1 in KYSE450 *in vivo*, we performed ChIP assays. In brief, protein-DNA complexes were immunoprecipitated with antibodies against c-Jun, c-Fos and Fra-1, and bound DNA fragments were recovered and subjected to PCR with specific primers for SPRR1A, SPRR2A, SPRR3, cystatin A, keratin 13 and involucrin promoters. c-Jun and c-Fos were almost involved in regulation of all differentiation-associated genes, whereas Fra-1 only regulated expression of SPRR2A (Figure 4-8B). The findings were consistent with previous evidence that c-Jun/AP-1 binding to the TRE is important for the transcription of differentiation-associated genes [Sark et al., 1998; Gibbs et al., 1990; Fischer et al., 1999; Takahashi et al., 1998; Lin et al., 2002].

The differentiation-associated genes were previously characterized as containing one or several AP-1 DNA binding sites in their promoter regions. c-Jun can play a dual role as both a transactivator and a transrepressor of target genes. To investigate the role of c-Jun involved in regulation of these genes, we performed transient co-transfections of KYSE450 cells with c-Jun and various promoter constructs. The results revealed that c-Jun/AP-1 activated all promoter constructs including SPRR1A, SPRR2A, SPRR3, cystatin A, keratin 13 and involucrin (Figure 4-8 C, D). Taken together, our data showed that c-Jun/AP-1 regulated differentiation-associated genes via MAPK pathways in KYSE450 cells (Figure 4-9).

Discussion

In this study, based on a common feature that all the promoter regions of the differentiation-associated genes contain AP-1 DNA binding sites, we speculate that transcription factor AP-1 may emerge in the process of altering expression of the differentiation-associated genes. Initially, we examined the endogenous expression of the components of transcription factor AP-1. The results showed that c-Jun was obviously down-regulated in ESCC. Normally, c-Jun is expressed broadly at low level in many cell types and its expression is transiently amplified by a number of different stimuli, classically involving MAPK pathways [Lohman et al., 1997; Lee et al., 2002]. It raises questions that whether the downregulation of c-Jun is associated with the decrease in the expression of the differentiation-associated genes.

The conventional position of c-Jun/AP-1 in signal transduction cascades has been closer to the end of MAPK pathways such as ERK, JNK, and p38. Among the AP-1 subunits, c-Jun is a main component and thus plays a key role in response to a variety of extracellular stimuli. TPA is known to effectively induce AP-1 transcriptional activity through activating the JNK pathway that phosphorylates and activates c-Jun in various cell lines. We selected KYSE450 cell line as a useful model to elucidate how MAPK cascades control gene expression via the transcriptional regulation of c-Jun. It was indicated that in response to TPA, c-Jun was directly phosphorylated by PKC/JNK signaling. The inhibitors of PKC (GF109203X) and JNK (SP600125) can obviously block the DNA-binding activity and transactivation activity of AP-1 in KYSE450 cell induced by TPA. Nonetheless, inhibition of MEK-ERK activity with PD98059 had little effect. It is well documented that JNK phosphorylates c-Jun more efficiently than ERKs *in vitro* [De′rijard et al., 1994; Minden et al., 1994]. The phosphorylated c-Jun with enhanced transactivation and DNA-binding activity mediated signaling transductions, which control many different, sometimes opposing cellular processes [Belguise et al., 2005]. Taken together, these results suggested that differentiation-associated genes were regulated by TPA-induced c-Jun/AP-1 mainly via a PKC/JNK pathway in esophageal cancer cell line KYSE450.

Activation of Jun N-terminal kinase leads to phosphorylation and thereby activation of transcription factor c-Jun/AP-1 and consequently alters gene expression. There are many nuclear Jun N-terminal kinase substrates including c-Jun, ATF-2, and Elk-1 proteins [Gupta et al., 1995; van Dam et al., 1995]. AP-1 proteins are the final target of signal-transducing kinase cascades and upon phosphorylation become transcriptionally active triggering the activity of AP-1-driven promoters and the expression of their corresponding regulated genes [Cavigelli et al., 1995]. We demonstrated that the transcription level and protein expression of a group of differentiation-associated genes keratin 4, keratin 13, SPRR3 and involucrin increased in response to TPA stimulation. Notably, though phosphorylation of ERKs and JNKs is induced by TPA, only SP600125 pretreatment attenuates TPA-induced keratin 4 protein expression, suggesting an important role of JNK in TPA-induced keratin 4 expression in KYSE450 cells. Keratin 4 is highly expressed in the esophageal squamous suprabasal cells and its expression is associated with the switch from proliferation to differentiation [Welter et al., 1995]. Keratin 4 homozygous null mice have a phenotype largely restricted to the esophagus and cornea where there is impairment of the normal differentiation program [Angel et al., 1987]. Involucrin is a major component of the cornified envelope and is considered to be an appropriate marker for terminal differentiation. Our results suggested that c-Jun/AP-1 target genes were essential for the differentiation program of esophageal squamous epithelium.

Positive, negative or even dual functions of individual subunits influence pivotal cellular processes, such as differentiation and cell-cycle regulation [Takahashi et al., 2001]. To explore whether the individual subunits c-Jun, c-Fos and Fra-1 directly regulate the expression of differentiation-associated genes, we performed ChIP assays. The results showed that c-Jun and c-Fos were involved in regulation of almost all the differentiation-associated genes, whereas Fra-1 only regulated expression of SPRR2A *in vivo*. Luciferase activity assays showed that c-Jun, as a transcriptional activator, activated SPRRs, keratins, cystatin A and involucrin in KYSE450. Previous studies showed that c-Jun/AP-1 positively regulated expression of the keratinocyte terminal differentiation markers SPRR1A and SPRR3. In contrast, expression of SPRR2 was negatively regulated by c-Jun [Sark et al., 1998; Fischer et al., 1999; Lohman et al., 1997]. c-Fos and c-Jun were involved in cystatin A expression, which was positively regulated via RAS/MEKK1/MKK7/JNK pathway [Takahashi et al., 1998; Takahashi et al., 2001]. Fra-1, JunB, and JunD activated human involucrin promoter transcription by binding to proximal and distal AP1 sites to mediate phorbol ester effects on promoter activity [Welter et al., 1995]. The inconsistency could result from the abundance of different AP-1 members within a given cell, as well as the cell lineage, differentiation stage, microenvironment and type of stimulus.

In conclusion, we found that c-Jun/AP-1 orchestrated the modulation of a group of TPA-induced differentiation-associated genes in ESCC through PKC/JNK signaling pathway.

5. Allelic Imbalance of Chromosome 1q in ESCC: A Novel Region of Allelic Loss and Significant Association with Differentiation

The above cDNA array work has shown that genes located in 1q21, i.e. chromosome 1 open reading frame 10 (C1orf10), small proline-rich proteins (SPRRs), calcium-binding S100 proteins (S100A8, S100A9, S100A14 etc) were coordinately down-regulated in ESCC, we are interested in exploring the mechanism which cause the coordinate down-regulation of the genes. The down-regulation of the genes could be caused by deletion, mutation and methylation, since many of the genes were clustered on chromosome 1q21 region, we firstly investigated whether loss of heterozygosity (LOH) on chromosome 1q21 occurred in ESCC, and if LOH occurred, the correlation of chromosome 1q21 LOH with clinical and histopathological parameters.

Like other cancers, ESCC is thought to arise from the stepwise accumulation of genetic changes that result in disruption of cell growth control and terminal differentiation [Hanahan et al., 2000]. In the past several years, loss of heterozygosity in several specific regions of genome has been identified, and the likely locations of tumor suppressors that may play roles in tumorigenesis of ESCC were characterized; these include 9p [Hu et al., 2004; Tokugawa et al., 2002], 9q [Lichun et al., 2004], 17p [Smeds et al., 2002], 13q [Li et al., 2003] and 18q [Karkera et al., 2000]. However, very little is known about the molecular genetic consequences of other chromosomal abnormalities in ESCC, especially chromosome 1q. Although a number of studies showed recently that chromosome 1q LOH is frequent in various tumors (breast cancer, thyroid carcinoma and parathyroid tumor [Gaki et al., 2000; Kitamura et al., 2000; Dwight et al., 2000; Teh et al., 1998] and 1q21 is one of the most common breakpoints on this chromosome [Plaat et al., 1999; Au et al., 2002; Nakagawa et al., 2003], the research of LOH on chromosome 1q in ESCC is rare so far.

Materials and Methods

Patients and Samples

For this study, sixty-one patients with esophageal squamous cell carcinoma diagnosed at Beijing Friendship Hospital Affiliate of Capital University of Medical Sciences and The First Affiliate Hospital of Anhui Medical University between 2002 and 2003 were collected. The characteristics of the patients including age, sex, stage, grade, lymph node metastasis, family history (i.e., one or more first-, second-, or third-degree relatives with cancer of the esophagus or gastric cardia), smoking history and alcohol consumption are summarized in Table 5-1. Esophageal and gastric cardia cancers were combined for family history because historically both were diagnosed based on dysphagia, and also because their coexistence at very high rates in the same geographic region which strongly suggests common environmental and genetic etiologies. Freshly excised tumor samples were frozen immediately in liquid nitrogen until use. None of the patients received medications before surgery. Histopathological examination was performed on representative sections of each tumor by experienced pathologists before DNA extraction, and only samples that were comprised of a majority of tumor cells were included in this study. Peripheral blood samples were collected before surgery followed by addition citrate (final concentration 0.2%) and stored at -20°C until DNA extraction.

DNA Extraction

Lymphocytes were isolated by dilution of 3-5ml peripheral blood with two volumes of sterile distilled water, and centrifugation at room temperature (3000rpm for 15 minutes). Leukocytes was collected and washed twice with phosphate buffered saline. Tumor tissues were pulverized in the present of liquid nitrogen. Pulverized tumors and leukocytes were digested with 50μg/ml proteinase K in 1% sodium dodecyl sulfate for 8-14 hours at 48°C and extracted twice with phenol/chloroform. DNA was precipitated by addition of two volumes of ethanol and 1/3 volume of 3M sodium acetate.

Microsatellite DNA Markers of Chromosome 1q

Seventeen loci spanning the chromosome arm 1q21-25 were detected by using primer pairs for 18 di- and tetranucleotide repeat markers (Table 5-2), which were ordered from Bioasia (Shanghai, China) based on the Genome Database (Internet address: http://www.gdb.org). Laboratory investigators were blinded to clinical information until all of the data had been collected and analyzed.

PCR

The template DNA of both tumor and leukocytes was amplified by PCR using a GeneAmp PCR system 9600 (PE Applied Biosystems, Foster City, CA) with the following conditions: after 95°C for 5 minutes, the reaction mixture were subjected to 35 cycles, each cycle consisted of denaturating at 95°C for 30 seconds, annealing at 55°C to 65°C for 30 seconds, and extension at 72°C for 30 seconds. There was a single final extension at 72°C for 4-7 minutes. Each reaction contained 2.0 μl of tumor DNA or normal control DNA (approximately 50-100 ng), 2.0 μl of 10×PCR buffer [100 mM Tris-HCl (pH8.3), 500 mM KCl], 1.5 mM $MgCl_2$, 0.4 μM of each primer, 2 units of Taq DNA polymerase (10U/μl; Takara, Dalian, China), and 0.2 mM deoxynucleotide triphosphate.

Table 5-1. Summary of clinical and histopathological information of patients

Sex/ number	Average age	Area		Stage				Grade			LNM [a]		Distant metastasis		Family history [b]		Smoking history		Alcohol consumption		Pickled vegetable consumption [b]	
		Beijing	Anhui	T_1	T_2	T_3	T_4	G1	G2	G3	Yes	No	Yes	No	Yes	No	Yes	No	Yes	No	Yes	No
Male/50	59.6±10.5	12	38	0	17	28	5	26	16	8	17	33	3	47	8	41	28	22	32	18	20	22
Female/11	57.8±9.4	3	8	0	3	6	2	4	3	4	4	7	1	10	2	9	2	9	2	9	4	5
Total/61	59.3 ±10.2	15	46	0	20	34	7	30	19	12	21	40	4	57	10	50	30	31	34	27	24	27

[a] LNM, lymph node metastasis.

[b] The family history and pickled vegetable consumption of patients 1 and 10 are unknown, respectively.

Table 5-2. Loss of heterozygosity of each microsatellite marker

Marker	Band	Position(Mb)	Allelic loss/ informative cases (%)
D1S2344	1q21.1	143.3	0/40(0)
D1S3466	1q21.2	147.4	24/40(60)
D1S498	1q21.3	148.5	3/35(8.57)
D1S2343	1q21.3	148.8	4/40(10)
D1S2346	1q21.3	150.4	11/49(22.45)
D1S305	1q21.3	151.5	8/41(19.51)
D1S2777	1q22	152.6	18/56(32.14)
D1S2140/ D1S1595	1q22	152.9	16/54(29.63)
D1S2624	1q23.1	153.8	1/41(2.44)
D1S1653	1q23.1	155.1	5/23(21.74)
D1S2635	1q23.2	156.4	9/41(21.95)
D1S1167	1q23.2	157.2	12/59(20.34)
D1S2878	1q23.3	162.6	16/57(28.07)
D1S2851	1q24.2	167.5	9/41(21.95)
D1S452	1q24.2	167.7	7/39(17.95)
D1S2790	1q25.1	170.3	12/58(20.69)
D1S158	1q25.3	181.4	15/52(28.85)

Analysis of Allelic Imbalance

After PCR, products were mixed with the same volume of loading buffer (98% formamide, 20 mM EDTA, 0.05%bromophenol blue, and 0.05% xylene cyanol), denatured at 95°C for 10 minutes, then chilled immediately on ice. Five or 7μl of the products was separated by electrophoresis through a 7 or 9% denaturing urea- polyacrylamide- formamide gel in 1×Tris-borate EDTA buffer. The results were visualized by silver staining. Informative cases (i.e. cases occurring when the length of two alleles vary and can be separated by polyacrylamide gel electrophoresis) were scored as LOH when a reduction of at least 50% in one allele compared to the normal allele occurred. Microsatellite instability (MSI) was defined as the presence of an extra band (allele) in the tumor DNA but not in the normal DNA. We considered the alleles were homozygously deleted (HD) when it was repeatedly found that visually evident complete or near complete loss of the both bands (alleles) in the tumor sample relative to corresponding normal DNA, provided that the same DNA samples gave normal PCR products with another primer set. All results were confirmed by at least two independent reviewers. Results were verified by independent PCR in cases with positive identification by either of the readers or PCR failure.

Statistic Analysis

Fisher's exact test, Chi-square test and *t* test were used to evaluate the relationship between alteration of given microsatellite marker and clinical/pathological parameters. All *P* values were two-sided and were considered statistically significant if $P<0.05$.

Results

Sixty-one ESCCs from 61 different patients were studied. The average age of the patients at surgery was 59.3 ±10.2 years; 82% were males, and 16.7% with family history. In 34.4% of the patients the ESCCs showed regional lymph node metastasis and in 6.6% showed distant metastasis (Table 5-1).

In the ESCCs from each of the 41 patients (patient # 1 to # 41), possible LOH on 17 loci spanning the chromosome 1q21-1q25 was investigated (Figure 5-1A). However, some of the loci gave very low rate of LOH and had almost no correlation between LOH and the clinical/pathological parameters. In the additional 20 tumors (patient # 42 to # 61), we only investigated 8 loci trending to have relatively frequent LOH or association with clinical/pathological parameters (Figure 5-1B). All of the loci assessed were informative in >60% of the ESCCs. One locus in 1q22 were tested by two makers, D1S2140 and D1S1595, because each of two markers showed low informative of <55% when it was used alone. Informative data from each individual tumor was >63% (Figure 5-1). Chromosome 1q LOH was identified in 46 of 61 tumors (75.4%). The frequency of LOH with different markers varied from 0 to 60% and the average is 22% (Figure 5-1). The most frequent LOH was observed with marker D1S3466 at 1q21.2 with a fractional ratio of LOH of 60%, followed by D1S2777 at 1q22 with a fractional ratio of LOH of 32.14%, D1S2140/D1S1595 (1q22, 29.63%), D1S158 (1q25.3, 28.85%), D1S2878 (1q23.3, 28.07%) (Table 5-2). Of the 46 patients with tumor samples showing chromosome 1q LOH, in 20 tumors LOH was detected in more than 3 different loci (Figure 5-1). Representative results of LOH at different loci are shown in Figure 5-2. Patient # 8 lost approximately half of chromosome 1q21-25, whereas three patients (# 14, 53 and 60) had only 1 locus of LOH (Figure 5-1). In this study, only 8 cases of microsatellite instability (MSI) and 4 cases of homozygous deletion (HD) were detected, indicating neither of them is frequent genetic alteration on chromosome 1q in ESCC (Figure 5-1).

To determine whether correlations between chromosome 1q LOH and carcinogenesis exist in ESCC, the presence or absence of chromosome 1q LOH was correlated with various clinical or pathological parameters reported to be associated with ESCC tumorigenesis. Of the 46 tumors with 1q LOH, 28 tumors showed histopathological grade ≥G2 (60.9%), whereas this occurred significantly less frequent (P=0.006) in tumors without 1q LOH (20%), with only 3 of 15 tumors showing histopathological grade ≥G2. No significant correlation was observed for the presence or absence of 1q LOH with age, gender, lymph node metastasis, distant metastasis, alcohol consumption, smoking history, pickled vegetable consumption or family history. Likewise, the group of patients suffering from multiple allelic loss (4 or more genetic alterations among the 8 loci that were investigated for all patients in this study) was statistically correlated with advanced histopathological grade (i.e., 11 of 12 tumors with multiple allelic loss showed histopathological grade ≥G2 (91.7%), whereas this occurred significantly less frequent (P=0.003) in tumors without multiple allelic loss on 1q (40.8%), with only 20 of 49 tumors showing histopathological grade ≥G2) and without any other clinicopathological parameters.

Marker	Patient Number 1	2	3	4	5	6	7	8	9	10	11	12	13	14	15	16	17	18	19	20	21	22	23	24	25	26	27	28	29	30	31	32	33	34	35	36	37	38	39	40	41
D1S2344	O	O	O	O	O	O	O	O	O	O	O	O	×	O	O	O	O	O	O	O	O	O	O	O	O	O	O	O	O	O	O	O	O	O	O	O	O	O	O	O	O
D1S3466	×	●	×	●	●	●	●	●	O	O	●	●	×	×	O	×	O	×	×	●	×	×	×	●	×	O	O	×	●	●	O	O	●	O	×	×	×	O	×	×	●
D1S498	O	O	O	O	O	O	×	O	O	O	O	O	O	O	O	O	O	O	O	×	O	O	O	×	×	O	O	O	O	×	●	×	●	●	O	O	O	O	O	O	O
D1S2343	●	O	O	O	O	O	O	O	O	●	O	O	O	O	●	MI	O	O	O	O	O	●	O	O	O	O	O	O	O	O	O	O	O	O	O	O	O	O	O	×	O
D1S2346	O	O	×	●	●MI	O	●	●	●	O	×	O	O	O	O	O	×	×	O	●	O	O	O	×	×	×	O	O	O	×	O	●	●	●	O	O	O	O	O	O	HD
D1S305	O	O	HD	O	O	O	O	O	O	●	O	●	MI	●	●	O	O	O	O	O	O	O	●	O	MI	MI	O	O	O	O	●	●	●	O	O	O	O	O	O	O	O
D1S2777	●	O	O	●	●	●	●	●	●	O	●	O	●	O	O	O	O	●	O	O	O	O	O	O	O	O	O	O	O	O	O	●	O	●	O	O	O	O	×	×	●
D1S2140/D1S1595	O	O	×	×	●	×	●	●MI	●MI	O	O	O	●	O	O	O	O	●	O	●	●	O	O	O	O	●	●	O	●	O	●MI	O	●	O	O	O	O	O	O	×	●MI
D1S2624	O	O	O	O	O	O	O	O	O	O	O	O	O	O	●	MI	O	O	O	O	O	O	O	O	O	O	O	O	O	O	O	O	O	O	O	O	O	O	O	O	O
D1S1653	O	×	—	×	●MI	×	O	O	O	×	●	×	×	O	●	O	O	O	×	●	×	O	●	O	×	×	×	O	O	O	O	O	O	×	×	×	O	×	×	—	O
D1S2635	O	●	●	O	O	O	O	●	●	O	●	O	O	O	●	O	O	O	O	O	●	O	O	O	●	●MI	O	O	O	O	O	O	O	O	O	O	O	O	O	O	O
D1S1167	O	O	O	O	O	O	×	●MI	●MI	O	O	O	●	O	●	O	O	O	O	O	●	O	O	O	O	O	O	O	O	O	O	O	●	O	●	O	O	O	O	O	O
D1S2878	●	●	O	O	×	O	O	●	●	O	●	O	●	O	O	O	O	O	O	O	O	O	O	O	●	●	O	O	O	O	O	O	O	O	●	O	O	O	O	×	●MI
D1S2851	O	O	O	●	O	●	O	O	O	O	O	●	O	O	O	O	O	O	O	●	O	O	O	●MI	●MI	●	O	O	O	O	O	O	O	O	●	O	O	O	O	O	●MI
D1S452	O	O	O	O	O	O	●	●	●	O	O	●	●	O	O	O	HD	O	O	O	O	HD	O	O	●MI	MI	O	O	O	O	O	O	●	O	×	O	O	×	O	O	O
D1S2790	×	●MI	O	O	×	O	●	●	●	O	O	O	O	O	O	O	×	O	O	O	●	O	O	O	O	●	●	O	O	O	O	O	O	O	O	O	O	O	O	O	●
D1S158	×	●	×	●	●	●	—	●	O	O	O	O	O	×	×	O	×	O	O	O	O	O	O	O	O	O	O	O	●	●	O	O	●	O	×	O	×	O	×	O	●

Figure 5-1. LOH for chromosome 1q in 61 patients (41 cases in A and 20 cases in B) with ESCC. The results for each patient with each marker are shown. Results at each locus assessed are expressed as LOH (●), retention of heterozygosity (O), noninformative (X), microsatellite instability (MI), homozygous deletion (HD),or data not available (—).

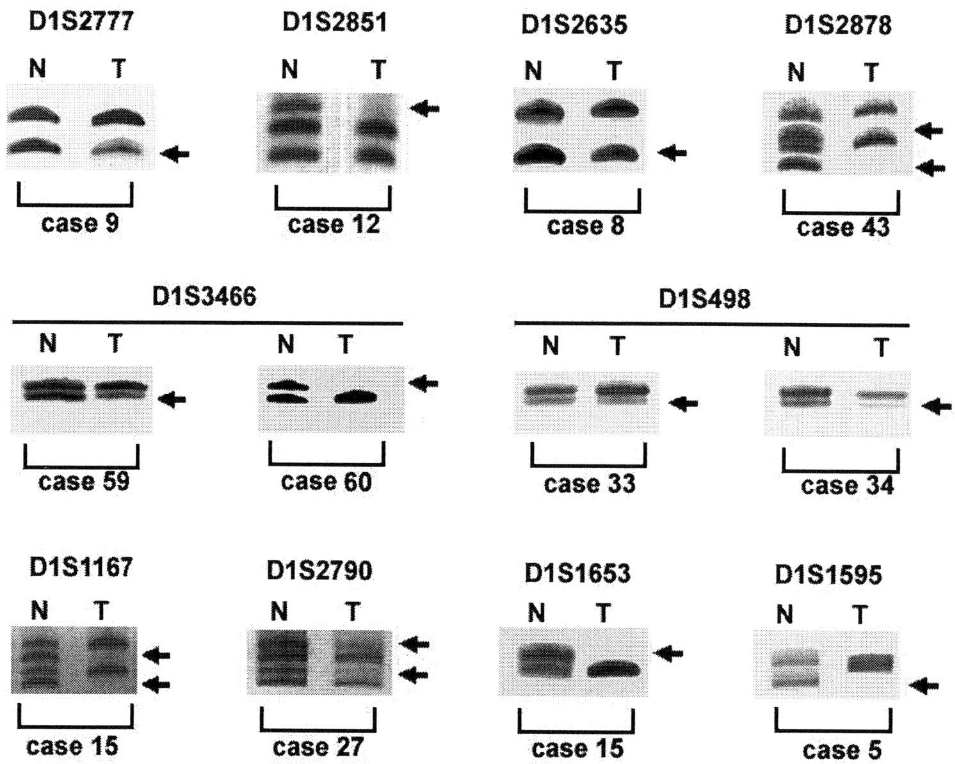

Figure 5-2. Representative results of LOH on chromosome 1q in tumor tissue (T) compared to peripheral blood leukocytes DNA (N). The patient (case) numbers are the same as those in Figure 5-1. The arrows show the LOH.

We have also examined a correlation of LOH at the 17 loci with clinicopathological parameters. A statistically significant association of LOH with histopathological grade was observed at the D1S3466, D1S2346, D1S2777 and D1S2878 locus (P=0.004, P=0.013, P=0.002 and P=0.038, respectively) (Table 5-3). Tumor samples showing poor and moderate differentiation exhibited more frequent allelic loss at these loci than well differentiated tumors. Another statistically significant correlation was found between LOH at D1S3466 and family history. Tumors with family history tend to have reduced frequencies of allelic loss at this locus (P=0.024; Table 5-4). We found a significant correlation of LOH at D1S2777 with smoking, as shown in Table 5-4 (P=0.035). No other significant correlation between LOH and clinicopathological parameters was detected.

Table 5-3. Correlation of LOH at D1S3466, D1S2346, D1S2777 and D1S2878 with histopathological grade

Histopathological grade	D1S3466		D1S2346		D1S2777		D1S2878	
	loss	retention	loss	retention	loss	retention	loss	retention
<G2	5	11	1	20	3	24	4	24
≥G2	19	5	10	17	15	14	12	17
P value	0.004		0.013		0.002		0.038	

Table 5-4. Correlation of LOH at D1S3466 with family history and D1S2777 with smoking

Locus	Parameter	Loss	Retention	*P* value
D1S3466	Family history			
	Yes	1	5	
	No	23	10	0.024
D1S2777	Smoking			
	Yes	13	16	
	No	5	22	0.035

Discussion

By extension, it is generally thought that the occurrence of nonrandom allele loss in cancer may attest to the presence of tumor suppressor genes involved in the development or progression of ESCC. Thus, many chromosome arms have been reported in ESCC as candidate sites for tumor suppressor genes, such as 3p, 5q, 9p, 9q, 13q, 17p, 17q and 18q [Hu et al., 2000; Hu et al., 1999; Ko et al., 2001]. Alterations of common tumor suppressor genes, such as p53 and Rb gene have been reported in ESCC. High frequencies of mutations and allelic loss of p53 gene were detected in ESCC patients [Hu et al., 2001], and there was a strong association between LOH of the Rb gene and alteration of pRb expression in human ESCC samples, suggesting LOH is a main event leading to Rb inactivation [Xing et al., 1999]. However, at present, no genes or other biomarkers have been identified that are sufficiently predictive to be clinically useful in determining the growth pattern of the tumor in an individual patient to allow early institution of antitumor treatments.

Recently, frequent LOH on chromosome 1q were detected in various human cancers and correlated with tumor growth, aggressive tumor behavior, or tumorigenesis [Chen et al., 2003; Borg et al., 1992; Kitamura et al., 2000; Teh et al., 1996]. These findings suggest that this chromosome arm harbors tumor suppressor genes. Studies in ESCC have given contradictory results [Wei et al., 2002; Hu et al., 2000; Yen et al., 2001; Mayama et al., 2000]. Moreover, the results were often limited by few cases or only small numbers of loci in chromosome 1q were examined. This study included a significant number of cases to allow correlations (61 cases), many loci (17 loci) spread evenly throughout chromosome 1q21-25 were investigated.

We have identified chromosome 1q LOH in 46 of 61 tumors (75.4%) in this study, the frequency of LOH with different markers varied from 0 to 60% and the average is 22%. This finding is similar to the average frequency of 24% from another study using similar methodology to ours [Hu et al., 2000]. However, a number of studies using comparative genomic hybridization showed chromosome 1q had no losses in ESCC [Wei et al., 2002; Yen et al., 2001; Mayama et al., 2000]. We speculate that this difference may caused by the presence of short intervals of LOH below the threshold of comparative genomic hybridization detection. In this study, only 8 cases of MSI were detected. This result agrees with the previous findings [Menin et al., 2000; Lam et al., 2002] and indicating that this type of genetic instability is uncommon and MSI probably represent random replication errors rather than being associated with DNA mismatch repair deficiency in ESCC [Hayashi et al., 2003]. MSI occurred relatively frequently on chromosome 1 in other tumors, such as breast cancer (14%) [Gaki et al., 2000] and prostate cancer (20%) [Karan et al., 2001], but no significant correlation with histological features was observed.

A number of studies have reported that the presence of chromosome 1q LOH has prognostic value in different tumors [Gaki et al., 2000; Chen et al., 2003; Borg et al., 1992; Kitamura et al., 2000]. In ESCC, other chromosome losses associating with lymph node metastasis [Harada et al., 2001; Ogasawara et al., 1995], family history [Hu et al., 1999], poor prognosis [Ikeguchi et al., 1999], and female gender [Aoki et al., 1994] have been reported. Our data revealed that ESCC with higher histopathological grade showed more frequent (60.9% vs 20.0 %, P=0.006) and multiple (91.7% vs 40.8%, P=0.003) allelic loss on chromosome 1q than those with lower grade. Furthermore, chromosome 1q LOH had a negative correlation with family history and a positive correlation with smoking history (P=0.024 and P=0.035, respectively). Although studies from others have shown the correlation between chromosome 1q LOH and the distant metastasis in gastrinomas [Chen et al., 2003], we did not find the similar association in ESCC in our study. It may due to the limited amount of cases with distant metastasis in our study (4 of 61 cases).

In our study, the presence of LOH at D1S3466 mapping to 1q21.2, D1S2346 (1q21.3), D1S2777 (1q22) and D1S2878 (1q23.3) all correlated with advanced histopathological grade (P=0.004, P=0.013, P=0.002 and P=0.038, respectively). These results suggest that 1q21-22 as well as 1q23.3 harbor genes that contribute to differentiation. Abnormalities in cellular differentiation are frequent occurrences in human cancers [Huang et al., 2001]. At present it is unclear which tumor suppressor genes in 1q21-22 or 1q23.3 are playing roles in cell differentiation in ESCC. Even if in 1q21.2, a region encompassing the peak LOH locus—D1S3466 and including many genes, such as VPS45A, CKIP-1, ANP32E, CA14, APH-1A and MRPS21, there are no obvious candidate tumor suppressor genes. However, the chromosome band 1q21 has been shown to harbor gene families involved in terminal differentiation of the human epidermis and the gene complex was named as epidermal differentiation complex (EDC) [Marenholz et al., 2001; Marshall et al., 2001]. Our previous study revealed that a number of genes involved in squamous cell differentiation were downregulated in ESCC, including small proline-rich proteins (SPRRs) and calcium-binding S100 protein (S100A8, S100A9) encoded by the EDC [Luo et al., 2004] and exactly overlapped with our defined regions which significantly correlated with differentiation. It suggested that allelic loss might be a partial molecular mechanism for defective expression of SPRRs and S100 proteins and result in abnormal differentiation. Further investigations will be required to verify whether the SPRR and S100 genes are target genes of allelic loss in these regions or other tumor suppressor gene(s) in these regions might be involved.

We also found LOH on 1q21.2 (D1S3466) may associate with family history and LOH on 1q22 (D1S2777) may associate with smoking history. The significant difference in the frequency of chromosome 1q LOH between patients with and without family history indicated that sporadic ESCC may be characterized by different patterns of chromosomal alterations from those present in ESCC with family history. On the other hand, previous studies have shown tobacco is one of the major risk factors for ESCC [Parkin et al., 2000;

Ribeiro et al., 1996]. Our results that LOH on 1q occurs more frequently in tumors of smokers than in those of nonsmokers suggested chromosome 1q LOH may have an association with smoking history and strengthened the notion that esophageal carcinogenesis as well as lung carcinogenesis is associated with smoking history.

This is the first report to identify this chromosome region as harboring important genes affecting tumorigenesis and differentiation for ESCC. At present it is unknown which tumor suppressor genes in 1q21-23 play important roles in tumorigenesis and differentiation in ESCC. Further studies using additional tumor samples will be needed to narrow the region to identify which genes may contribute to the development and differentiation in ESCC, and confirm the finding of our study.

In conclusion, our study demonstrates that chromosome 1q LOH frequently occurs in ESCC. Moreover, the LOH on 1q correlated significantly with histopathological grade, family history and smoking. The results should be useful in at least two ways. First, assessment of 1q LOH might be useful as a molecular prognosticator to identify the subset of patients whose tumors show poor differentiation. Second, the find of the frequent allelic loss in chromosome 1q can serve as a starting point for future studies to identify which genes may be responsible for the carcinogenesis and cancer development of the ESCC [Li et al., 2005].

Recent data indicate that activity of *EDC* genes may be controlled by highly coordinated changes of local chromatin structure, specific spatial distribution of chromatin regions, and temporally and spatially ordered transcriptional regulation [Williams et al., 2002; Elder et al., 2002; Mahy et al., 2002]. These data show that in interphase nuclei of keratinocytes (but not lymphoblasts) the EDC is positioned external to chromosome 1 [Williams et al., 2002]. Interestingly, that suppression of histone deacetylases led to the "open" conformation of the EDC and increased expression of some genes. Silencing of genes in this locus was connected with CpG methylation [Elder et al., 2002]. Such coordinated changes of conformation may be tightly coordinated with transcriptional control [Kimchi et al., 2005]. However, the exact mechanism of the coordinate down-regulation of these genes still needs further investigation.

Functional Study of the Differentially Expressed Genes in ESCC

By cDNA array analysis, we got many of the genes with altered expression in ESCC, we then studied the function of these genes in ESCC development. Following are two of the genes, Stefin A and SLP-2.

Overexpression of Stefin A in human ESCC Cells Inhibits Tumor Cell Growth, Angiogenesis, Invasion and Metastasis

Stefin A, also named as Cystatin A, was the first inhibitor of cysteine proteases reported to be associated with malignant tumors. Stefin A is decreased in many cancer tissues and cells. For example, in breast cancer patients, an inverse correlation of stefin A expression with metastatic potential and relapse-free period was suggested. Lowered stefin A had an impact on total cystatins activity and was associated with cathepsin B activity [Sloane et al., 1990].

An inverse correlation between increased cathepsin B expression and decreased stefin A level has been demonstrated in a variety of human tumors [Sinha et al., 2002; Hirai et al., 1999; Strojnik et al., 2000; Luo et al., 2004].

Evidence is accumulating that an inverse correlation exists between the level of stefin A and malignant progression. The above cDNA microarray results demonstrated a 6-7-fold decrease in stefin A levels and a 3-fold increase in cathepsin B levels in human ESCC tissue. To investigate the role of stefin A in human ESCC cells, we stably transfected sense-oriented stefin A cDNA into EC9706 or KYSE150 esophageal squamous cell carcinoma cells. Our study showed that overexpression of stefin A not only obviously reduced the tumor growth, angiogenesis and invasion but also significantly inhibited the lung metastasis in xenograft mice from EC9706 cells.

The most important characteristic of malignant tumors is their ability to invade surrounding tissues and metastasize. Tumor invasion and metastasis are usually the major causes that lead cancer patients to death. Although many advances have been made in cancer therapy, most cancer deaths still result from metastatic disease. Tumor invasion and metastasis are aggressive processes during which tumor cells invade, migrate and proliferate. Penetration and degradation of components of extracellular matrix (ECM) and basement membrane are key steps in the metastatic cascade of cancer cells. In the metastatic process, a variety of proteolytic enzymes are triggered to be expressed, secreted and activated. Production of proteolytic enzymes of different classes is a property strongly associated with the metastatic state of cancer cell [Mignatti et al., 1993]. These enzymes include matrix metalloproteinases (MMPs), serine proteases (uPA, plasmin, etc)□and lysosomal cysteine proteinase cathepsin B.

Cathepsin B has been implicated in the progression of tumors from a premalignant to a malignant state and in various disease states, including tumor growth, angiogenesis, invasion and metastasis [Coulibaly et al., 1999; Sloane et al., 1990; Szpaderska et al., 2001], rheumatoid arthritis, cholestatic liver injury, and pancreatitis. It is widely held that invasion is facilitated by a membrane or secreted form of cathepsin B that acts outside the cell to degrade ECM components at or adjacent to the surface of the invading cell [Frosch et al., 1999]. Cathepsin B has a broad pH optimum and is able to degrade the components of the ECM and basement membrane either directly or indirectly by activating other proteases like pro-uPA [Kobayashi et al., 1991]. Overexpression of cathepsin B mRNA, increased cathepsin B staining and elevated cathepsin B activity have been found in different human cancers [Yan et al., 1998]. An amplicon at 8p22-23 resulting in cathepsin B gene amplification and overexpression supports an important role for cathepsin B in esophageal adenocarcinoma and possibly in other tumors [Hughes et al., 1998]. Many investigators have demonstrated a correlation between increased activity of cathepsin B and increased metastatic capability of animal tumors or the malignancy of human tumors [Szpaderska et al., 2001; Khan et al, 1998; Bervar et al., 2003; Nagai et al., 2003; Werle et al., 1999]. These increases in cathepsin B activity correspond in part to increased amount of cathepsin B mRNA transcripts and in part to reduced regulation by endogenous low-molecular-weight cysteine proteinase inhibitors [Sloane et al., 1990]. The major means of regulation of mature cysteine proteinases is by its endogenous inhibitors of the cystatin superfamily (stefins, cystatins and kininogens) and thyropins. Intracellularly, the most abundant inhibitors are stefin A (cystatin A) and stefin B (cystatin B), whereas extracellularly, cystatin C is the most widely distributed inhibitor. Quantitatively different combinations of cystatins are the major constituents of the inhibitory potential against cathepsin B in squamous cell lung cancer and normal lung tissue [Krepela et al., 1998]. Cathepsin B and its endogenous inhibitors may facilitate proteolysis by the hepatoma cells and thereby contribute to the invasive phenotype of this type of cancer [Calkins et al., 1998].

Materials and Methods

Cell Culture and Reagent

Human EC9706 esophageal squamous cell carcinoma cell line was established and maintained in our laboratory [Han et al., 2002], two other human esophageal squamous cell carcinoma cell lines KYSE150 and KYSE510 [Shimada et al., 1992] were generous gifts from Dr. Shimada Y, First Department of Surgery, Faculty of Medicine, Kyoto University, Japan. All the three cell lines were grown in RPMI-1640 medium supplemented with 10% FBS, 100μg/ml streptomycin and 100μg/ml penicillin (pH 7.2-7.4) in a humidified incubator containing 5% CO_2 at 37°C. CA074Me (Sigma, St. Louis, MO), a highly selective cathepsin B inhibitor, which is readily cell-permeant, was dissolved in dimethyl sulfoxide (DMSO).

Preparation of Constructs and Transfection

Human stefin A cDNA was subcloned into an eukaryotic expression vector pcDNA3.1(+) (Invitrogen, San Diego, CA) in the sense orientation. EC9706 and KYSE150 cells were transfected with either the stefin A cDNA construct or the pcDNA3.1(+) empty vector alone by using lipofectamine™ 2000 (Invitrogen, San Diego, CA). After G418 (400μg/ml, Life Technologies, Rockville, MD) selection, the stable clones were identified by Western blot for stefin A expression. Two stable clones (A1, A2 from EC9706 cells or S1, S2 from KYSE150 cells, respectively) expressing higher stefin A and one empty vector clone (Vector from EC9706 cells or Vector(150) from KYSE150 cells, respectively) for transfected EC9706 cells or KYSE150 cells were expanded and used for the following studies.

Western Blot Analysis

Western blot analysis was carried out as described previously [Zhu et al., 1999]. In brief, equal amounts of cell lysates (20-100μg) were loaded on 10% or 15% SDS-PAGE and transferred onto PVDF membranes. After blocking, the membranes were incubated with the appropriate diluted primary antibody. Proteins were detected by the ECL detection system. Mouse monoclonal antibodies against stefin A (1:200; Axxora, LLC, San Diego, CA), cathepsin B (1:800; Axxora, LLC, San Diego, CA) and β-actin (1:5000; Sigma, St. Louis, MO) as primary antibodies, horseradish peroxidase–conjugated anti-mouse immunoglobulins (IgGs) as secondary antibodies were used. The signal of target protein was detected by the ECL detection system (Amersham Life Science, Braunschweig, Germany) and recorded on film in the linear detectable range. Protein quantification was performed on the relevant band by Leica QWin image analysis and image processing software.

Growth Kinetics Assay

Cells (10×10^4 cells/well) were seeded in 6-well plates and allowed to grow for different times. The growth rate was determined by cell number counting on a hemacytometer in triplicate on every day of culture up to the seventh day.

Cathepsin B Activity Assay

Cathepsin B activity assay was performed using MR-Cathepsin B detection kit (B-Bridge International Inc., Sunnyvale, CA) according to the manufacture's instructions with some modifications. Briefly, cells were seeded onto sterile coverslip in RPMI-1640 media supplemented with 10% FBS. After reaching about 60% confluence, cells were washed twice with PBS, 520 μl of media containing 20 μl of 26 × MR-cathepsin solution was then added. After incubation for 1 h at 37°C in a CO_2 incubator, cells were rinsed twice with PBS, and MR-Cathepsin B-stained cells were observed for cathepsin B activity using a fluorescence microscope. For observation of cathepsin B inhibitor, cells were treated with 10 μM CA074Me for 12 h before cathepsin B activity assay was performed. The cell fluorescence intensity was assessed by Leica QWin image analysis and image processing software. The mean cell fluorescence intensity for cathepsin B activity in untreated EC9706 or KYSE150 cells was considered as 100%.

Motility and Invasion Assay

AP48 invasion chamber (Neuro Probe Inc., Gaithersburg, MD) was used for in vitro motility and invasion assays according to the manufacture's instructions. For treatment with cathepsin B inhibitor, cells were treated with 10 μM CA074Me for 12 h. Cells (6×10^4) in serum-free media were plated in the upper chamber and incubated in a humidified CO_2 incubator at 37°C. 10% FBS was added to RPMI-1640 media in the lower chamber as a chemoattractant. For invasion assay, the upper side of the filter was covered with 300 μg/ml of Matrigel (BD Biosciences, Bedford, MA). After 8 h for motility assay or 24 h for invasion assay, cells on the upper side of the filter were mechanically removed. Cells migrated to the lower side were fixed with 100% methanol and stained with 0.25% crystal violet. The number of cells on the lower side was counted under microscope. The counting was performed for five different fields in each sample. Experiments were carried out in triplicate, and the results were shown as mean ± SD of three independent experiments. The mean cell counting in untreated EC9706 or KYSE150 cells was considered as 100%.

Experiments in Nude Mice

Single cell suspensions of each of the transfectants and parental cells were trypsinized and collected. The cell viability was > 95% as determined by trypan blue staining. Cells (2×10^6) in a 0.1-ml volume of PBS were inoculated s.c. into the right flank of 4-5-week-old female BALB/c nude mice (eight for each group). Once palpable tumors were established, tumor volume measurements were taken once a week using calipers along two major axes. Tumor volume was calculated as follows: $V = (4/3)\pi R1^2 R2$, where R1 is radius 1 and R2 is radius 2 and R1 < R2. At the end of 3 months, all mice were sacrificed and the tumor volume and weight were measured. All the tumors were fixed in 4% polyformaldehyde and cut into 6 μm sections. For observation of metastasis, all the lungs, livers and brains were excised, fixed in 4% polyformaldehyde and cut into consecutive 6 μm sections. All these sections above were stained with haematoxylin and eosin (H and E), observed under a microscope. The presence of lung metastasis was determined and confirmed by the pathologists in cancer hospital, Chinese Academy of Medical Sciences. The number of lung metastases for each group was counted.

Factor VIII Staining and Microvessel Quantitation

Tumor tissue sections were analyzed by immunohistochemical staining using PV-9000 detection kit (Golden Bridge International Inc., Suite G Lynnwood, WA) according to manufacture's instructions with some modifications. Briefly, Sections were deparaffinized, quenched with 3% hydrogen peroxide in deionized water for 20 minutes, microwaved 10 minutes in 0.01 M sodium citric acid buffer (pH 6.0), and blocked for 30 minutes with PBS containing 3% goat serum, 0.1% Triton X-100 and 0.05% Tween20. Slides were incubated with rabbit antihuman factor VIII IgGs (1:80; Chemicon International Inc., Temecula, CA) overnight at 4°C. The Poly HRP-anti-Mouse/Rabbit IgGs was used as the secondary antibody. Sections were exposed to diaminobenzidine peroxidase substrate (Sigma, St. Louis, MO) for 2-3 minutes and counterstained with hematoxylin. Corresponding tissue sections without primary antibody served as negative controls. Microvessel density (MVD) was quantified by examining areas of vascular hotspots as previously described by Weidner et al [Weidner et al., 1991] with some modifications. Sections were scanned at low magnification (40× and 100×) for the localiztion of vascular hotspots. The three most vascular areas of the tumor, not containing necrosis, were determined and then counted in the high power field (HPF) (200×). The values of the three sections were averaged and the results were analyzed. Branching structures were counted as a single vessel as previously shown [Mukherjee et al., 1999].

Statistical Analysis

Results were showed as mean ± SD. Student's t test was used for comparison unless particular test was notified. $P < 0.05$ was considered statistically significant.

Results

High Invasive EC9706 Cells Show Low Stefin A Protein Level

We grew EC9706, KYSE150 and KYSE510 cells to 90-95% confluence and assessed the stefin A protein levels by Western blot (Figure 6I-1A and 6I-1B) as well as the invasiveness using invasion assay (Figure 6I-1C and 6I-1D). In the three cancer cell lines, the most invasive EC9706 cells showed the lowest stefin A protein expression; KYSE510 cells with the highest protein level were found to be least invasive; KYSE150 cells were less invasive with less protein. Densitometric quantification verified that the amount of stefin A protein was significantly 4-5 times higher in KYSE150 cells, and 17-18 times higher in KYSE510 cells, than that in EC9706 cells (Figure 6I-1B; $P < 0.01$). In Figure 6I-1A, cathepsin B was detected in two forms (about 31kDa and 26kDa), which was consistent with the previous study [Konduri et al., 2002]. To investigate the role of stefin A in human esophageal squamous cell carcinoma cells, we chose EC9706 cells for transfection and further study.

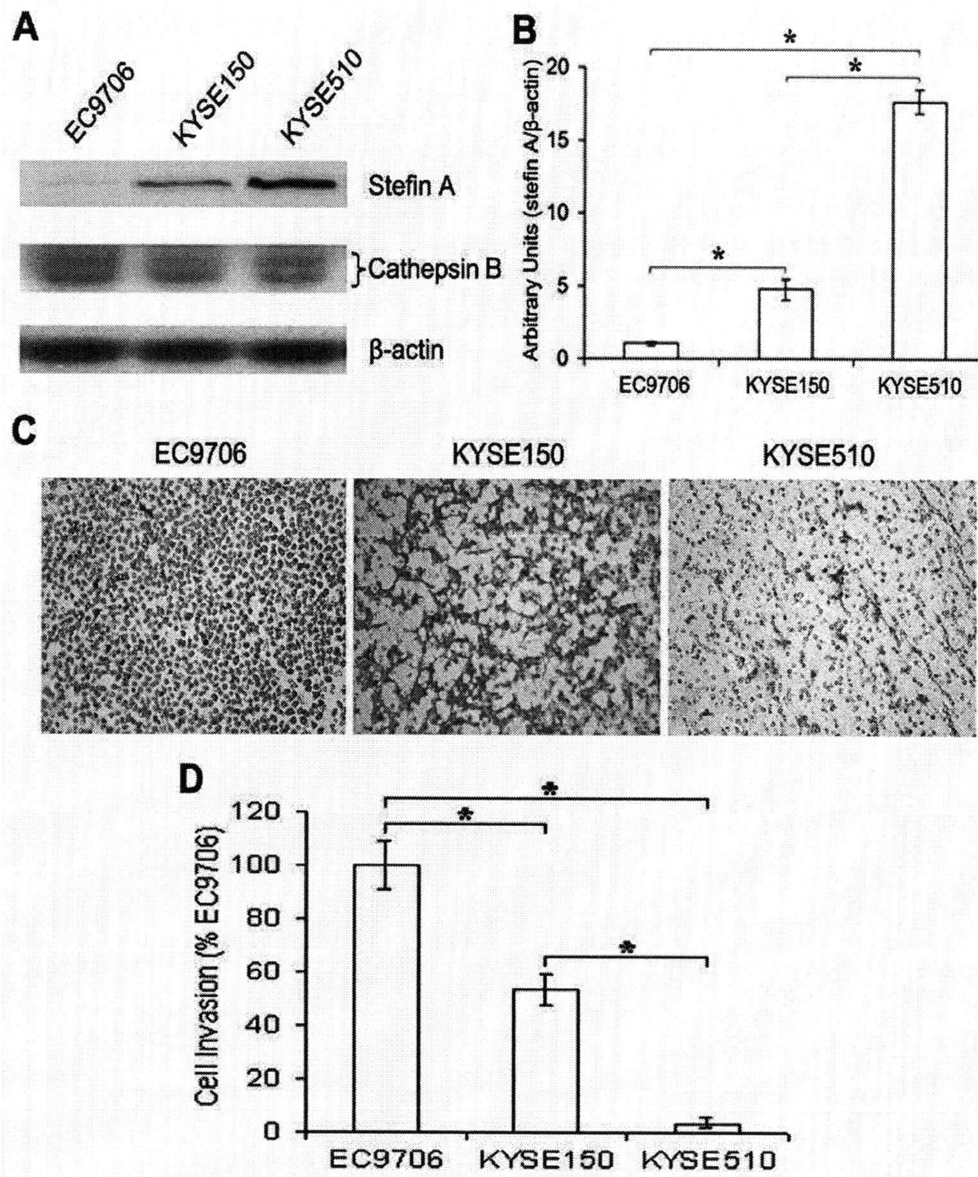

Figure 6I-1. Expression of stefin A and invasiveness of three different human ESCC cell lines(EC9706, KYSE150 and KYSE510). *A*, Western blot results of stefin A and cathepsin B; β-actin was used as internal control. *B*, ratios of stefin A quantification: corresponding β-actin quantification from (*A*) were calculated and represented the relative expression level of stefin A protein. Protein quantification was obtained by densitometric analysis of the protein optical density × resulting band area. For comparative analysis, ratios of stefin A quantification: corresponding β-actin quantification in EC9706 cells were expressed as 1 arbitrary unit. Data are mean ±SD of three independent experiments. *, $P < 0.01$. *C*, Matrigel invasion of three cell lines. Similar results were obtained by three independent experiments. *D*, histogram of invasion from (*C*) was expressed as % of EC9706 cells. Data were shown as mean ± SD of three independent experiments. *, $P < 0.01$.

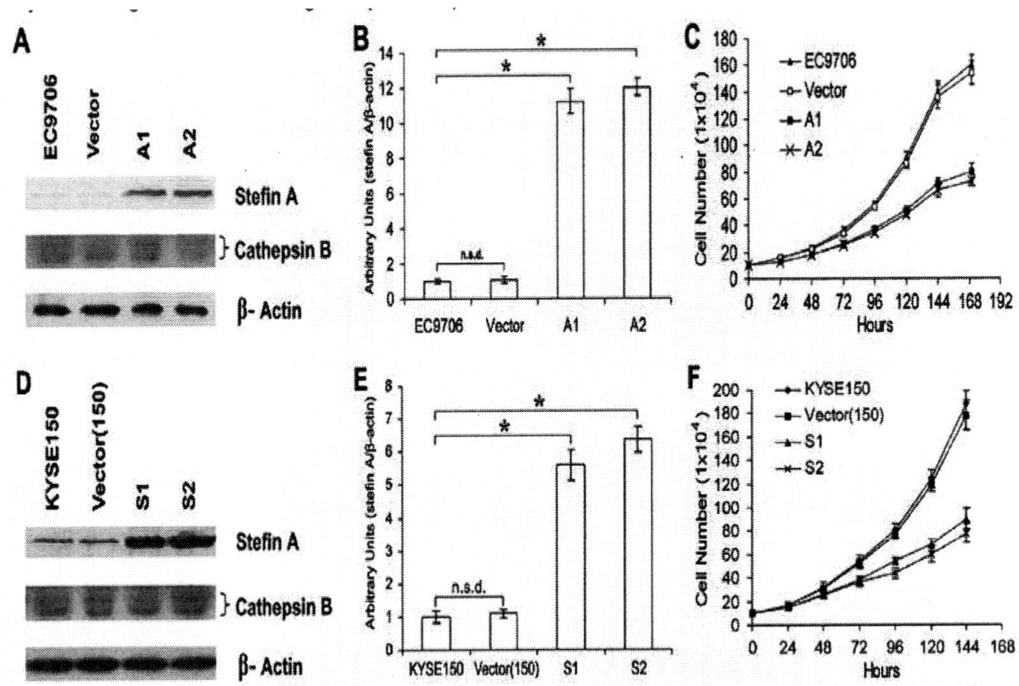

Figure 6I-2. Effect of stefin A overexpression on the growth of EC9706 (*A*, *B* and *C*) or KYSE150 (*D*, *E* and *F*) cells. *A* and *D*, Western blot results. Higher expression of stefin A was obtained in A1 and A2 clones from EC9706 cells (*A*) and in S1 and S2 clones from KYSE150 cells (*D*). *B* and *E*, ratios of stefin A quantification: corresponding β-actin quantification from (*A*) and (*D*) were calculated and represented the relative expression level of stefin A protein. Protein quantification was obtained by densitometric analysis of the protein optical density × resulting band area. For comparative analysis, ratios of stefin A quantification: corresponding β-actin quantification in EC9706 or KYSE150 cells were expressed as 1 arbitrary unit. Data are mean ±SD of three independent experiments. *, $P < 0.001$; n.s.d., no significant difference. *C* and *F*, growth kinetics of stefin A-transfected EC9706 (*C*) or KYSE150 (*F*) cells. Growth curves were plotted for cell number versus time of incubation of tumor cells *in vitro*. Values are expressed as mean ± SD of three independent experiments.

Transfection with Sense-stefin A Construct Inhibits Cell Growth in vitro

EC9706 cells were transfected with sense-stefin A construct or empty vector for stable clones. As shown in Figure 6I-2A, two stable cell clones (A1 and A2) expressed higher stefin A protein. Densitometric quantification indicated that the amount of stefin A protein was significantly 11-12 times higher in A1 or A2 cells than that in EC9706 cells ($P < 0.01$) (Figure 6I-2B). No significant difference in the amount of stefin A protein was observed between EC9706 cells and Vector cells (Figure 6I-2A and 6I-2B). Therefore, A1, A2 and Vector cells were identified and chosen for the following studies.

As shown in Figure 6-2C, the growth rates of A1 or A2 cells were slower than that of EC9706 or empty vector cells under normal culture conditions. There was no significant difference in growth rates between EC9706 and empty vector cells. Therefore, overexpression of stefin A protein inhibited cell proliferation. Overexpression of stefin A in KYSE150 cells also inhibited cell proliferation (Figure 6I-2D, 6I-2E and 6-2F). No significant apoptosis was found in stefin A-transfected EC9706 or KYSE150 cells I (data not shown).

Stefin A-Transfected ESCC Cells Show Reduced Cathepsin B Activity and Reduced Invasiveness in vitro

Previous studies confirmed that stefin A was one of cathepsin B inhibitors. Using MR-Cathepsin B detection kit, we evaluated the effect of overexpression of stefin A on cathepsin B activity in EC9706 cells. As shown in Figure 6I-3A and Figure 6I-3B, cathepsin B activities were reduced either by transfection with stefin A or by treatment of CA074Me. Transfection of stefin A reduced cathepsin B activity in EC9706 cells by 87-92%. CA074Me at a concentration of 10 μM reduced cathepsin B activity by 80-83%. Combination of both further reduced cathepsin B activity by 95-96%.

Many studies confirmed that cathepsin B was involved in invasion and metastasis of tumor [Szpaderska et al., 2001; Werle et al., 1999; Krepela et al., 1998]. As stefin A was endogenously a cathepsin B inhibitor, we examined whether transfection with stefin A inhibited cell invasion. Using AP48 invasion chamber coated with or without Matrigel, we investigated the effects of transfection with stefin A combined with or without CA074Me on cell motility and invasion. There was no significant difference between EC9706 or empty vector cells and A1 or A2 cells in cell motility; Treatment with cathepsin B inhibitor didn't affect cell motility either (data not shown). However, as shown in Figure 6I-3C, transfection with stefin A reduced cell invasion by 78-83% while treatment with 10 μM CA074Me reduced invasion by 65-68%. The most significant reduction in invasiveness was observed in combination of both above and cell invasion was reduced by 88-95%. Reduced cathepsin B activity and reduced invasiveness were also observed in stefin A-transfected KYSE150 cells (Figure 6I-3D and 6I-3E).

Stefin A-transfected Cells Show Reduced Tumor Growth in vivo

Our data showed that the tumors from stefin A-transfected cells in nude mice grew more slowly than that from EC9706 or empty vector cells (Figure 6I-4A, 6I-4B and 6I-4C), which was consistent with the cell growth curve above. At the end of 3 months, the weights of tumors from stefin A-transfected cells were 5 times less than that of tumors from EC9706 or empty vector cells. As shown in Figure 6I-4D and 6I-4E, stefin A transfection in KYSE150 cells also inhibited tumor growth and reduced tumor weight.

Stefin A-transfected EC9706 Cells Show a Significant Inhibition of Lung Metastasis in Nude Mice

At the end of 3 months, all the mice were sacrificed. All the lungs, livers and brains were cut to consecutive 6-μm sections. The metastases were carefully observed on serial microscopic sections of whole specimens in all mice. Lung metastases were seen in two out of eight mice from EC9706 cells without the help of a microscope. By a microscope, parental or empty vector EC9706 cells metastasized to the lungs in half of nude mice, resulting in a total number of 27 or 26 lung tumor colonies, respectively. In contrast, no lung metastasis was seen in nude mice from stefin A-transfected EC9706 cell clones (A1 and A2) even with the help of microscope (Figure 6I-5A and 6I-5B; Table 6I-1). Overexpression of stefin A inhibited the lung metastasis in nude mice (P = 0.038, Fisher's exact test). Neither liver metastases nor brain metastases were found in all mice. Nevertheless, no lung metastasis was observed in nude mice from parental KYSE150 cells or transfected KYSE150 cells.

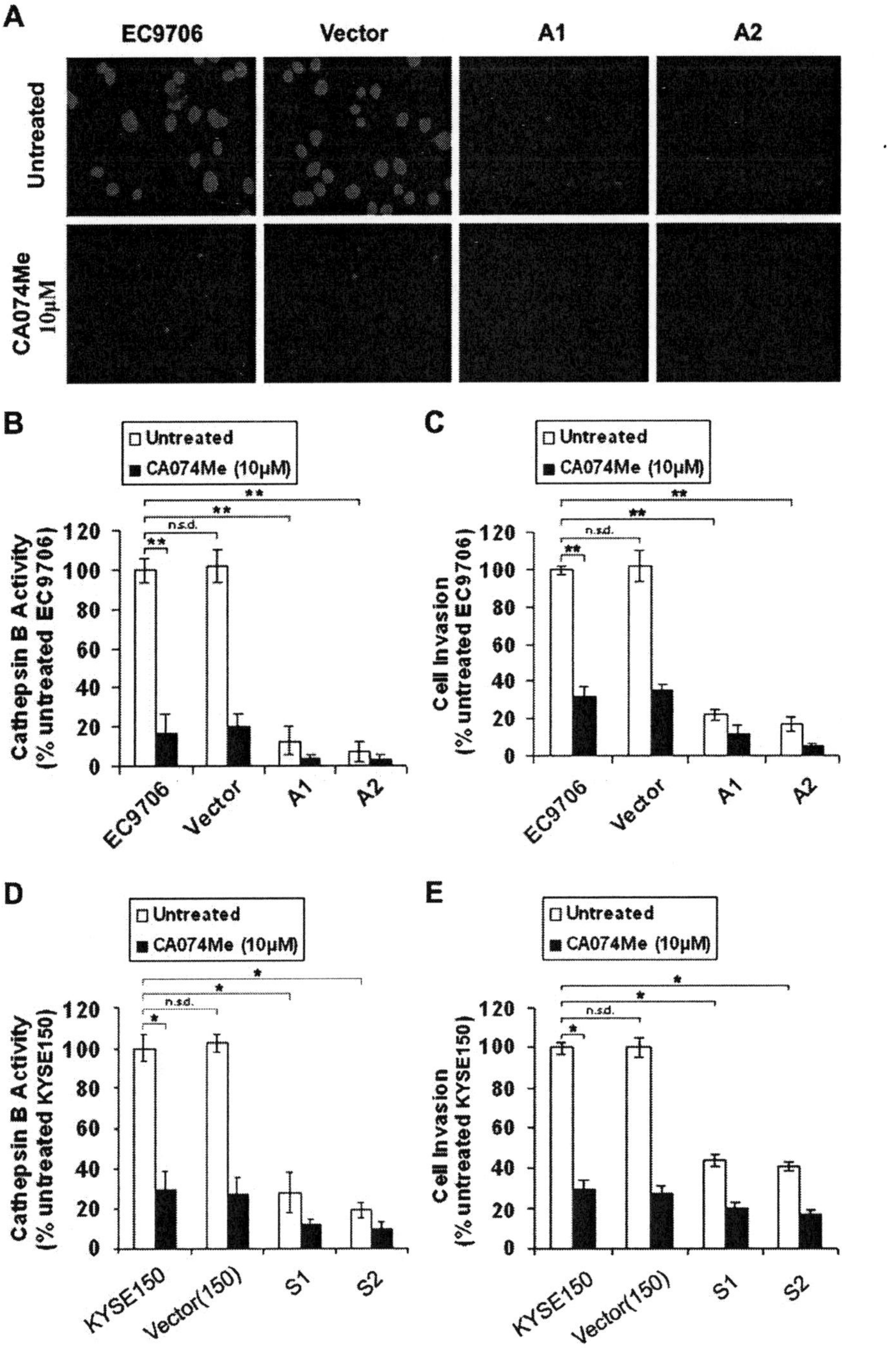

Figure 6I-3. Effect of transfection with stefin A or treatment with 10 μM CA074Me on cathepsin B activity and invasion in EC9706 (*A*, *B* and *C*) or KYSE150 (*D* and *E*) cells. *A*, cathepsin B activity assay: photographic results from a fluorescence microscope. *B* and *D*, histogram showing inhibition of cathepsin B activity expressed as % of untreated EC9706 (*B*) or KYSE150 (*D*) cells according to "Materials and Methods". Results were shown as mean ± SD of three independent experiments. *, $P < 0.05$; **, $P < 0.001$; n.s.d., no significant difference. *C* and *E*, histogram showing inhibition of invasion expressed as % of untreated EC9706 (*C*) or KYSE150 (*E*) cells according to "Materials and Methods". Results were shown as mean ± SD of three independent experiments. *, $P < 0.05$; **, $P < 0.001$; n.s.d., no significant difference.

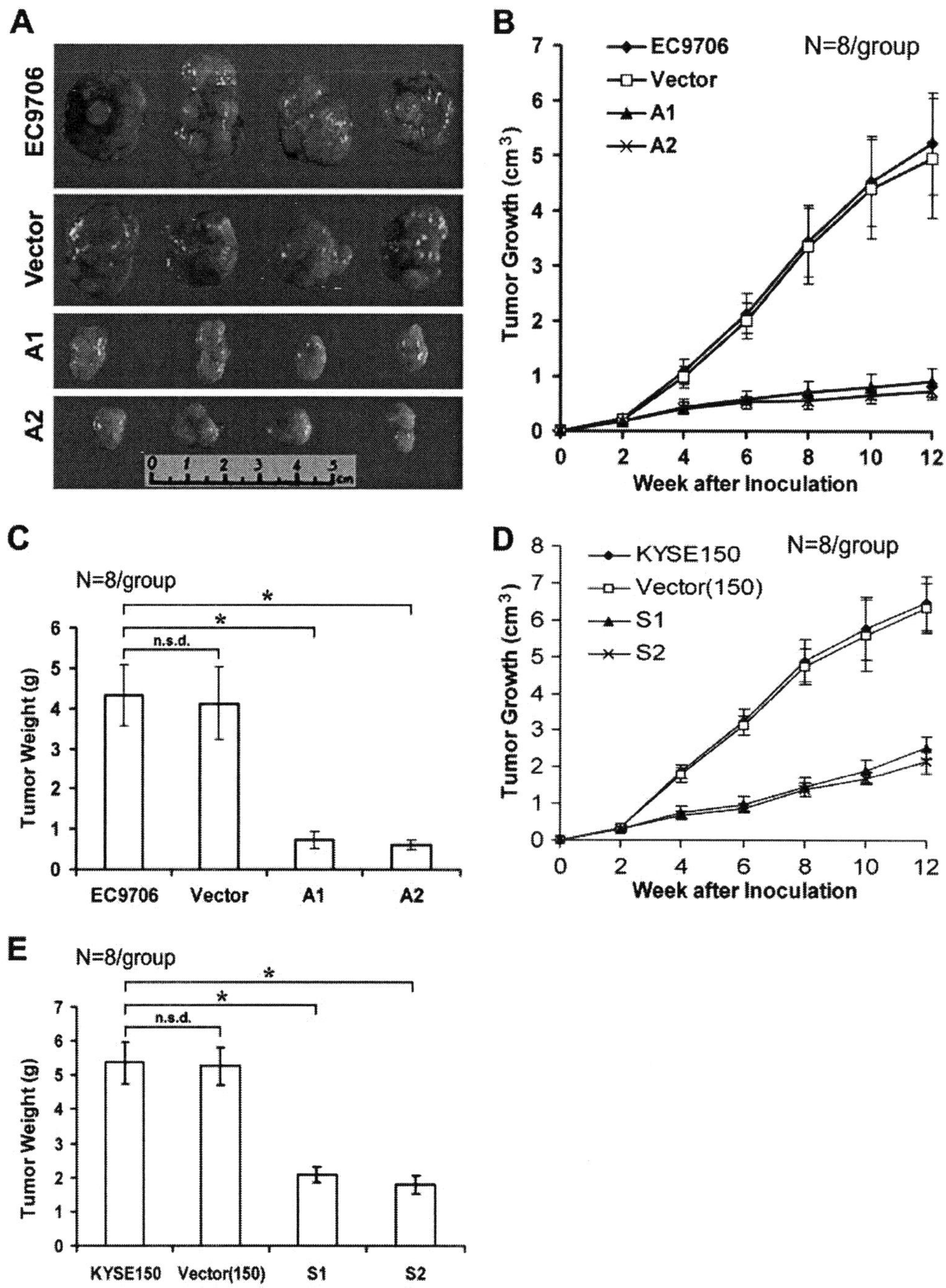

Figure 6I-4. The growth of tumors from stefin A-transfected EC9706 (*A*, *B* and *C*) or KYSE150 (*D* and *E*) clones in nude mice. *A*, representative gross pictures of the tumors from EC9706 cells 3 months after the inoculation. *B* and *D*, tumor growth curve from EC9706 (*B*) or KYSE150 (*D*) cells. Cross-sectional tumor diameters were measured externally, and the approximate tumor volume was calculated as described in "Materials and Methods". *C* and *E*, the weight of the tumors from EC9706 (*C*) or KYSE150 (*E*) cells 3 months after the inoculation. Values are expressed as mean ± SD of eight developed tumors in the experimental mice. Similar results were obtained in two additional experiments. *, $P < 0.001$; n.s.d., no significant difference.

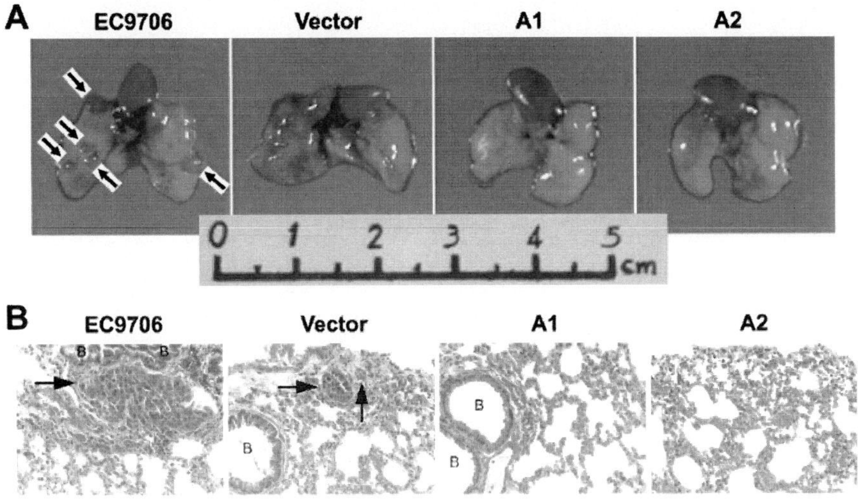

Figure 6I-5. Transfection with stefin A inhibited lung metastasis in nude mice inoculated with EC9706 cells. *A*, representative gross pictures of the mouse lungs from parental, empty vector and stefin A-transfected EC9706 cells 3 months after s.c. injection of cells. Tumor colonies are showed by arrows. *B*, representative lung sections stained with H and E under a microscope. Metastatic tumors (arrows) deposit in the lungs of the mice from EC9706 cells or empty vector transfectants. The deposit is adjacent to bronchioles (B, bronchioles). Magnification, × 200.

Stefin A-transfection Inhibits Tumor Angiogenesis

As shown in the tumor sections stained with H and E, the tumors from stefin A-transfected EC9706 cells showed obvious necrosis compared with the tumors from parental or empty vector cells (Figure 6I-6A). In this study, we assumed that tissue necrosis was due to inhibition of tumor angiogenesis. Immunohistochemical staining with factor VIII antibody was employed. Transfection with stefin A showed a dramatic reduction of factor VIII staining (Figure 6I-6B). As shown in Figure 6I-6C, MVD was significantly decreased in the tumors from EC9706 cells or empty vector cellls (54 ± 7.3 / HPF or 57 ± 5.5 / HPF, respectively) compared with the tumors from A1 or A2 cells (17 ± 3.8 / HPF or 12 ± 3.6, respectively; $P < 0.001$), which supported the assumption. Similar inhibition of tumor angiogenesis was obtained in the tumor sections from stefin A-transfected KYSE150 cells (Figure 6I-6D).

Discussion

Cathepsin B is believed to participate in tumor progression from a premalignant to a malignant state. Regulation of cathepsin B expression levels by molecular genetic techniques led to changes of malignant phenotype of tumor cell. Stable transfection [Coulibaly et al., 1999] or transient transfection [Szpaderska et al., 2001] of human cathepsin B cDNA into tumor cells resulted in an increase in cathepsin B activity and an increase in invasiveness. Conversely, Down-regulation of cathepsin B expression reduces cathepsin B activity and inhibits Matrigel invasiveness of tumor cells by stable transfection with an antisense cathepsin B cDNA [Szpaderska et al., 2001; Krueger et al., 1999; Krueger et al., 2000; Mohanam et al., 2001; Gondi et al., 2004] or by RNA interfering [Gondi et al., 2004; Lakka et al., 2004]. Because invasiveness through Matrigel is often correlated with metastatic potential *in vivo* [Bonfil et al., 1989; Axelrod et al., 1989; Dedhar et al., 1993; Albini et al.,

1998], these data provide further evidence for the relationship between increased levels of cathepsin B and malignancy.

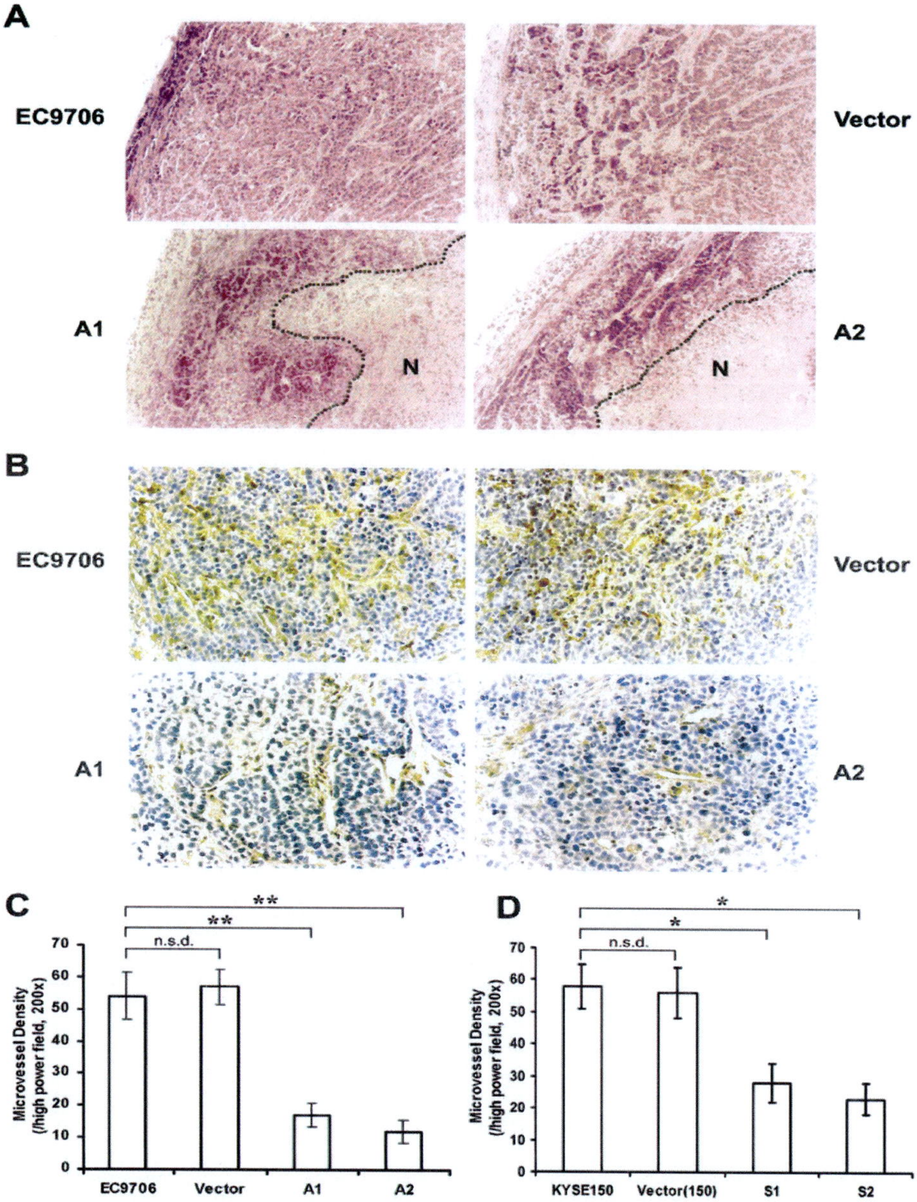

Figure 6I-6. Stefin A transfection inhibited angiogenesis of the tumors from EC9706 (*A*, *B* and *C*) or KYSE150 (*D*) cells. *A*, tumor section stained with H and E from EC9706, empty vector, A1 and A2 cells. Sections from A1 and A2 cells show obvious necrosis. N, necrosis; magnification, X100. *B*, immunohistochemical analysis of tumor tissue sections stained with anti-human factor VIII antibody. Obviously decreased staining of factor VIII was observed in stefin A-transfected tumor sections compared with high intense staining in the tumor sections from EC9706 cells and empty vector transfectants. Magnification, × 200. *C* and *D*, microvessel density analysis in the tumor sections from EC9706 (*C*) or KYSE150 (*D*) cells was performed according to "Materials and Methods". Values are expressed as mean ± SD. *, $P < 0.01$; **, $P < 0.001$; n.s.d., no significant difference.

In normal cells or tissues, there is a balance between cathepsin B and its endogenous inhibitors. Alterations in the balance have been postulated to contribute to malignant progression. An inverse correlation between increased cathepsin B expression and decreased levels of its endogenous inhibitors has been demonstrated in a variety of malignant tumors [Sinha et al., 2002; Hirai et al., 1999; Strojnik et al., 2000; Luo et al., 2004]. Up-regulated expression of the endogenous inhibitors in tumor cells improved the malignant phenotype. Stable transfection of cystatin C into B16F10 melanoma cells with low cystatin C levels inhibited motility and *in vitro* invasiveness by 50% in both stimulated (autocrine motility factor, laminin) and unstimulated cells [Sexton et al., 1997]. Sense-cystatin C-transfected SNB19 cells were markedly less invasive than control cells in a Matrigel invasion assay and in a coculture assay of SNB19 spheroids and fetal rat brain aggregates, and sense-cystatin C-transfected SNB19 cells did not form tumors in nude mice upon intracerebral injection [Konduri et al., 2002]. Larger changes in invasiveness were observed in murine SCC-VII squamous cell carcinoma cells by heterologous expression of cathepsin B and cystatin C than were seen in the previous cathepsin B overexpression study [Coulibaly et al., 1999]. Recently, overexpression of cystatin M, a potent endogenous protein inhibitor of lysosomal cysteine proteases, in human MDA-MB-435S breast carcinoma cells significantly suppressed *in vitro* cell proliferation, migration, and Matrigel invasion and showed significantly delayed primary tumor growth and lower metastatic burden in the lungs and liver in mice when compared with mock controls [Shridhar et al., 2004; Zhang et al., 2004].

In our study, we observed that the relative amounts of stefin A decreased as the invasion ability increased. The human EC9706 esophageal squamous cell carcinoma cell line is normally highly invasive and produces very little stefin A protein. Introducing a sense-stefin A construct into EC9706 cells inhibited the *in vitro* invasiveness by 78-83% in a Matrigel invasion assay and lung metastasis in nude mice significantly. The reason why the metastasis was significantly inhibited might be that the invasiveness of EC9706 cells was reduced to some extent which prevented metastasis. The fact that metastasis was only found in lungs but not in livers and brains indicate lung is the most important target for metastasis of human esophageal squamous cell carcinoma cells in nude mice, which is in agreement with our assumption. This is the first observation for overexpression of endogenous cathepsin B inhibitor in experimental metastasis of esophageal carcinoma cells. Previous study showed that treatment of a liver-homing Lewis lung carcinoma subline H-59 cells with E-64, an exogenous cathepsins inhibitor, inhibited experimental liver metastases formation by up to 90% [Navab et al., 1997].

Conflicting results about the contributions of cysteine proteinases to cell motility were reported in different studies. Some reports demonstrated that E-64 [Redwood et al., 1992; Boike et al., 1992], stefins [Boike et al., 1992] and transfection with cystatin C [Sexton et al., 1997] cystatin M [Shridhar et al., 2004] or antisense cathepsin B [Krueger et al., 1999] reduced the cell motility. Some studies showed no effect of cysteine proteinases on cell motility [Coulibaly et al., 1999; Szpaderska et al., 2001; Bjornland et al., 1996]. In our study, both transfection of stefin A cDNA and treatment with CA074Me did not affect the motility of human EC9706 esophageal sqamous carcinoma cells but significantly reduced Matrigel invasion. CA074Me, a membrane-permeable proinhibitor for intracellular cathepsin B, which becomes active following internalization and conversion to CA074 [Buttle et al., 1992], is currently the reagent of choice to inactivate cathepsin B within living cells. As stefin A is an endogenous inhibitor of cathepsins B, H, L and S, here we employed CA074Me to evaluate the inhibition of cathepsin B by stefin A. In PC3M cells, CA-074Me at a concentration of 10 μM completely abolished intracellular cathepsin B activity and produced a 45–75% inhibition of Matrigel invasion, suggesting an important intracellular function for cathepsin B in matrix degradation [Szpaderska et al., 2001]. In our study, CA-074Me at a concentration of 10 μM inhibited cathepsin B activity by 80-83% and Matrigel invasion by 65-68%. Stefin A-transfection inhibited cathepsin B activity by 87-92% and Matrigel invasion by 78-83%. Combined applications of both resulted in the largest inhibition of both cathepsin B activity by 95-96% and Matrigel invasion by 88-95%. Taken together, transfection with stefin A and treatment with cathepsin B inhibitor contributed synergistically to inhibit cancer invasion by inhibiting cathepsin B activity. And we speculate that cathepsin B serves as a very important role in the invasion of EC9706 cells. In EC9706 cells, stefin A may be the potent inhibitor for intracelllualr cathepsin B activity. Confocal results supported the interaction of cathepsin B and stefin A proteins in cells. Stefin A was reported to inhibit cathepsin B by a two-step mechanism, involving an initial weak interaction followed by a conformational change [Pavlova et al., 2000]. The reason why the invasion inhibition of CA074Me was not strong as other study [Szpaderska et al., 2001] may be due to the variation of the cell lines.

It has been reported that cathepsin B could promote cell proliferation by activating growth factors [Guo et al., 2002] or liberating them from the extracellular matrix where they are sequestered [Taipale et al., 1997]. Cathepsin B inhibitor was reported to inhibit cell proliferation [Xing et al., 1998]. In our study, transfection with stefin A also inhibited both *in vitro* cell proliferation and *in vivo* tumor growth. In addition, we observed a noteworthy phenomenon that the tumors from stefin A-transfected EC9706 cells were prone to necrosis compared with the tumors from parental or empty vector cells. Cathepsin B is believed to participate in tumor growth and angiogenesis. Inhibition of cathepsin B and MMP-9/ uPAR gene expression in glioblastoma cell line via RNA interference reduced tumor growth and angiogenesis [Gondi et al., 2004; Lakka et al., 2004]. Adenovirus-mediated expression of antisense urokinase plasminogen activator receptor and antisense cathepsin B also inhibited tumor growth, invasion, and angiogenesis in gliomas [Gondi et al., 2004]. Therefore, we assume that transfection with stefin A might inhibit tumor angiogenesis and consequently cause tumor necrosis. Immunohistochemical analysis with factor VIII antibody, an indicator, which is widely used for measure of angiogenesis, supported our hypothesis. Moreover, inhibition of tumor angiogenesis induced by transfection with stefin A must contribute in part to inhibit tumor growth and metastasis. Therefore we have come to two conclusions: one is that both inhibition of cell proliferation and tumor angiogenesis induced by stefin A transfection may contribute to inhibition of tumor growth; the other is that both reduced invasiveness and inhibition of tumor angiogenesis caused by stefin A transfection may contribute to inhibiting tumor metastasis.

In summary, transfection of stefin A cDNA into human EC9706 esophageal squamous cell carcinoma cells inhibits tumor growth, angiogenesis, invasion and metastasis, and this is mainly through the inhibiting of cathepsin B activity. Similar results were observed in stefin A-transfected KYSE150 cells, another human esophageal squamous cancer cell line (not all data shown). Therefore, our study provides evidence of the role of stefin A expression in human esophageal squamous cell carcinoma cells and may shed a light of a novel strategy for cancer therapy [Li et al., 2005].

Table 6I-1 Experimental lung metastasis of human EC9706 esophageal squamous cancer cells [parental or transfected with empty vector or sense-stefin A (A1 or A2)] in nude mice

Nude Mice	No. of mice with lung metastasis	Total No. of lung tumor nodules
EC9706 (n=8)	4	27
Vector (n=8)	4	26
A1 (n=8)	0^a	0
A2 (n=8)	0^a	0

$^a P < 0.05$ (Fisher's Exact Test), clone A1 or A2 group versus EC9706 or vector group.

II Stomatin-like Protein 2 is Overexpressed in ESCC and Involved in Regulating Cell Growth and Cell Adhesion in Human ESCC

In the above cDNA array work, we found a gene named Stomatin-like protein 2 (SLP-2) was significantly overexpressed over 6 times in ESCC compared with their normal counterparts [Luo et al., 2004; Zhang et al., 2004; Zhang et al., 2005]. SLP-2 is a novel and unusual stomatin homologue of unknown functions. It has been implicated in interaction with erythrocyte cytoskeleton and presumably other integral membrane proteins but not directly with the membrane bilayer. We show here the involvement of SLP-2 in human ESCC, and the effects of SLP-2 on ESCC cells. KYSE450 cells transfected with antisense SLP-2 showed decreased cell growth, proliferation, tumorigenecity and cell adhesion. SLP-2 was first identified as a novel caner-related gene overexpressed in human ESCC, decreased cell growth, cell adhesion and tumorigenesis in the antisense transfectants revealed that SLP-2 may be important in ESCC tumorigenesis.

SLP-2 is a novel and unusual member of the stomatin gene superfamily [Wang et al., 2000; Seidel et al., 1998; Sprenger et al., 2004]. The vertebrate homologues identified are SLP-1, SLP-2 and SLP-3 [Wang et al., 2000; Seidel et al., 1998; Goldstein et al., 2003]. SLP-1 is most abundant in brain and shares many similarities with UNC-24 [Wang et al., 2000; Seidel et al., 1998]. SLP-3 is specifically expressed in olfactory sensory neurons [Wang et al., Goldstein et al., 2003]. All of these proteins as well as the stomatin from other species share a characteristic NH2-terminal hydrophobic domain as well as a consensus stomatin signature sequence that defines the stomatin gene family [Wang et al., 2000]. Similar to other family members, SLP-2 shares the cognate stomatin signature sequence. However, it is the first member of this family to be recognized that lacks an NH2-terminal hydrophobic domain [Wang et al., 2000]. These features distinguish it from stomatin. SLP-1 and SLP-3. SLP-2 may link stomatin or other integral membrane proteins to the peripheral cytoskeleton and play a role in regulating ion channel conductance or the organization of sphingolipid and cholesterol-rich lipid rafts [Wang et al., 2000]. However no relationship between SLP-2 and tumorigenesis was reported yet.

In this study, the overexpression of SLP-2 in ESCC was evaluated by semi-quantitative RT-PCR, Western blot and immunohistochemistry. To investigate the role of SLP-2 in tumorigenesis, antisense-oriented SLP-2 was stably transfected into KYSE450 cells. Antisense transfection of SLP-2 in KYSE450 cells not only obviously reduced cell growth and proliferation *in vitro*, but also inhibited tumor formation and growth *in vivo*. The identification of SLP-2 as a novel caner-related gene may have implications for understanding tumorigenesis.

Materials and Methods

Tissue Specimens

Fresh tissue specimens were taken from patients presented to Cancer Hospital of Chinese Academy of Medical Sciences (CAMS) (Beijing, China) and the First Affiliated Hospital of Anhui Medical University (Anhui province, China) immediately after surgery and stored at –80°C until use. None of the patients had received radio- or chemotherapy before surgery. Tumor tissues were dissected from the resected specimens and the normal tissue block was taken from the distal resection margin. In addition, the sections used for immunohistochemistry were fixed in 4% polyformaldehyde and embedded in paraffin.

Cell Culture

ESCC cell lines YES2, KYSE30, KYSE70, KYSE140, KYSE150, KYSE180, KYSE410, KYSE450, KYSE510, COLO-680N, TE12 and EC9706 were previously established from primary human ESCC patients [Shimada et al., 1992; Kanda et al., 1994; Han et al., 2002]. All the cell lines were grown in RPMI-1640 medium supplemented with 10% fetal bovine serum (FBS), 100μg/ml streptomycin and 100μg/ml penicillin (pH 7.2-7.4) in a humidified incubator containing 5% CO_2 at 37°C. KYSE series were generous gifts from Dr. Shimada Y at Kyoto University, Japan [Shimada et al., 1992].

Semi-quantitative RT-PCR

Total RNA was extracted from frozen tissues using TRIzol reagent (Invitrogen) according to the manufactures' instructions. The quality of the RNA was assessed by 1% denaturing agarose gel electrophoresis and spectrophotometry. Five micrograms of total RNAs of each sample were reverse-transcribed to the first strand of cDNA primed with oligo-$(dT)_{12\text{-}18}$ using Transcriptase SuperScript II™ Preamplification System for First Strand cDNA kit (Invitrogen). Then 0.5~1 μl aliquots of the cDNA was used as template to amplify SLP-2 fragments with primers: sense 5'-GTGACTCTCGACAATGTAAC-3'; antisense: 5'-TGATCTCATAACGGAGGCAG-3' and fibronectin 1 fragments with primers: sense 5'-AAGGAGAAGACCGGACCAAT-3'; antisense: 5'-GGCTTGATGGTTCTCTGGAT-3' both at annealing temperature of 57°C for 27 cycles. The expression of the housekeeping gene, glyceraldehyde-3-phosphate dehydrogenase (GAPDH), was used as an internal control.

Antibody production and Western blot analysis

Amino acid sequences of SLP-2 were obtained from NCBI Protein database, and subjected to analysis of secondary structure, hydrophilicity profile and antigenecity using the program Protean^{TM} of DNAStar software (DNAStar, Inc.). The polypeptide composed of 13 amino acids (ASLDEELDRVKMS) at the C-terminal of SLP-2 (corresponding to amino acids 344 aa-356 aa) was chosen and synthesized, coupled with KLH at the end of the N-terminal, and emulsified in complete Freund's adjuvant used as antigens for immunization with New Zealand White rabbit and chicken respectively. Immune response was enhanced with antigens emulsified in incomplete Freund's adjuvant after 14 days and 21 days of immunization, respectively. Antibodies were recovered after being subjected to enzyme-linked immunosorbent assay (ELISA) by Zhuhai Bioinforbody Inc. (Zhuhai, China). The rabbit SLP-2 antibody was then purified by protein A while the chicken SLP-2 antibody was purified by coupling to the antigen.

For Western blot analysis, cells or tissues were lysed with the buffer (1% SDS, 10 mM Tris-Cl, pH 7.6, 20 μg/ml aprotenin, 20 μg/ml leupeptin and 1 mM AEBSF). The protein concentrations were determined using the BCA Protein Assay kit (Pierce). Ten micrograms of protein were separated on 12 % of SDS-PAGE gels and transferred to polyvinylidene difluoride (PVDF) membranes. After blocking, the membranes were incubated with the appropriate primary antibody: anti-SLP-2 antibody (1:1,000 dilution) or mouse monoclonal antibody against fibronectin 1 (1:1,000 dilution; Santa Cruz Biotechnology Inc.) at 4°C overnight. After washing, the membranes were incubated with secondary antibody at a dilution of 1:3,000 at room temperature for 1 hour. Proteins were detected with the enhanced chemiluminescence (ECL) kit (Amersham Pharmacia Biotechnology Inc.), and anti-β-actin antibody (Sigma) was used as loading control.

Immunohistochemical Staining

Tissue sections were dewaxed with xylene and rehydrated through gradient ethanol into water. After endogenous peroxidase activity was quenched with 3% H_2O_2 for 30 min, sections were digested with 0.1% trypsin at 37°C for 30 min. After PBS washes, nonspecific antibody binding was blocked by preincubating slides in 10% normal goat non-immune serum at 37°C for 30 min. Sections were incubated with the polyclonal primary antibody against SLP-2 at 1:200 dilution or mouse monoclonal antibody against fibronectin at 1:1000 dilution (Santa Cruz Biotechnology Inc.) for 2 hours at 37°C or overnight at 4°C. For chicken SLP-2 antibody, sections were incubated with polymer helper at 37°C for 30 min. After PBS washes again, sections were incubated with secondary antibody at 1:200 dilution for 30 min at 37°X. After PBS washes, sections were developed using diaminobenzidine tetrahydrochloride (DAB, Sigma). As for the rabbit SLP-2 antibody or mouse fibronectin 1 antibody, sections were incubated with biotinylated link secondary antibody at 1:200 dilution for 30 min and then with horseradish peroxidase (HRP)-labeled streptavidin for 30 min. After PBS washes, sections were developed using DAB. Sections were washed in running tap water and lightly counterstained with hematoxylin, followed by dehydration and coverslip mounting. Negative controls were obtained by omitting the primary antibody.

SLP-2 and fibronectin 1 expression was evaluated as described previously [Sarbia et al., 1999; Hao et al., 2001]. The percentage of SLP-2 positive tumor cells was determined semi-quantitatively by assessing the entire tumor section. Each sample was assigned to one of the following categories: 0 (0–4%), 1 (5–24%), 2 (25–49%), 3 (50–74%), or 4 (75–100%). The intensity of immunostaining was determined as 0 (negative), 1+, (weak), 2+, (moderate) or 3+, (strong). A final immunoreactive score between 0 and 12 was calculated by multiplying the percentage of positive cells with the staining intensity score. All slides were blind evaluated for immunostaining without any knowledge of the clinical outcome of other clinical or pathological data.

Preparation of Constructs and Transfection

Full-length of SLP-2 was amplified with the primers 5'-GGCGGTTGGGAAATGCTGGCGCG-3' and 5'- GTCCCCAGACTCCCTGGCC -3' by PCR using cDNA template reverse transcribed from normal esophageal epithelia, and was subcloned into an eukaryotic expression vector pcDNA3.1/myc-His(-) (Invitrogen) in the antisense orientation. Human ESCC cell line KYSE450 were transfected with either the antisense construct or the empty vector using lipofectamine^{TM} 2000 (Invitrogen). After G418 (400μg/ml, Life Technologies, Rockville, MD) screening, the stable clones were identified by semi-quantitative RT-PCR and Western blot.

MTT Assay

KYSE450 cells transfected with either antisense SLP-2 expression vector or empty vector were digested with trypsin and inoculated in 96-well plates at a concentration of 1×10^3 cells/well after counting. After incubation at 37°C in a humidified incubator containing 5% CO_2 for 1, 2, 3, 4, 5, 6, 7 days, 3-(4,5-dimethylthiazol-2-yl)-2,5-diphenyltetazolium bromide (MTT) dissolved in RPMI1640 at the final concentration of 0.5 mg/ml was added in a 96-well plate each day. The plates were incubated for an additional 4 h, then detected at 570 nm on Bio-Kinetics Reader (Bio-Rad) after 200 μl of dimethyl sulfoxide (DMSO) instead of RPMI-FBS was added to each well to solubilize the formazan crystals.

Clonogenecity Assay

KYSE450 cells transfected with either antisense SLP-2 expression vector or empty vector were digested with trypsin and seeded in six-well plates at a concentration of 5×10^2 cells/well after counting. The plates were incubated at 37°C in a humidified incubator containing 5% CO_2. When the colonies became visible (2-4 weeks), cells were fixed with methanol, stained with Gimesa and counted.

Flow cytometry Assay

Flow cytometry assay was performed by propidium iodide staining. KYSE450, empty vector and antisense SLP-2-transfected cells were grown to 80-90% confluence, then digested with trypsin, washed twice with PBS and fixed overnight at 4°C in 70% ethanol. After washed twice with PBS, cells were incubated with 5 μg/ml propidium iodide and 50 μg/ml RNase A in PBS for 1 h at room temperature. Flow activated cell sorter analysis was carried out using a FACSCalibur flow cytometer (Becton Dickson, Mountain View, CA) with CELLQUEST software. A total of 10,000 events were measured per sample.

Experiments in Nude Mice

Single cell suspensions of each of the transfectants and parental cells were digested with trypsin and collected. The cell viability was >95% as determined by trypan blue staining. Cells (2×10^6) in a 0.1-ml volume of PBS were inoculated s.c. into the right flank of 4-week-old male BALB/c nude mice (ten for each group). Once palpable tumors were established, tumor volume measurements were taken once a week using calipers along two major axes. At the end of 8 weeks, all mice were sacrificed and the tumor weights were measured.

Cell Attachment Determination

Cell attachment assay was measured by the CellTiter 96® AQueous One Solution Assay (Promega, Madison, WI) according to the manufacturer's instructions. Briefly, aliquots of 2×10^4 cells/well were distributed in 96-well plates coated with Matrigel (BD Bioscience, Bedford, MA) in 100 μl of medium. Cell attachment was subsequently determined at 2-, 4-, 6-, 12- and 24- hour time points after inoculation. At each time point, 20 μl of [3-(4, 5-dimethylthiazol-2-yl)-5-(3-carboxymethoxyphenyl)-2-(4-sulfophenyl)-2H- tetrazolium, inner salt], MTS, and an electron-coupling reagent (phenazine methosulfate, PMS) were added to the medium of the cultured cells and incubated for an additional 1 hour. The relative cell viability was determined with a 490-nm filter. Each experiment was performed in triplicate and repeated a minimum of three times. Relative attachment ratio was calculated as the following, where attachment ratio of parental KYSE450 was considered as 100%:

$$\text{relative attachment ratio(\%)} = \frac{\text{OD490 of calibrated experimental group}}{\text{OD490 of calibrated parental KYSE450 cells}} \times 100\%$$

Confocal Imaging

Cells were plated on cover slips, fixed in 4% paraformaldehyde/phosphate-buffered saline (PBS), permeabilized with 0.5% Triton X-100 and blocked for 30 min with 3% bovine serum albumin. After simultaneous overnight incubation at 4°C with rabbit anti-SLP-2 IgGs (1:100) and mouse fibronectin 1 IgGs (1:100, Santa Cruz Biotechnology Inc.) as primary antibodies, the slides were washed in PBS and incubated with the secondary antibodies: FITC-conjugated goat anti-mouse IgGs (1:50; Molecular Probes, Eugene, OR) and TRITC-conjugated goat anti-rabbit IgGs (1:50; Molecular Probes, Eugene, OR). Nuclei were counterstained with 1 μg/ml DAPI (Sigma, St. Louis, MO). Slides were mounted with mowoil and examined with Leica TCS SP2 confocal microscope (Leica Microsystems, Wetzlar, Germany). Series of images were processed and analyzed with the accompanying software package.

Statistical Analysis

Statistical analysis was performed using the SPSS statistical software (SPSS Inc.). The correlation between SLP-2 expression and clinicopathological characteristics or parameters was analyzed using Spearman's correlation analysis. Chi-square test was performed for comparison unless particular test was notified. Data from MTT assay, clonogenecity assay and tumorigenecity assay were presented as means±SD. $P < 0.05$ was considered statistically significant.

Results

Overexpression of SLP-2 in ESCC

The above cDNA microarry work indicated that SLP-2 was overexpressed in ESCC, we further evaluated the expression by semi-quantitative RT-PCR in 54 pairs of ESCC samples. As it was shown in Figure 6II-1A, overexpression of SLP-2 was observed in 74% (40/54) of ESCC samples.

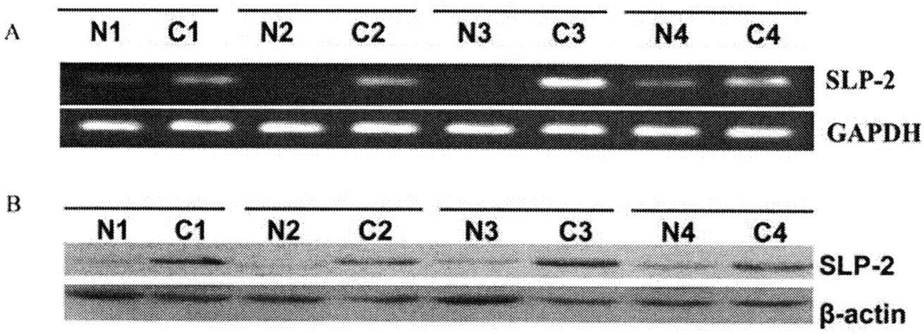

Figure 6II-1. Overexpression of SLP-2 in ESCC detected by semi-quantitative RT-PCR (A) and Western blot (B). N: normal tissue; C: patient-matched ESCC tissue. GAPDH and β-actin served as an internal control.

To confirm the overexpression of SLP-2 in cancer tissues at the protein level, amino acid sequences of SLP-2 obtained from NCBI Protein database were analyzed using the program Protean™, secondary structure and hydrophilicity profile results were the same as reported [Wang et al., 2000]. Polypeptide composed of 13 amino acids (ASLDEELDRVKMS□ was chosen as antigens for immunization with rabbit and chicken respectively. SDS-PAGE and ELISA indicated that both antibodies against SLP-2 were free of contaminating proteins (purity >90%) and high of titer (1:1,000-1:10,000). Immunoblot analysis of SLP-2 protein expression in patient-matched normal and tumor epithelium from different individuals with ESCC was performed, and rabbit polyclonal antibody against SLP-2 was fit for Western blotting. Using the available antibody, SLP-2 was also found to be up-regulated in 72% of ESCC tissues (13/18) (Figure 6II-1B). Western blotting with the commercially available antibody against SLP-2 (Cat No. 10348-1-AP, Proteintech Group Inc., Chicago, IL) also got the similar results (data not shown).

Immunohistochemistry was also carried out to reveal the expression of SLP-2 in ESCC, and both antibodies could be used for detection of the antigen in tissues. However, the specificity of the chicken polyclonal antibody is better than the rabbit one. Using chicken polyclonal antibody, SLP-2 was found up-regulated in 92% of ESCC samples (86 of 93, $P \leq 0.0001$), compared with their normal counterparts. Strong positive staining was presented in plasma membrane and cytoplasm of ESCC (Figure 6II-2). However, no correlation was found between SLP-2 expression and the pathological characteristics, including differentiation grade, lympho node metastasis, and invasion.

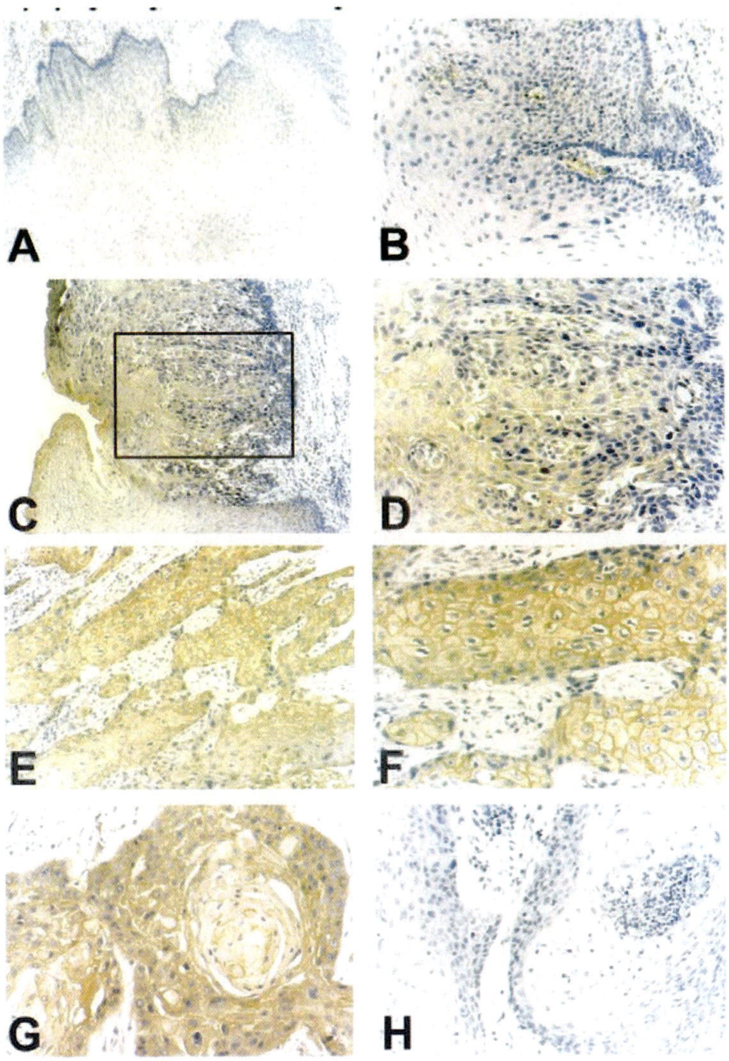

Figure 6II-2. Immunohistochemical staining analysis of SLP-2 expression in normal epithelia, mild dysplasia, severe dysplasia and squamous cell carcinoma of esophagus. A: Negative staining of SLP-2 in normal esophageal epithelia (×100); B: Positive staining was seen in mild dysplasia of esophagus (×200); C and D: Strong positive cell membrane staining was found in severe dysplasia of esophagus (C:×100; D:×200); E, F and G: Strong positive staining of SLP-2 was found in plasma membrane and cytoplasm of ESCC (E:×100; F:×200) and keratinpearl (G:×200); H: Negative control. The primary antibody was replaced by non-immune serum (×200). All sections were counterstained with hematoxylin.

Overexpression of SLP-2 in Premalignant Lesions of ESCC Development

To determine whether the overexprssion of SLP-2 occurred in premalignant lesions of ESCC, immunohistochemical staining was performed with mild dysplasia, severe dysplasia and squamous cell carcinoma of esophagus. The expression of SLP-2 in normal epithelial cells was always negative, whereas positive staining was found in mild dysplasia, severe dysplasia and squamous cell carcinoma including keratinpearl, which indicated that overexpression of SLP-2 is an early event in esophageal cancer development. Moreover, its expression was significantly increased as ESCC tumorigenesis progressed (Figure 6II-2).

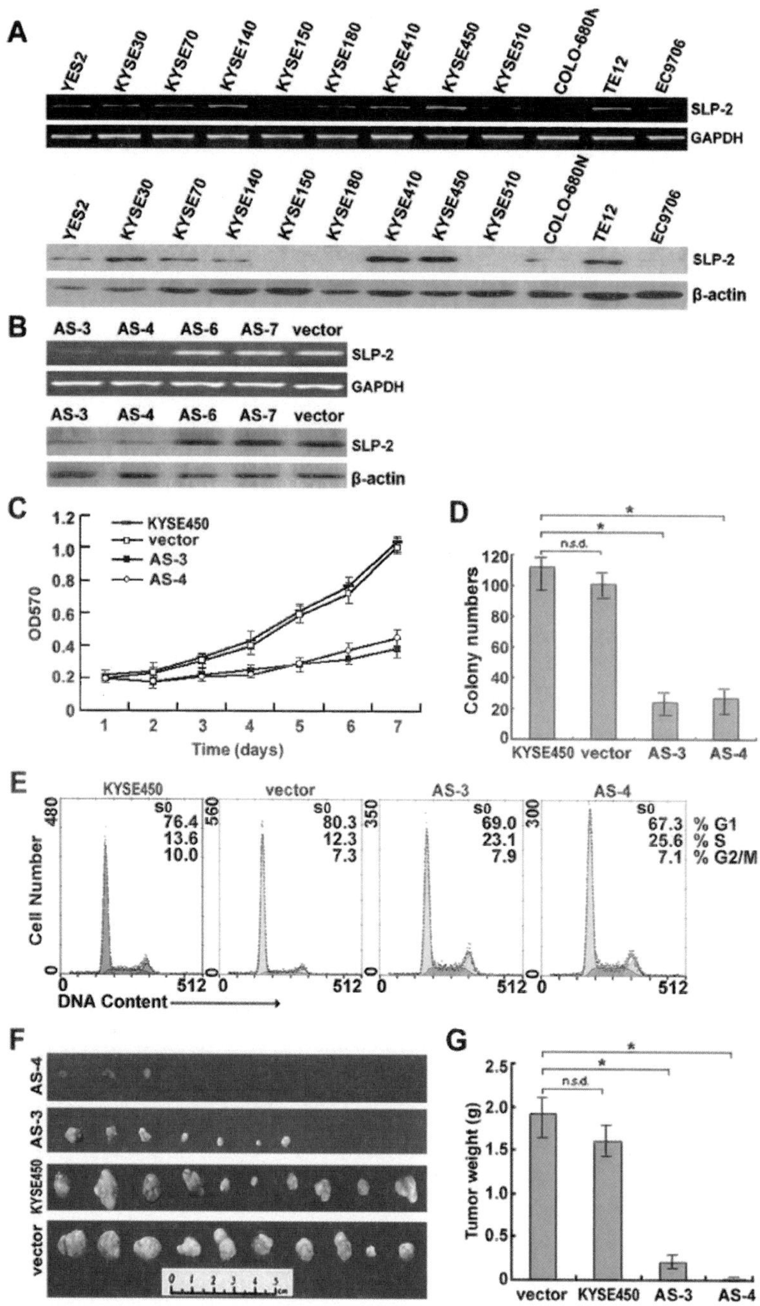

Figure 6II-3. Expression of SLP-2 in ESCC cell lines and the effects of SLP-2 on tumor growth and tumorigenecity. A: ESCC cell lines with endogenous SLP-2 expression were screened by semi-quantitative RT-PCR and Western blot; B: Positive clones transfected with antisense SLP-2 were identified by semi-quantitative RT-PCR and Western blot (AS-3, AS-4, AS-6 and AS-7 were different individual transfectants); C: Growth curve by MTT assay; D: Clonogenecity assay; E: Flow cytometry analysis; F, G: Tumorigenecity of antisense clones AS-3 and AS-4 compared with empty vector in nude mice. Representative gross pictures of the tumors 2 months after the inoculation (F) and tumor weight from KYSE450 cells, empty vector transfectants and cell clones AS-3 and AS-4 cells (G). Values were shown as mean ± SD of three independent experiments. *, $P < 0.001$; n.s.d., no significant difference.

Screening of ESCC Cell Lines and Positive Transfectants

Semi-quantitative RT-PCR and Western blot analysis of 12 ESCC cell lines showed that KYSE30, KYSE410, KYSE450 and TE12 had higher expression of endogenous SLP-2 (Figure 6II-3A). To investigate the role of SLP-2 gene in human ESCC cells, KYSE450 cells were chosen for antisense transfection and further study. After G418 screening, two antisense stable clones (AS-3 and AS-4) and one empty vector clone were identified and chosen for the following studies (Figure 6II-3B).

Transfectants with Antisense SLP-2 Inhibits Cell Growth in vitro and Tumor Growth in vivo

MTT assay and clonogenecity assay showed that cell growth and proliferation of antisense stable clones (AS-3 and AS-4) was slower than that of KYSE450 or empty vector cells under normal culture conditions (Figure 6II-3C, 6II-3D). As inhibition of cell growth and proliferation can be the consequence of an induction either of apoptosis or of cell cycle arrest, flow cytometry assay was performed. As shown in Figure 6II-3E, no apoptosis was detected in antisense SLP-2-transfected cells, while antisense SLP-2 constructs resulted in an arrest at S phase by approximately 11-13% compared with empty vector controls (Figure 6II-3E). In addition, we have previously observed that antisense transfection of SLP-2 caused S-phase arrest in another ESCC cell line TE12 [Zhang et al., 2005].

Four groups of nude mice inoculated with antisense SLP-2, empty vector or KYSE450 cells were sacrificed at the end of 8 weeks. The data showed that tumors from antisense SLP-2 transfectants in nude mice grew more slowly than that from KYSE450 or empty vector cells (Figure 6II-3F, 6II-3G), which was consistent with the cell proliferation results *in vitro*. The weights of tumors from antisense SLP-2 transfectants were significantly less than that from KYSE450 or empty vector cells, whereas no obvious difference was found between parental cells and the vector control cells.

Transfectants with Antisense SLP-2 Inhibits Cell Attachment

Cell *attachment* was determined by the CellTiter 96® AQueous One Solution Assay. The result showed that cell adhesion of antisense transfectants (AS-3 and AS-4) was decreased sharply than that of parental KYSE450 or empty vector cells after 4 hours inoculation (P<0.05) whereas there was no significant difference between parental KYSE450 cells and empty vector cells. With time prolongation the adhesive ability of antisense transfectants increased, though adhesive force of antisense transfectants was still weaker than that of parental KYSE450 or empty vector cells (data not shown).

Cell adhesion played important roles in maintaining normal functions of cells including cellular organization and structure, proliferation, survival, metabolism and gene expression. Alterations in cell adhesion were in the control of cell behavior during invasion and metastasis of malignant cancer cells. Several cellular adhesion molecules such as fibronectin 1, laminin B1, β-catenin, E-cadherin, ICAM-1 and E-selectin were detected in antisense transfectants (AS-3 and AS-4) and parental KYSE450 or empty vector cells by semi-quantitative RT-PCR and Western blot. Only fibronectin 1 was found drastically decreased in antisense transfectants compared with parental KYSE450 cells and empty vector cells (Figure 6II-4A), whereas no differences were found for other adhesive molecules we tested.

Fibronection 1 and SLP-2 expression was evaluated simultaneously in 25 pairs of patient-matched ESCC tissues by semi-quantitative PCR and in 18 pairs of ESCC tissues which were from the above 25 paired samples by Western blot (Figure 6II-4B). After band quantification carried out by Leica QWin image analysis and image processing software, correlation between fibronection 1 and SLP-2 expression was performed using Pearson's correlation analysis. The results suggested that there was a good correlation between fibronection 1 and SLP-2 expression (r_p=0.661, P=0.000). Immunohistochemical staining was also performed to analyze the correlation between fibronection 1 and SLP-2 expression in ESCC tissue arrays including 50 pairs of patient-matched samples. Correlation analysis using Spearman's correlation analysis indicated that there was a good correlation between fibronection 1 and SLP-2 expression (r_s=0.297□p=0.004). Confocal imaging was further performed to analyze the localization of SLP-2 and fibronectin 1, and the result showed that the two proteins co-localized at the membrane (Figure 6II-4C).

Discussion

Tumorigenesis is a complex and multistage process with many genes involved. To develop rational approaches to the diagnosis and treatment of cancer depends on identifying and understanding the molecular mechanisms that underlie tumor formation and progression. In this aspect, studies that seek to identify dysregulated genes and proteins in neoplasm are critical. Thus, research efforts aimed at systematically identifying the gene and protein expression profiles of normal and tumor cells are critically needed. The above cDNA microarry work identified the first time that SLP-2 was overexpressed in ESCC tissues [Luo et al., 2004]. The present study is the first, to our knowledge, to identify SLP-2 as a novel cancer-related gene and to examine the potential role of SLP-2 in tumorigenesis.

Our findings demonstrate that SLP-2 was associated with the different stages of tumor progression from normal tissue to premalignant and malignant lesions of esophagus. Usually, squamous epithelial dysplasia occurs about 10 years before the appearance of ESCC. Overexpression of SLP-2 in the early stage of ESCC suggests that SLP-2 may be fundamentally important in human tumorigenesis. The elucidation of mechanisms governing malignant changes from dysplasia to ESCC at the molecular level may provide a further understanding of tumorigenesis, as well as new approaches to the strategy of early prevention and treatment of ESCC. In addition, an effective non-surgical therapy for patients with early ESCC-endoscopic laser therapy-has been designed and may play a role in early ESCC treatment. The overexpression of SLP-2 in premaligant lesions indicated that it might serve as a marker for early detection of ESCC.

SLP-2 was first identified to be overexpressed in human ESCC, as well as lung cancer, laryngeal cancer and endometrial adenocarcinoma, which demonstrated that SLP-2 overexpression was very common in cancer development [Zhang et al., 2006]. However, the status of SLP-2 expression in other kinds of human tumors still remained unclear. The overexpression of SLP-2 in many kinds of human tumors and the inhibition of tumorigenesis of antisense construct made it to be more likely a potential oncogene. We further transfected sense-oriented SLP-2 to human fibroblast cell NIH3T3. Unfortunately, no malignant transformation was found. As we know, oncogenes would confer to tumorigenic potential in athymic nude mice. However, in many cases, a single oncogene was weakly sufficient for

malignant transformation of immortalized NIH3T3 cells. Co-transfection of one oncogene with another would demonstrate a synergistic action between the two oncogenes in the transformation of NIH3T3 fibroblasts [Velasco et al., 1994]. Although there were many oncogenes identified by now, this study will add more information on understanding the molecular mechanisms of tumorigenesis and will shed a light on development of better treatment and better diagnostic and preventive approaches to cancer.

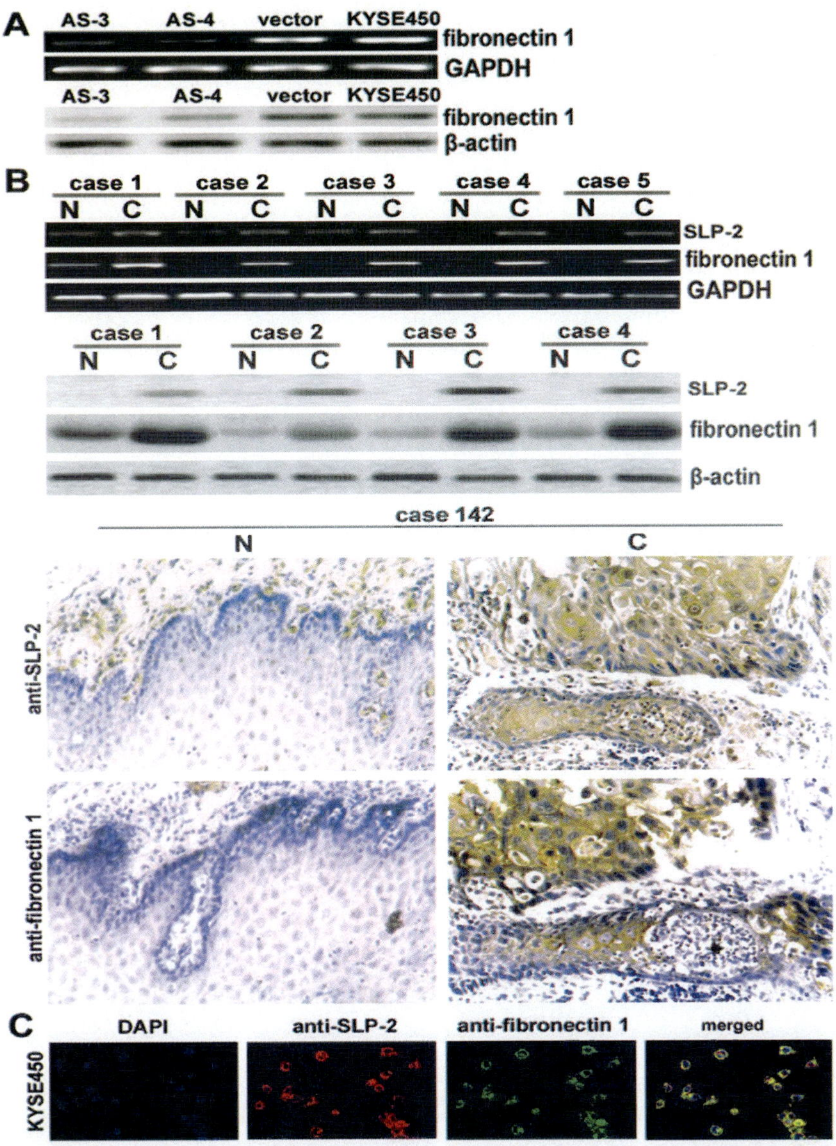

Figure 6II-4. Correlation analysis between fibronectin 1 and SLP-2 expression. A: Expression of fibronectin 1 in different individual transfectants detected by semi-quantitative RT-PCR and Western blot; B: Expression of fibronection 1 and SLP-2 in patient-matched ESCC tissues by semi-quantitative PCR, Western blot and immunohistochemistry; C: Co-localization of SLP-2 and fibronectin 1 by confocal imaging. AS-3, AS-4: different individual transfectants with antisense SLP-2; vector: transfectants with empty vector; KYSE450: parental KYSE450 cells. N: normal tissue; C: patient-matched tumor tissue.

Most cells adhere to their neighbors and to the extracellular matrix (ECM), a fibrillar meshwork surrounding or underlying most cells in the body. During embryological development, cell adhesion is important for the correct movements of cells modeling the embryo. In the adult, appropriate cell adhesion is necessary for numerous physiological processes and can be deranged in many diseases, including cancer. During cell culture we found that cells transfected with antisense SLP-2 were more easily to be digested by trypsin than parental KYSE450 cells and empty vector cells [Zhang et al., 2005]. We then speculated that SLP-2 may have some relation with cell adhesion, and subsequently carried out the cell attachment assay to determine cell adhesion ability by the CellTiter 96® AQueous One Solution Assay. This assay was a nonradioactive procedure that measured metabolic function that directly correlated with living cell numbers. Transfectants with antisense SLP-2 reduced cell attachment revealed that SLP-2 might be involved in cell adhesion. To further characterize the cell adhesion molecules involved in the process, the expression changes of some adhesion molecules, such as fibronectin 1, laminin B1, β-catenin, E-cadherin, ICAM-1 and E-selectin, were tested by RT-PCR and Western blot analysis. The results showed that one particularly important extracellular matrix protein, fibronectin 1, was sharply down-regulated in transfectants with antisense SLP-2 and co-localized at membrane with SLP-2. Fibronectins comprised a group of closely related proteins all encoded by a single gene and they promoted cell adhesion and cell migration and affected many other cellular processes [Briggs et al., 2005]. Co-localization of SLP-2 and fibronectin 1 at membrane indicated that SLP-2 might serve as transmembrane linkers between the extracellular matrix outside and the cytoskeleton and signaling systems inside cells.

The etiology of increased SLP-2 expression in human cancers is unknown. Possible mechanisms include point mutation, gene amplification, gene rearrangement and insertion of strong promoter or enhancer. Epigenetic modifications including demethylation and deacetylation may also be responsible. Mutation detection of SLP-2 exons with 30 patient-matched ESCC tissues did not reveal any mutation within the ORF of SLP-2 (data not shown). Further studies to investigate the mechanism of SLP-2 overexpression are currently underway. Antisense transfection of SLP-2 gene led to the arrest of S phase and the inhibition of cell growth, proliferation and tumor growth. These findings revealed that up-regulation of SLP-2 causes the cancer cells to become hyperproliferative and overexpression of SLP-2 may contribute to the malignant phenotype of ESCC.

Conclusion

A Mendelian autosomal recessive major gene may contribute to genetic susceptibility to ESCC in a moderate high-incidence area of Northern China;

A group of genes involved in cell differentiation were coordinately down-regulated in human ESCC, most of them encoded Ca^{2+}-binding or Ca^{2+}-modulating proteins that constitute the cell envelope. ESCC is associated with suppression of the genes involved in epidermal differentiation, including genes in 1q21 loci and corresponding to the epidermal differentiation complex;

c-Jun/AP-1 activated the expression of the differentiation-associated genes via MAPK pathways. In addition, GKLF was also able to regulate the expression of differentiation-associated genes SPRR1A, SPRR2A and KRT4 in human ESCC;

A significant association was found between chromosome 1q LOH and histopathological grade, which suggests this region harbor putative tumor suppressor gene(s) contributing to tumorigenesis and differentiation in ESCC;

Stefin A plays an important role in the growth, angiogenesis, invasion and metastasis of human ESCC;

A new candidate oncogene, Stomatin-like Protein 2 (SLP-2) is over-expressed and involved in regulating cell growth and cell adhesion in human ESCC.

Acknowledgements

This work was co-supported by China Key Program on Basic Research (2004CB518604), National Science Foundation of China (30425027), and the Chinese Hi-tech R and D program (2004AA231041).

References

Albini A. Tumor and endothelial cell invasion of basement membranes. The matrigel chemoinvasion assay as a tool for dissecting molecular mechanisms. *Pathol Oncol Res.* 1998; 4: 230-241.

Ameyar M, Wisniewska M, Weitzman JB. A role for AP-1 in apoptosis: the case for and against. *Biochimie.* 2003;85:747-752.

Angel P, Imagawa M, Chiu R, et al. Phorbol ester-inducible genes contain a common cis element recognized by a TPA-modulated trans-acting factor. *Cell.* 1987; 49:729-739

Angel P, Karin M. The role of Jun, Fos and the AP-1 complex in cell-proliferation and transformation. *Biochim Biophys Acta.* 1991;1072: 129-157.

Aoki T, Mori T, Du X, et al. Allelotype study of esophageal carcinoma, *Genes Chromosomes Cancer.* 1994; 10: 177-182.

Arai K, Yamada T, Nozawa R. Immunohistochemical investigation of migration inhibitory factor-related protein (MRP)-14 expression in hepatocellular carcinoma. *Med Oncol.* 2000;17:183-188.

Arnould T, Kim E, Tsiokas L,et al. The polycystic kidney disease 1 gene product mediates protein kinase C alpha-dependent and c-Jun N-terminal kinase-dependent activation of the transcription factor AP-1. *J Biol Chem.*1998;273:6013-6018.

Au W, Gascoyne R, Viswanatha D, et al. Cytogenetic analysis in mantle cell lymphoma: a review of 214 cases. *Leuk Lymphoma.* 2002; 43:783-791.

Axelrod JH, Reich R, Miskin R. Expression of human recombinant plasminogen activators enhances invasion and experimental metastasis of H-ras-transformed NIH 3T3 cells. *Mol Cell Biol.* 1989; 9: 2133-2141.

Banks-Schlegel SP, Quintero J. Growth and differentiation of human esophageal carcinoma cell lines. *Cancer Res.*1986; 46:250-258.

Baudier J, Delphin C, Grunwald D, et al. Characterization of the tumor suppressor protein p53 as a protein kinase C substrate and a S100b-binding protein. *Proc Natl Acad Sci* USA .1992; 89:11627-11631

Belguise K, Kersual N, Galtier F, et al. FRA-1 expression level regulates proliferation and invasiveness of breast cancer cells. *Oncogene.* 2005; 24:1434-1444.

Bervar A, Zajc I, Sever N, et al. Invasiveness of transformed human breast epithelial cell lines is related to cathepsin B and inhibited by cysteine proteinase inhibitors. *Biol Chem.* 2003; 384: 447-455.

Bikle DD, Ng D, Tu CL, et al. Calcium- and vitamin D-regulated keratinocyte differentiation. Mol Cell Endocrinol. 2001; 177:161-171.

Bjornland K, Buo L, Kjonniksen I, et al. Cysteine proteinase inhibitors reduce malignant melanoma cell invasion in vitro. *Anticancer Res.*1996; 16: 1627-1631.

Boike G, Lah T, Sloane BF, et al. A possible role for cysteine proteinase and its inhibitors in motility of malignant melanoma and other tumour cells. *Melanoma Res.* 1992; 1: 333-340.

Bonfil RD, Reddel RR, Ura H, et al. Invasive and metastatic potential of a v-Ha-ras-transformed human bronchial epithelial cell line. *J Natl Cancer Inst.* 1989; 81: 587-594.

Borg A, Zhang QX, Olsson H, et al. Chromosome 1 alterations in breast cancer: allelic loss on 1p and 1q is related to lymphogenic metastases and poor prognosis. *Genes Chromosomes Cancer.* 1992; 5: 311-320.

Briggs J, Chamboredon S, Castellazzi M, et al. Transcriptional upregulation of SPARC, in response to c-Jun overexpression, contributes to increased motility and invasion of MCF7 breast cancer cells. *Oncogene*, 2002; 21:7077-7091.

Briggs SL. The role of fibronectin in fibroblast migration during tissue repair. *J Wound Care.* 2005; 14:284-287.

Broome AM, Ryan D, Eckert RL. S100 protein subcellular localization during epidermal differentiation and psoriasis. *J Histochem Cytochem.* 2003; 51:675-685.

Brown LM, Hoover R, Silverman D, et al. Excess incidence of squamous cell esophageal cancer among US Black men: role of social class and other risk factors. *Am. J. Epidemiol.* 2001; 153:114-122.

Buttle DJ, Murata M, Knight CG, et al. CA074 methyl ester: a proinhibitor for intracellular cathepsin B. *Arch Biochem Biophys.* 1992; 299: 377-380.

Cajone F, Sherbet GV. Stathmin is involved in S100A4-mediated regulation of cell cycle progression. *Clin Exp Metastasis.* 1999;17:865-871

Calkins CC, Sameni M, Koblinski J, et al. Differential localization of cysteine protease inhibitors and a target cysteine protease, cathepsin B, by immuno-confocal microscopy. *J Histochem Cytochem.* 1998; 46: 745-751.

Carter CL, Hu N, Wu M, et al. Segregation analysis of esophageal cancer in 221 high-risk Chinese families. *J Natl Cancer Inst.* 1992; 84: 771–776 .

Cavigelli M, Dolfi F, Claret FX, et al. Induction of c-fos expression through JNK-mediated TCF/Elk-1 phosphorylation. *EMBO J.* 1995; 14: 5957-5964.

Chang L, Karin M. Mammalian MAP kinase signalling cascades. *Nature.* 2001;410: 37-40.

Chang S, Ge K, Zhai F, et al. Dietary intake of some micronutrients in different provinces of China. *Wei Sheng Yan Jiu.* 1999; 28:364-366.

Chang-Claude J, Becher H, Blettner M, et al. Familial aggregation of oesophageal cancer in a high incidence area in China. *Int J Epidemiol.* 1997; 26:1159–1165.

Chen B, Wang M, Cai Y, et al. Transglutaminase-3, an esophageal cancer-related gene. *Int J Cancer.* 2000; 88:862-865.

Chen B, Wang M, Cai Y, et al. Decreased expression of SPRR3 in Chinese human oesophageal cancer. *Carcinogenesis.* 2002; 21:2147-2150

Chen YJ, Vortmeyer A, Zhuang Z, et al. Loss of heterozygosity of chromosome 1q in gastrinomas: occurrence and prognostic significance. *Cancer Res.* 2003; 63:817-823.

Claus EB. The genetic epidemiology of cancer. *Cancer Surv.* 1995; 25:13–26.

Coulibaly S, Schwihla H, Abrahamson M, et al. Modulation of invasive properties of murine squamous carcinoma cells by heterologous expression of cathepsin B and cystatin C. *Int J Cancer.* 1999; 83: 526-531.

Davis R J. Signal transduction by the JNK group of MAP kinases, Cell. 2000; 103:239-252.

Day NE, Varghese C. Oesophageal cancer. *Cancer Surv.* 1994; 19–20:43–54.

Dayne, N.E. and Munoz, N. (1982). Esophagus. In: Schottenfeld D, Fraumeni J (eds) *Cancer epidemiology and prevention.* pp 596–623. Saunders, Philadelphia.

Deng T, Karin M. c-Fos transcriptional activity stimulated by H-Ras-activated protein kinase distinct from JNK and ERK. *Nature.* 1994; 371:171-175.

Dérijard B, Hibi M, Wu I, et al. JNK1: a protein kinase stimulated by UV-light and Ha-Ras that binds and phosphorylates the c-Jun activation domain. *Cell.* 1994; 76:1025-1037.

Donato R. S100: a multigenic family of calcium-modulated proteins of the EF-hand type with intracellular and extracellular functional roles. *Int J Biochem Cell Biol.* 2001; 33:637-668.

Dorsey MJ, Tae HJ, Sollenberger KG, et al. B-ATF: a novel human bZIP protein that associates with members of the AP-1 transcription factor family. *Oncogene.* 1995; 11: 2255-2265.

Dotto GP. Signal transduction pathways controlling the switch between keratinocyte growth and differentiation. *Crit Rev Oral Biol Med.* 1999; 10: 442-457.

Draghici S, Khatri P, Martins RP, et al. Global functional profiling of gene expression. *Genomics.* 2003; 81: 98–104.

Dwight T, Twigg S, Delbridge L, et al. Loss of heterozygosity in sporadic parathyroid tumours: involvement of chromosome 1 and the MEN1 gene locus in 11q13. *Clin. Endocrinol.* (Oxf) 2000; 53: 85-92.

Elder JT, Zhao X. Evidence for local control of gene expression in the epidermal differentiation complex. *Exp Dermatol.* 2002; 11: 406–412.

El-Rifai W, Moskaluk CA, Abdrabbo MK, et al. Gastric cancers overexpress S100A calcium-binding proteins. *Cancer Res.* 2002; 62: 6823-6826.

Fischer DF, Sark MW, Lehtola MM, et al. Structure and evolution of the human SPRR3 gene: implications for function and regulation. *Genomics.* 1999; 55: 88-99.

Frey BM, Reber BF, Vishwanath BS, et al. Annexin I modulates cell functions by controlling intracellular calcium release. *FASEB J.* 1999; 13: 2235-2245.

Frosch BA, Berquin I, Emmert-Buck MR, et al. Molecular regulation, membrane association and secretion of tumor cathepsin B. *Apmis*. 1999; 107: 28-37.

Gaki V, Tsopanomichalou M, Sourvinos G, et al. Allelic loss in chromosomal region 1q21-23 in breast cancer is associated with peritumoral angiolymphatic invasion and extensive intraductal component. *Eur. J. Surg. Oncol.* 2000; 26: 455-460.

Garrett SC, Varney KM, Weber DJ, et al. S100A4, a mediator of metastasis. *J Biol Chem.* 2006 ; 281: 677-680.

Gebhardt C, Breitenbach U, Tuckermann JP, et al. Calgranulins S100A8 and S100A9 are negatively regulated by glucocorticoids in a c-Fos-dependent manner and overexpressed throughout skin carcinogenesis. *Oncogene*. 2002; 21: 4266-4276.

Gendler SJ, Cohen EP, Craston A, et al. The locus of the polymorphic epithelial mucin (PEM) tumour antigen on chromosome 1q21 shows a high frequency of alteration in primary human breast tumours. *Int J Cancer.* 1990; 45: 431-435.

Gibbs S, Lohman F, Teubel W, et al. Characterization of the human spr2 promoter: induction after UV irradiation or TPA treatment and regulation during differentiation of cultured primary keratinocytes. *Nucleic Acids Res.* 1990; 18: 4401-4407.

Goldstein BJ, Kulaga HM, Reed RR. Cloning and characterization of SLP-3: a novel member of the stomatin family expressed by olfractory receptor neurons. *J Assoc Otolaryngol.* 2003; 4: 74-82.

Gomes LI, Esteves GH, Carvalho AF, et al. Expression profile of malignant and nonmalignant lesions of esophagus and stomach: differential activity of functional modules related to inflammation and lipid metabolism. Cancer Res. 2005; 65: 7127-7136.

Gondi CS, Lakka SS, Dinh DH, et al. RNAi-mediated inhibition of cathepsin B and uPAR leads to decreased cell invasion, angiogenesis and tumor growth in gliomas. *Oncogene.* 2004; 23: 8486-8496.

Gondi CS, Lakka SS, Yanamandra N, et al. Adenovirus-mediated expression of antisense urokinase plasminogen activator receptor and antisense cathepsin B inhibits tumor growth, invasion, and angiogenesis in gliomas. *Cancer Res.* 2004; 64: 4069-4077.

Grigorian M, Andresen S, Tulchinsky E, et al. Tumor suppressor p53 protein is a new target for the metastasis-associated Mts1/S100A4 protein: functional consequences of their interaction. *J Biol Chem.* 2001; 276: 22699-22708.

Guo W, Blot WJ, Li JY, et al. A nested case-control study of oesophageal and stomach cancers in the Linxian nutrition intervention trial. *Int J Epidemiol.* 1994; 23: 444–450.

Guo M, Mathieu PA, Linebaugh B, et al. Phorbol ester activation of a proteolytic cascade capable of activating latent transforming growth factor-beta a process initiated by the exocytosis of cathepsin B. *J Biol Chem.* 2002; 277: 14829-14837.

Gupta S, Campbell D, De'rijard B, et al. Transcription factor ATF2 regulation by the JNK signal transduction pathway. *Science*. 1995 ;267: 389–393.

Han Y, Wei F, Xu X, et al. Establishment and comparative genomic hybridization analysis of human esophageal carcinomas cell line EC9706. *Zhonghua Yi Xue Yi Chuan Xue Za Zhi* 2002; 19: 455-457.

Hanahan D, Weinberg RA. *The hallmarks of cancer.* Cell. 2000; 100: 57-70.

Hanaoka K, Qian F, Boletta A, et al. Co-assembly of polycystin-1 and -2 produces unique cation-permeable currents. *Nature*. 2000; 408 :990-994.

Hao XP, Pretlow TG, Rao JS, et al. Beta-catenin expression is altered in human colonic aberrant crypt foci. *Cancer Res*. 2001; 61: 8085-8088.

Harada H, Uchida N, Shimada Y, et al. Polymorphism and allelic loss at the AS3 locus on 13q12-13 in esophageal squamous cell carcinoma. *Int. J. Oncol.* 2001; 18: 1003-1007.

Hayashi M, Tamura G, Jin Z, et al. Microsatellite instability in esophageal squamous cell carcinoma is not associated with hMLH1 promoter hypermethylation. *Pathol Int.* 2003; 53: 270-276.

Heizmann CW, Fritz G, Schafer BW. S100 proteins: structure, functions and pathology. *Front Biosci*. 2002; 7: d1356-1368.

Hermani A, Hess J, De Servi B, et al. Calcium-binding proteins S100A8 and S100A9 as novel diagnostic markers in human prostate cancer. *Clin Cancer Res.* 2005; 11: 5146-5152.

Hirai K, Yokoyama M, Asano G, et al. Expression of cathepsin B and cystatin C in human colorectal cancer. *Hum Pathol.* 1999; 30: 680-686.

Hitomi J, Kimura T, Kusumi E, et al. Novel S100 proteins in human esophageal epithelial cells: CAAF1 expression is associated with cell growth arrest. *Arch Histol Cytol.* 1998; 61: 163-178.

Hitomi J, Yamaguchi K, Kikuchi Y, et al. A novel calcium-binding protein in amniotic fluid, CAAF1: its molecular cloning and tissue distribution. *J Cell Sci.* 1996; 109:805-815.

Hitomi K, Presland RB, Nakayama T. Analysis of epidermal-type transglutaminase (transglutaminase 3) in human stratified epithelia and cultured keratinocytes using monoclonal antibodies. *J Dermatol Sci.* 2003; 32: 95–103.

Hohl D, de Viragh PA, Amiguet-Barras F, et al. The small proline-rich proteins constitute a multigene family of differentially regulated cornified cell envelope precursor proteins. *J Invest Dermatol.* 1995; 104: 902-909.

Hu N, Dawsey SM, Wu M, et al. Familial aggregation of oesophageal cancer in Yangcheng County, Shanxi Province, China. *Int J Epidemiol.* 1992; 21:877–882.

Hu N, Roth MJ, Emmert-Buck MR, et al. Allelic loss in esophageal squamous cell carcinoma patients with and without family history of upper gastrointestinal tract cancer. *Clin Cancer Res* .1999; 5: 3476–3482.

Hu N, Roth MJ, Polymeropolous M, et al. Identification of novel regions of allelic loss from a genomewide scan of esophageal squamous-cell carcinoma in a high-risk Chinese population. *Genes Chromosomes Cancer.* 2000; 27: 217-228.

Hu N, Huang J, Emmert-Buck MR, et al. Frequent inactivation of the TP53 gene in esophageal squamous cell carcinoma from a high-risk population in China. *Clin Cancer Res.* 2001; 7: 883-891.

Hu N, Goldstein AM, Albert PS, et al. Evidence for a familial esophageal cancer susceptibility gene on chromosome 13. *Cancer Epidemiol Biomarkers Prev.* 2003; 12: 1112-1115.

Hu N, Flaig MJ, Su H, et al. Comprehensive Characterization of Annexin I Alterations in Esophageal Squamous Cell Carcinoma. Clinical Cancer Research. 2004; 10: 6013-6022.

Hu N, Wang C, Han XY, et al. Evaluation of BRCA2 in the genetic susceptibility of familial esophageal cancer. *Oncogene.* 2004; 23: 852-858.

Hu N, Wang C, Su H,et al. Goldstein A. High frequency of CDKN2A alterations in esophageal squamous cell carcinoma from a high-risk Chinese population. *Genes Chromosomes Cancer.* 2004; 39:2 05-216.

Hu YC, Lam KY, Law S, et al. Identification of differentially expressed genes in esophageal squamous cell carcinoma (ESCC) by cDNA expression array: overexpression of Fra-1, Neogenin, Id-1, and CDC25B genes in ESCC. *Clin Cancer Res.* 2001; 7: 2213-2221.

Huang EY, Madireddi MT, Gopalkrishnan RV, et al. Genomic structure, chromosomal localization and expression profile of a novel melanoma differentiation associated (mda-7) gene with cancer specific growth suppressing and apoptosis inducing properties. *Oncogene.* 2001; 20: 7051-7063.

Hughes SJ, Glover TW, Zhu XX, et al. A novel amplicon at 8p22-23 results in overexpression of cathepsin B in esophageal adenocarcinoma. *Proc Natl Acad Sci U S A.* 1998; 95: 12410-12415.

Hunter MJ, Chazin WJ. High-level expression and dimer characterization of the S100 EF-hand proteins, migration inhibitory factor related proteins 8 and 14. *J Biol Chem.* 1998; 273: 12427–12435.

Ikeguchi M, Unate H, Maeta M, et al. Detection of loss of heterozygosity at microsatellite loci in esophageal squamous-cell carcinoma. *Oncology.* 1999; 56: 164-168.

Ikura M, Yap KL. Where cancer meets calcium--p53 crosstalk with EF-hands. *Nat Struct Biol.* 2000; 7: 570-574.

Imazawa M, Hibi K, Fujitake S, et al. S100A2 overexpression is frequently observed in esophageal squamous cell carcinoma. *Anticancer Res.* 2005; 25:1247-1250.

Jameson BA, Wolf H. The antigen index: a novel algorithm for predicting antigenic determinatants. *Comput Appl Biosci.* 1998; 4: 181-186.

Jankowski J, Hopwood D, Dover R, et al. Development and growth of normal, metaplastic and dysplastic oesophageal mucosa: biological markers of neoplasia. *Eur J Gastroenterol Hepatol.* 1993; 5: 235-246.

Jenkins TD, Opitz OG, Okano J, et al. Transactivation of the human keratin 4 and Epstein-Barr virus ED-L2 promoters by gut-enriched Kruppel-like factor. *J Biol Chem.* 1998; 273:10747-10754.

Ji J, Zhao L, Wang X, et al. Differential Expression of S100 Gene Family in Human Esophageal Squamous Cell Carcinoma. *Journal of Cancer Research and Clinical Oncology.* 2004; 130: 480-486.

Jones SJ, Dicker AJ, Dahler AL, et al. E2F as a regulator of keratinocyte proliferation: implications for skin tumor development. *J Invest Dermatol.* 1997; 109: 187-193.

Kadowaki Y, Fujiwara T, Fukazawa T, et al. Induction of differentiation-dependent apoptosis in human esophageal squamous cell carcinoma by adenovirus-mediated p21sdi1 gene transfer. *Clin Cancer Res.* 1999; 5: 4233-4241.

Kalinin AE, Kajava AV, Steinert PM. Epithelial barrier function: assembly and structural features of the cornified cell envelope. *Bioessays.* 2002; 24: 789–800.

Kanda Y, Nishiyama Y, Shimada Y, et al. Analysis of gene amplification and overexpression in human esophageal-carcinoma cell lines. *Int J Cancer.* 1994; 58: 291-297.

Karin M, Liu Z, and Zandi E. AP-1 function and regulation. *Curr. Opin. Cell Biol.* 1997; 9: 240-246.

Karan D, Schmied BM, Dave BJ, et al. Decreased androgen-responsive growth of human prostate cancer is associated with increased genetic alterations. *Clin. Cancer Res.* 2001; 7: 3472-3480.

Karkera J, Ayache S, Ransome R, et al. Refinement of regions with allelic loss on chromosome 18p11.2 and 18q12.2 in esophageal squamous cell carcinoma. *Clin Cancer Res.* 2000; 6: 3565-3569.

Katz JP, Perreault N, Goldstein BG, et al. The zinc-finger transcription factor Klf4 is required for terminal differentiation of goblet cells in the colon. *Development.* 2002; 129: 2619-2628.

Kazumori A, Takumi T, Ryushi N, et al. Immunohistochemical investigation of S100A9 expression in pulmonary adenocarcinoma: S100A9 expression is associated with tumor differentiation. *Oncology Reports.* 2001; 8: 591-596.

Kerkhoff C, Klempt M, Sorg C. Novel insights into structure and function of MRP8 (S100A8) and MRP14 (S100A9). *Biochim Biophys Acta.* 1998; 1448: 200-211.

Khan A, Krishna M, Baker SP, et al. Cathepsin B expression and its correlation with tumor-associated laminin and tumor progression in gastric cancer. *Arch Pathol Lab Med.* 1998; 122: 172-177.

Kimchi ET, Posner MC, Park JO, et al. Progression of Barrett's metaplasia to adenocarcinoma is associated with the suppression of the transcriptional programs of epidermal differentiation. *Cancer Res.* 2005; 65: 3146-3154.

Kitamura Y, Shimizu K, Tanaka S, et al. Association of allelic loss on 1q, 4p, 7q, 9p, 9q, and 16q with postoperative death in papillary thyroid carcinoma. *Clin. Cancer Res.* 2000; 6: 1819-1825.

Ko JM, Wong CP, Tang CM, et al. Frequent loss of heterozygosity on multiple chromosomes in Chinese esophageal squamous cell carcinomas. *Cancer Lett.* 2001; 170:131-138.

Kobayashi H, Schmitt M, Goretzki L, et al. Cathepsin B efficiently activates the soluble and the tumor cell receptor-bound form of the proenzyme urokinase-type plasminogen activator (Pro-uPA). *J Biol Chem.* 1991; 266: 5147-5152.

Koch PJ, de Viragh PA, Scharer E, et al. Lessons from loricrin-deficient mice: compensatory mechanisms maintaining skin barrier function in the absence of a major cornified envelope protein. *J Cell Biol.* 2000; 151: 389-400.

Konduri SD, Yanamandra N, Siddique K, et al. Modulation of cystatin C expression impairs the invasive and tumorigenic potential of human glioblastoma cells. *Oncogene.* 2002; 21: 8705-8712.

Kouzarides, T, Ziff E. The role of the leucine zipper in the fos–jun interaction. *Nature.*1988; 336: 646-651.

Krepela E, Prochazka J, Karova B,et al. Cysteine proteases and cysteine protease inhibitors in non-small cell lung cancer. *Neoplasma.*1998; 45:318-331.

Krueger S, Haeckel C, Buehling F, et al. Inhibitory effects of antisense cathepsin B cDNA transfection on invasion and motility in a human osteosarcoma cell line. *Cancer Res.* 1999; 59: 6010-6014.

Krueger S, Haeckel C, and Buehling F. Antisense inhibition of cathepsin B in a human osteosarcoma cell line. *Adv Exp Med Biol.* 2000; 477: 439-444.

Kustikova O, Kramerov D, Grigorian M, et al. Fra-1 induces morphological transformation and increases in vitro invasiveness and motility of epithelioid adenocarcinoma cells. *Mol Cell Biol.* 1998; 18: 7095-7105.

Kyriazanos ID, Tachibana M, Dhar DK, et al. Expression and prognostic significance of S100A2 protein in squamous cell carcinoma of the esophagus. *Oncol Rep.* 2002; 9: 503-510.

Lagergren J, Bergstrom R, Lindgren A, et al. Symptomatic gastroesophageal reflux as a risk factor for esophageal adenocarcinoma. *N Engl J Med.* 1999; 340: 825-831.

Lakka SS, Gondi CS, Yanamandra N, et al. Inhibition of cathepsin B and MMP-9 gene expression in glioblastoma cell line via RNA interference reduces tumor cell invasion, tumor growth and angiogenesis. *Oncogene.* 2004; 23: 4681-4689.

Lam KY, Tsao SW, Zhang D, et al. Prevalence and predictive value of p53 mutation in patients with oesophageal squamous cell carcinomas: a prospective clinico-pathological study and survival analysis of 70 patients. *Int J Cancer.* 1997; 74: 212-219.

Lam CT, Tang CM, Lau KW, et al. Loss of heterozygosity on chromosome 11 in esophageal squamous cell carcinomas. *Cancer Lett.* 2002; 178: 75-81.

Lee W, Mitchell P, Tjian R. Purified transcription factor AP-1 interacts with TPA-inducible enhancer elements. *Cell.* 1987; 49: 741-752.

Lestou VS, Ludkovski O, Connors JM, et al. Characterization of the recurrent translocation t(1;1)(p36.3;q21.1-2) in non-Hodgkin lymphoma by multicolor banding and fluorescence in situ hybridization analysis. *Genes Chromosomes Cancer.* 2003; 36: 375-381.

Li, G.H. and He, L.J. (1986). A survey on the familial aggregation of esophageal cancer in Yangcheng county, Shanxi Province. In: Wu M, Nebert DW (eds) *Genes and diseases: proceedings of the first Sino-American human genetics workshop.* pp 43-47. Beijing: Science Press.

Li G, Hu N, Goldstein AM, et al. Allelic loss on chromosome bands 13q11-q13 in esophageal squamous cell carcinoma. *Genes Chromosomes Cancer.* 2001; 31: 390-397.

Li JY. Epidemiology of esophageal cancer in China. *Natl Cancer Inst Monogr.* 1982; 62: 113-120.

Li J, Liu Z, Wang Y, et al. Allelic imbalance of chromosome 1q in esophageal squamous cell carcinomas from China: a novel region of allelic loss and significant association with differentiation. *Cancer Letters.* 2005; 220: 221-230.

Li JY, Ershow AG, Chen ZJ, et al. A case-control study of cancer of the esophagus and gastric cardia in Linxian. *Int J Cancer.* 1989; 43: 755-761.

Li W, Lu W, Zhu M, et al. Determination of copper, zinc, iron and calcium in wheat and maize and three nitrogen compounds in high and low risk areas of esophageal cancer. *Wei Sheng Yan Jiu.* 1998; 27: 69-71.

Li W, Wang X, Zhang C, et al. Genetic epidemiological survey on esophageal carcinoma in part of population of Yangquan City. *Natl Med J China.* 1998; 78: 203-206.

Li W, Hu N, Su H, et al. Allelic loss on chromosome 13q14 and mutation in deleted in cancer 1 gene in esophageal squamous cell carcinoma. *Oncogene.* 2003; 22: 314-318.

Li W, Ding F, Zhang L, et al. Overexpression of Stefin A in Human Esophageal Squamous Carcinoma Cells Inhibits Tumor Cell Growth, Angiogenesis, Invasion and Metastasis. *Clin Cancer Res.* 2005; 11: 8753-8762.

Lichun Y, Ching Tang CM, Wai Lau K, et al. Frequent loss of heterozygosity on chromosome 9 in Chinese esophageal squamous cell carcinomas. *Cancer Lett.* 2004; 203: 71-77.

Liew CC, Hwang DM, Wang RX, et al. Construction of a human heart cDNA library and identification of cardiovascular based genes (CVBest). *Mol Cell Biochem.* 1997; 172: 81-87.

Lin G, Xiao J, Zhao S, et al. Identification of AP1 cis-element and transcriptional effect on cytokeratin 13 gene expression. *Zhonghua Yi Xue Yi Chuan Xue Za Zhi.* 2002;19: 479-483.

Lin J, Blake M, Tang C, et al. Inhibition of p53 transcriptional activity by the S100B calcium-binding protein. *J Biol Chem.* 2001; 276: 35037-35041.

Liu, B.Q. and Li, B. (1984). Epidemiology of carcinoma of the esophagus in China. In: Huang GJ, Wu YK (ed.), *Carcinoma of the esophagus and gastric cardia.* pp1–24. Berlin: Springer Verlag.

Liu D, Rudland PS, Sibson DR, et al. Expression of calcium-binding protein S100A2 in breast lesions. *Br J Cancer.* 2000; 83: 1473-1479.

Lohman FP, Gibbs S, Fischer DF, et al. Involvement of c-JUN in the regulation of terminal differentiation genes in normal and malignant keratinocytes. *Oncogene.* 1997;14:1623-1627.

Lu J, Liu Z, Xiong M, et al. Gene expression profile changes in initiation and progression of squamous cell carcinoma of esophagus. *Int J Cancer.* 2001; 91: 288-294.

Luo A, Kong J, Hu G, et al. Discovery of Ca^{2+}-relevant and differentiation-associated genes downregulated in esophageal squamous cell carcinoma using cDNA microarray. *Oncogene.* 2004; 23: 1291-1299.

Ma S, Rao L, Freedberg IM, et al. Transcriptional control of K5, K6, K14, and K17 keratin genes by AP-1 and NF-kappaB family members. *Gene Expr.* 1997; 6: 361-370.

Marenholz I, Volz A, Ziegler A, et al. Genetic analysis of the epidermal differentiation complex (EDC) on human chromosome 1q21: chromosomal orientation, new markers, and a 6-Mb YAC contig. *Genomics.* 1996; 37: 295-302.

Marenholz I, Zirra M, Fischer DF, et al. Identification of human epidermal differentiation complex (EDC)-encoded genes by subtractive hybridization of entire YACs to a gridded keratinocyte cDNA library. *Genome Res.* 2001; 11: 341-355.

Marenholz I. Identification of human epidermal differentiation complex (EDC)-encoded genes by subtractive hybridization of entire YACs to a gridded keratinocyte cDNA library. *Genome Res.* 2001; 11: 341-355.

Marshall D, Hardman MJ, Nield KM, et al. Differentially expressed late constituents of the epidermal cornified envelope. *Proc. Natl. Acad. Sci. USA.* 2001; 98: 13031-13036.

Matsuki M, Yamashita F, Ishida-Yamamoto A, et al. Defective stratum corneum and early neonatal death in mice lacking the gene for transglutaminase 1 (keratinocyte transglutaminase). *Proc Natl Acad Sci.* 1998; 95: 1044-1049.

Mayama T, Fukushige S, Shineha R, et al. Frequent loss of copy number on the long arm of chromosome 21 in human esophageal squamous cell carcinoma. *Int. J. Oncol.* 2000; 17: 245-252.

Menin C, Santacatterina M, Zambon A, et al. Anomalous transcripts and allelic deletions of the FHIT gene in human esophageal cancer. Cancer Genet Cytogenet 2000; 119: 56-61.

Mignatti P, Rifkin DB. Biology and biochemistry of proteinases in tumor invasion. *Physiol Rev.* 1993; 73: 161-195.

Minden A, Lin A, McMahon M, et al. Differential activation of ERK and JNK mitogen-activated protein kinases by Raf-1 and MEKK. *Science.* 1994; 266: 1719-1723.

Minden A, Lin A, Smeal T, et al. C-Jun N-terminal phosphorylation correlates with activation of the JNK subgroup but not the ERK subgroup of mitogen-activated protein kinase. *Mol Cell Biol.* 1994; 14: 6683-6688.

Mischke D, Korge BP, Marenholz I, et al. Genes encoding structural proteins of epidermal cornification and S100 calcium-binding proteins form a gene complex ("epidermal differentiation complex") on human chromosome 1q21. *J Invest Dermatol.* 1996; 106: 989-992.

Mohanam S, Jasti SL, Kondraganti SR, et al. Down-regulation of cathepsin B expression impairs the invasive and tumorigenic potential of human glioblastoma cells. *Oncogene.* 2001; 20: 3665-3673.

Montesano R, Hollstein M, Hainaut P. Genetic alterations in esophageal cancer and their relevance to etiology and pathogenesis: a review. *Int J Cancer.* 1996; 69: 225-235.

Moore BW. A soluble protein characteristic of the nervous system. *Biochem Biophys Res Commun.* 1965; 19: 739-744.

Moore RC, Durso NA, Cyr RJ. Elongation factor-1alpha stabilizes microtubules in a calcium/calmodulin-dependent manner. *Cell Motil Cytoskeleton.* 1998; 41: 168-180.

Mueller A, Schafer BW, Ferrari S, et al. The calcium-binding protein S100A2 interacts with p53 and modulates its transcriptional activity. *J Biol Chem.* 2005; 280: 29186-29193.

Mukherjee P, Sotnikov AV, Mangian HJ, et al. Energy intake and prostate tumor growth, angiogenesis, and vascular endothelial growth factor expression. *J Natl Cancer Inst.* 1999; 91: 512-523.

Nagai A, Terashima M, Harada T, et al. Cathepsin B and H activities and cystatin C concentrations in cerebrospinal fluid from patients with leptomeningeal metastasis. *Clin Chim Acta.* 2003; 329: 53-60.

Nagasawa S, Onda M, Sasajima K, et al. Cyclin D1 overexpression as a prognostic factor in patients with esophageal carcinoma. *J Surg Oncol.* 2001; 78: 208-214.

Nagy N, Brenner C, Markadieu N, et al. S100A2, a putative tumor suppressor gene, regulates in vitro squamous cell carcinoma migration. *Lab Invest.* 2001; 81: 599-612.

Nakagawa Y, Sawanobori M, Amaya H, et al. Clinical implications of abnormalities of chromosomes 1 and 13 in multiple myeloma, *Acta. Haematol.* 2003; 109: 129-136.

Nateri AS, Riera-Sans L, Costa C, et al. The ubiquitin ligase SCFFbw7 antagonizes apoptotic JNK signaling. *Science.* 2004; 303: 1374-1378.

National Cancer Control Office (1980) Investigation of cancer mortality in China. pp 65–96. Beijing: People's Health Publishing House.

Navab R, Mort JS, Brodt P. Inhibition of carcinoma cell invasion and liver metastases formation by the cysteine proteinase inhibitor E-64. *Clin Exp Metastasis. 1997*; 15: 121-129.

Neglia M, Bertoni L, Zoli W, et al. Amplification of the pericentromeric region of chromosome 1 in a newly established colon carcinoma cell line. *Cancer Genet Cytogenet.* 2003; 142: 99-106.

Ness SL, Edelmann W, Jenkins TD, et al. Mouse keratin 4 is necessary for internal epithelial integrity. *J Biol Chem.* 1998; 273: 23904-23911.

Ninomiya I, Ohta T, Fushida S, et al. Increased expression of S100A4 and its prognostic significance in esophageal squamous cell carcinoma. *Int J Oncol.* 2001; 18: 715-720.

Ogasawara S, Maesawa C, Tamura G, et al. Frequent microsatellite alteration on chromosome 3p in esophageal squamous cell carcinoma. *Cancer Res.* 1995; 55: 891-894.

Ohuchida K, Mizumoto K, Ishikawa N, et al. The role of S100A6 in pancreatic cancer development and its clinical implication as a diagnostic marker and therapeutic target. *Clin Cancer Res.* 2005; 11: 7785-7793.

Opitz OG, Jenkins TD, Rustgi AK. Transcriptional regulation of the differentiation- linked human K4 Promoter is dependent upon esophageal-specific nuclear factors. *J Biol Chem.* 1998; 273: 23912-23921.

Papavassiliou AG, Treier M, Bohmann D. Intramolecular signal transduction in c-Jun. *EMBO J.* 1995; 14: 2014-2019.

Parker C, Lakshmi MS, Piura B, et al. Metastasis-associated mts1 gene expression correlates with increased p53 detection in the B16 murine melanoma. *DNA Cell Biol.* 1994; 13: 343-351.

Parker C, Whittaker PA, Usmani BA, et al. Induction of 18A2/mts1 gene expression and its effects on metastasis and cell cycle control. *DNA Cell Biol.* 1994; 13: 1021-1028.

Parkin DM, Pisani P, Ferlay J. Estimates of the worldwide incidence of eighteen major cancers in 1985. *Int J Cancer.* 1993; 54: 594-606.

Pavlova A, Krupa JC, Mort JS, et al. Cystatin inhibition of cathepsin B requires dislocation of the proteinase occluding loop. Demonstration By release of loop anchoring through mutation of his110. *FEBS Lett.* 2000; 487: 156-160.

Pietas A, Schluns K, Marenholz I, et al. Molecular cloning and characterization of the human S100A14 gene encoding a novel member of the S100 family. *Genomics.* 2002; 79: 513-522.

Plaat B, Molenaar W, Mastik M, et al. Computer-assisted cytogenetic analysis of 51 malignant peripheral-nerve-sheath tumors: sporadic vs. neurofibromatosis-type-1-associated malignant schwannomas, *Int. J. Cancer.* 1999; 83: 171-178.

Pulverer BJ, Kyriakis JM, Avruch J, et al. Phosphorylation of c-jun mediated by MAP kinases. *Nature.* 1991; 353: 670-674.

Raingeaud J, Gupta S, Rogers JS, et al. Pro-inflammatory cytokines and environmental stress cause p38 mitogen-activated protein kinase activation by dual phosphorylation on tyrosine and threonine. *J Biol Chem.* 1995; 270: 7420-7426.

Redwood SM, Liu BC, Weiss RE, et al. Abrogation of the invasion of human bladder tumor cells by using protease inhibitor(s). *Cancer.* 1992; 69: 1212-1219.

Ribeiro U, Posner MC, Safatle-Ribeiro AV, et al. Risk factors for squamous cell carcinoma of the oesophagus. *Br. J. Surg.* 1996; 83: 1174-1185.

Robinson NA, Lapic S, Welter JF, et al. S100A11, S100A10, annexin I, desmosomal proteins, small proline-rich proteins, plasminogen activator inhibitor-2, and involucrin are components of the cornified envelope of cultured human epidermal keratinocytes. *J Biol Chem.* 1997; 272: 12035-12046.

Ross KF, Herzberg MC. Calprotectin expression by gingival epithelial cells. *Infect Immun.* 2001; 69: 3248-3254.

Roth J, Burwinkel F, van den Bos C, et al. MRP8 and MRP14, S-100-like proteins associated with myeloid differentiation, are translocated to plasma membrane and intermediate filaments in a calcium-dependent manner. *Blood.* 1993; 82: 1875-1883.

Rustandi RR, Baldisseri DM, Weber DJ. Structure of the negative regulatory domain of p53 bound to S100B(betabeta). *Nat Struct Biol.* 2000; 7: 570-574.

S.A.G.E. (1997) Statistical Analysis for Genetic Epidemiology, Release 3.0. Computer program package available from the Department of Epidemiology and Biostatistics. Cleveland: Case Western Reserve University.

Sakaguchi M, Miyazaki M, Inoue Y, et al. Relationship between contact inhibition and intranuclear S100C of normal human fibroblasts. *J Cell Biol.* 2000; 149:1193-1206.

Sark MW, Fischer DF, de Meijer E, et al. AP-1 and ets transcription factors regulate the expression of the human SPRR1A keratinocyte terminal differentiation marker. *J. Biol. Chem.* 1998; 273: 24683-24692.

Sato N, Hitomi J. S100P expression in human esophageal epithelial cells: Human esophageal epithelial cells sequentially produce different S100 proteins in the process of differentiation. *Anat Rec.* 2002; 267:60-69.

Schafer BW, Wicki R, Engelkamp D, et al. Isolation of a YAC clone covering a cluster of nine S100 genes on human chromosome 1q21: rationale for a new nomenclature of the S100 calcium-binding protein family. *Genomics.* 1995;25:638-643.

Scotto C, Deloulme JC, Rousseau D, et al. Calcium and S100B regulation of p53-dependent cell growth arrest and apoptosis. *Mol Cell Biol.* 1998, 18:4272-4281.

Seeliger S, Vogl T, Engels IH, et al. Expression of calcium-binding proteins MRP8 and MRP14 in inflammatory muscle diseases. *Am J Pathol.* 2003;163:947-956.

Seery JP, Watt FM. Asymmetric stem-cell divisions define the architecture of human oesophageal epithelium. *Curr Biol.* 2000;10:1447-1450.

Seery JP. Stem cells of the oesophageal epithelium. J Cell Sci. 2002;115:1783-1789.

Segre JA, Bauer C, Fuchs E. Klf4 is a transcription factor required for establishing the barrier function of the skin. *Nat Genet.* 1999; 22: 356-360.

Seidel G, Prohaska R. Molecular cloning of hSLP-1, a novel human brain-specific member of the band 7/MEC-2 family similar to Caenorhabditis elegans UNC-24. *Gene.* 1998; 225: 23-29.

Semov A, Moreno MJ, Onichtchenko A, et al. Metastasis-associated protein S100A4 induces angiogenesis through interaction with Annexin II and accelerated plasmin formation. *J Biol Chem.* 2005; 280: 20833-20841.

Sexton PS and Cox JL. Inhibition of motility and invasion of B16 melanoma by the overexpression of cystatin C. *Melanoma Res.* 1997; 7: 97-101.

Shaheen N, Ransohoff DF. Gastroesophageal reflux, Barrett esophagus and esophageal cancer. *JAMA*. 2002; 287:1972–1981.

Shimada Y, Imamura M, Wagata T, et al. Characterization of 21 newly established esophageal cancer cell lines. *Cancer*. 1992; 69: 277-284.

Shridhar R, Zhang J, Song J, et al. Cystatin M suppresses the malignant phenotype of human MDA-MB-435S cells. *Oncogene*. 2004; 23: 2206-2215.

Sinha AA, Quast BJ, Wilson MJ, et al. Prediction of pelvic lymph node metastasis by the ratio of cathepsin B to stefin A in patients with prostate carcinoma. *Cancer*. 2002; 94: 3141-3149.

Sloane BF, Moin K, Sameni M, et al. Membrane association of cathepsin B can be induced by transfection of human breast epithelial cells with c-Ha-ras oncogene. *J Cell Sci.* 1994; 107: 373-384.

Sloane, BF. Cathepsin B and cystatins: evidence for a role in cancer progression. *Semin Cancer Biol.* 1990; 1: 137-152.

Smeal T, Angel P, Meek J, et al. Different requirements for formation of Jun:Jun and Jun:Fos complexes. *Genes Dev.* 1989;3: 2091-2100.

Smeal T, Binetruy B, Mercola D, et al. Oncoprotein-mediated signaling cascade stimulates c-Jun activity by phosphorylation of serines 63 and 73. *Mol. Cell Biol.* 1992; 12: 3507-3513.

Smeds J, Berggren P, Ma X, et al. Genetic status of cell cycle regulators in squamous cell carcinoma of the oesophagus: the CDKN2A (p16(INK4a) and p14(ARF)) and p53 genes are major targets for inactivation. *Carcinogenesis.* 2002; 23: 645-655.

Sprenger RR, Speijer D, Back JW, et al. Comparative proteomics of human endothelial cell caveolae and rafts using two-dimensional gel electrophoresis and mass spectrometry. *Electrophoresis.* 2004; 25: 156-172.

Stein HJ, Siewert JR. Barrett's esophagus: pathogenesis, epidemiology, functional abnormalities, malignant degeneration, and surgical management. *Dysphagia.* 1993; 8:276-288.

Steven AC, Steinert PM. Protein composition of cornified cell envelopes of epidermal keratinocytes. *J Cell Sci.* 1994; 107:693-700.

Strojnik T, Zajc I, Bervar A, et al. Cathepsin B and its inhibitor stefin A in brain tumors. *Pflugers Arch.* 2000; 439: R122-123.

Stulik J, Osterreicher J, Koupilova K,et al. The analysis of S100A9 and S100A8 expression in matched sets of macroscopically normal colon mucosa and colorectal carcinoma: the S100A9 and S100A8 positive cells underlie and invade tumor mass. *Electrophoresis.* 1999; 20:1047-1054.

Su H, Hu N, Shih J, et al. Gene expression analysis of esophageal squamous cell carcinoma reveals consistent molecular profiles related to a family history of upper gastrointestinal cancer. *Cancer Res.* 2003; 63: 3872-3876.

Suzuki H, Abo S, Kitamura M, et al. Gene amplification of int-2 and erbB in human esophageal cancer: relationship to clinicopathological variables. *Cancer Invest.* 1997; 15:411-415.

Szpaderska, AM and Frankfater, A. An intracellular form of cathepsin B contributes to invasiveness in cancer. *Cancer Res.* 2001; 61: 3493-3500.

Taipale, J and Keski-Oja, J. Growth factors in the extracellular matrix. *Faseb J.* 1997; 11: 51-59.

Takahashi H, Asano K, Kinouchi M, et al. Structure and transcriptional regulation of the human cystatin A gene. *J. Biol. Chem.* 1998; 273:17375-17380.

Takahashi H, Honma M, Ishida-Yamamoto A, et al. Expression of human cystatin A by keratinocytes is positively regulated via the Ras/MEKK1/MKK7/JNK signal transduction pathway but negatively regulated via the Ras/Raf-1/MEK1/ERK pathway. *J Biol Chem.* 2001; 276: 36632-36638.

Tamoto E, Tada M, Murakawa K, et al. Gene-expression profile changes correlated with tumor progression and lymph node metastasis in esophageal cancer. *Clin Cancer Res.* 2004; 10:3629-3638.

Teh BT, Farnebo F, Kristoffersson U, et al. Autosomal dominant primary hyperparathyroidism and jaw tumor syndrome associated with renal hamartomas and cystic kidney disease: linkage to 1q21-q32 and loss of the wild type allele in renal hamartomas. *J Clin. Endocrinol Metab.* 1996; 81: 4204-4211.

Teh BT, Farnebo F, Twigg S, et al. Familial isolated hyperparathyroidism maps to the hyperparathyroidism-jaw tumor locus in 1q21-q32 in a subset of families. *J Clin Endocrinol Metab.* 1998; 83: 2114-2120.

Thiel C, Osborn M, Gerke V. The tight association of the tyrosine kinase substrate annexin II with the submembranous cytoskeleton depends on intact p11- and Ca(2+)-binding sites. *J Cell Sci.* 1992; 103:733-742.

Tokugawa T, Sugihara H, Tani T, et al. Modes of silencing of p16 in development of esophageal squamous cell carcinoma, *Cancer Res.* 2002;62 : 4938-4944.

Turksen K, Troy TC. Permeability barrier dysfunction in transgenic mice overexpressing claudin 6. *Development.* 2002; 129:1775-1784.

van Dam H, Wilhelm D, Herr I, et al. ATF-2 is preferentially activated by stress-activated protein kinases to mediate c-jun induction in response to genotoxic agents. *EMBO J.* 1995;14:1798-1811.

van Dam H, Castellazzi M, Distinct roles of Jun: Fos and Jun: ATF dimers in oncogenesis, *Oncogene.* 2001; 20: 2453-2464.

van Dorst EB, van Muijen GN, Litvinov SV, et al. The limited difference between keratin patterns of squamous cell carcinomas and adenocarcinomas is explicable by both cell lineage and state of differentiation of tumour cells. *J Clin Pathol.* 1998; 51:679-684.

Velasco JA, Castro R, Avila MA, et al. cph, a novel oncogene which cooperates with H-ras in the transformation of NIH3T3 fibroblasts. *Oncogene.* 1994; 9:2065-2069.

Wei F, Ni J, Wu S, et al. Cytogenetic studies of esophageal squamous cell carcinomas in the northern Chinese population by comparative genomic hybridization. *Cancer Genet Cytogenet.* 2002; 138: 38-43.

Werle, B, Lotterle H, Schanzenbacher U, *et al.* Immunochemical analysis of cathepsin B in lung tumours: an independent prognostic factor for squamous cell carcinoma patients. *Br J Cancer* 1999; 81: 510-519.

Vimalachandran D, Greenhalf W, Thompson C, et al. High nuclear S100A6 (Calcyclin) is significantly associated with poor survival in pancreatic cancer patients. *Cancer Res.* 2005;65:3218-3225.

Wang GQ, Dawsey SM, Li JY, et al. Effects of vitamin/mineral supplementation on the prevalence of histological dysplasia and early cancer of the esophagus and stomach: results from the General Population Trial in Linxian, China. *Cancer Epidemiol Biomarkers Prev.* 1994; 3:161–166.

Wang Q, Lu J, Yang C, et al. CASK and its target gene Reelin were co-upregulated in human esophageal cancer. *Cancer Letters.* 2002; 179:71-77.

Wang Q, Yang C, Zhou J, et al. Cloning and characterization of EC45 gene which encoded human ribosomal protein L15 and overexpressed in esophageal cancer. *Gene.* 2001; 263:205-209.

Wang Y, Morrow JS. Identification and characterization of human SLP-2, a novel homologue of stomatin (Band 7.2b) present in erythrocytes and other tissues. *J Biol Chem.* 2000; 275: 8062-8071.

Wang YP, Han XY, Su W, et al Esophageal cancer in Shangxi Province, People's Republic of China: a case-control study in high and moderate risk areas. *Cancer Causes Control.* 1992; 3:107–113.

Warner-Bartnicki AL, Murao S, Collart FR, et al. Regulated expression of MRP8 and MRP14 in human promyelocytic leukemic HL-60 cells treated with the differentiation-inducing agents mycophenolic acid and 1,25-dihydroxyvitamin D3. *Exp Cell Res.* 1993; 204: 241-246.

Watson PH, Leygue ER, Murphy LC. Psoriasin (S100A7). *Int J Biochem Cell Biol.* 1998; 30:567-571.

Werle B, Lotterle H, Schanzenbacher U, et al. Immunochemical analysis of cathepsin B in lung tumours: an independent prognostic factor for squamous cell carcinoma patients. *Br J Cancer* 1999; 81: 510-519.

Welter JF, Crish JF, Agarwal C, et al. Fos-related antigen (Fra-1), junB, and junD activate human involucrin promoter transcription by binding to proximal and distal AP1 sites to mediate phorbol ester effects on promoter activity. *J. Biol. Chem.* 1995; 270: 12614-12622.

Wertz IE, O'Rourke KM, Zhang Z, et al. Human De-etiolated-1 regulates c-Jun by assembling a CUL4A ubiquitin ligase. *Science.* 2004; 303:1371-1374.

Weterman MA, Wilbrink M, Dijkhuizen T, et al. Fine mapping of the 1q21 breakpoint of the papillary renal cell carcinoma-associated (X;1) translocation. *Hum Genet.* 1996; 98:16-21

Williams RR, Broad S, Sheer D, et al. Subchromosomal positioning of the epidermal differentiation complex (EDC) in keratinocyte and lymphoblast interphase nuclei. *Exp Cell Res* 2002; 272:163-175.

Wong N, Chan A, Lee SW, et al. Positional mapping for amplified DNA sequences on 1q21-q22 in hepatocellular carcinoma indicates candidate genes over-expression. *J Hepatol.* 2003; 38:298-306.

Wu, M., Hu, N. and Wang, X. (1989). Genetic factors in the etiology of esophageal cancer and the strategy for its prevention in high-incidence areas in northern China. In: Lynch HT, Hirayama T (ed.), *Genetic Epidemiology of Cancer.* Boca Raton, Florida: CRC Press.

Xia SH, Hu LP, Hu H, et al. Three isoforms of annexin I are preferentially expressed in normal esophageal epithelia but down-regulated in esophageal squamous cell carcinomas. *Oncogene*. 2002; 21: 6641-6648.

Xing EP, Yang GY, Wang LD, et al. Loss of heterozygosity of the Rb gene correlates with pRb protein expression and associates with p53 alteration in human esophageal cancer. *Clin. Cancer Res.* 1999; 5:1231-1240.

Xing R, Wu F, Mason RW. Control of breast tumor cell growth using a targeted cysteine protease inhibitor. *Cancer Res.* 1998; 58: 904-909.

Xu L, Hui L, Wang S, et al. Expression profiling suggested a regulatory role of liver-enriched transcription factors in human hepatocellular carcinoma. *Cancer Res.* 2001;61:3176-3181.

Xu Y, Selaru FM, Yin J, et al. Artificial neural networks and gene filtering distinguish between global gene expression profiles of Barrett's esophagus and esophageal cancer. *Cancer Res.* 2002; 62:3493-3497.

Yamaguchi K, Lantowski A, Dannenberg AJ, et al. Histone deacetylase inhibitors suppress the induction of c-Jun and its target genes including COX-2. *J. Biol. Chem.* 2005; 280:32569-32577.

Yamazaki M, Ishidoh K, Kominami E, et al. Genomic structure of human cystatin A. *DNA Seq.* 1997; 8:71-76.

Yan, S, Sameni, M, and Sloane, BF. Cathepsin B and human tumor progression. *Biol Chem* 1998; 379: 113-123.

Yen CC, Chen YJ, Chen JT, et al. Comparative genomic hybridization of esophageal squamous cell carcinoma: correlations between chromosomal aberrations and disease progression/prognosis. *Cancer*. 2001; 92: 2769-2777.

Yangquan Cancer Control Office. (1990) Report of cancer mortality investigation of Yangquan City. Cancer Prev.

Yu X, Luo A, Zhou C, et al. Differentiation-associated genes regulated by TPA-induced c-Jun expression via a PKC/JNK pathway in KYSE450 cells. *Biochem Biophys Res Commun.* 2006; 342:286-292.

Zhang LY, Ding F, Liu ZM, et al. Effect of stomatin-like protein 2 (SLP-2) Gene on growth and proliferation Esophageal Sqaumous Cell Carcinoma Cell Line TE12. *Ai Zheng.* 2005;24:155-159.

Zhang LY, Wang T, Ding F, et al. Differential expression and bioinformatics analysis of SLP-2 gene in esophageal squamous cell carcinoma. *World Chin J Digestol.* 2004; 12:1517-1521.

Zhang L, Ding F, Cao W, et al. Antisense Stomatin-like protein 2 (SLP-2) showed decreased cell growth, tumorigenecity and cell adhesion in human esophageal squamous cell carcinoma. *Clin Cancer Res.* 2006; 12:1639-1646.

Zhang S, Wang G, Liu D, et al. The C-terminal region of S100A4 is important for its metastasis-inducing properties. *Oncogene*. 2005; 24: 4401-4411.

Zhang W, Bailey-Wilson JE, Li W, et al. Segregation analysis of esophageal cancer in a moderately high-incidence area of northern China. *Am J Hum Genet.* 2000; 67:110-119.

Zhang J, Shridhar R, Dai Q, et al. Cystatin M: a novel candidate tumor suppressor gene for breast cancer. *Cancer Res.* 2004; 64: 6957-6964.

Zhi H, Zhang J, Hu G, et al. The deregulation of arachidonic acid metabolism-related genes in human esophageal squamous cell carcinoma. *Int J Cancer.* 2003;106:327-333.

Zhou H, Lin A, Gu Z, et al. 12-O-tetradecanoylphorbol-13-acetate (TPA)-induced c-Jun N-terminal kinase (JNK) phosphatase renders immortalized or transformed epithelial cells refractory to TPA-inducible JNK activity. *J Biol Chem.* 2000; 275:22868-22875.

Zhu, GH, Wong, BC, Eggo, MC, et al. Pharmacological inhibition of protein kinase C activity could induce apoptosis in gastric cancer cells by differential regulation of apoptosis-related genes. *Dig Dis Sci.* 1999; 44: 2020-2026.

In: Esophageal Cancer Research Developments
Editor: Henry K. Kinner, pp. 99-112
ISBN: 1-60021-309-X

Chapter II

Epidemiology of Esophageal Cancer

Eduardo De Stefani[1*], Hugo Deneo-Pellegrini[1], Alvaro L. Ronco[2], Paolo Boffetta[3] and Enrique Barrios[4,5]

[1]Grupo de Epidemiología, Departamento de Anatomía Patológica, Hospital de Clínicas, Facultad de Medicina, Montevideo, Uruguay.
[2]División de Epidemiología, Sección de Radiología, Hospital Pereira Rossell, Montevideo, Uruguay.
[3]International Agency for Resarch on Cancer, Lyon, France.
[4]Registro Nacional de Cáncer, Montevideo, Uruguay.
[5]Departamento de Biofísica, Facultad de Medicina, Montevideo, Uruguay.

Abstract

Esophageal carcinoma is characterized by sharp differences in incidence among different countries and , even, in short distances. Tobacco smoking and alcohol drinking are important risk factors in most areas in the World. Nevertheless, countries at high risk are characterized by lack of these effects and diet is suspected as playing an important role in esophageal carcinogenesis. More precisely, diet poor in fresh vegetables and citrus fruits appears to be a major determinant in the etiology of esophageal cancer. Perhaps nitrosamines are the major initiators in the process of esophageal carcinogenesis. Since the eighties adenocarcinoma of the esophagus have experimented a sharp increase in incidence. Obesity, body mass index and esophagogastric reflux have play a an importan role in the development of this histologic type.

* *Correspondence:* Dr. Eduardo De Stefani, Avenida Brasil 3080 dep. 402, 11300 Montevideo, Uruguay. Tel.: +598 2 708 23 14; Fax: +598 2401 91 89; E-Mail: estefani@adinet.com.uy

Introduction

According to Parkin et al [1], esophageal carcinoma is the ninth most common cancer in the World, with about 300,000 cases diagnosed every year. In developing countries esophageal cancer ranks in the fifth place. Its epidemiological behaviour, marked by large differences in incidence within small geographic areas and sharp changes in incidence over time suggests a a predominant role for external environmental risk factors [2]. In this Chapter we will outline the major characteristics of esophageal cancer.

Descriptive Epidemiology

Esophageal cancer is particulary frequent in the so-called high-risk areas [2]. They include certains areas of northern China, the zones of Iran borderdering with the Caspian sea, neighboring countries of the former Soviet Union, northern Afghanistan and Siberia. These countries and areas have been known as the Asian esophageal cancer belt [2]. In the province of Gonbad (Iran) have been registered rates of 174 per 100, 000 women. In high risk areas tobacco and alcohol are play a negligible role in the etiology of the disease [2].

Uruguay, a developing country located in the temperate zone of South America, display exteremelly high incidence and mortality rates for cancer. It has been suggested that Uruguay is characterized by the existence of two subpopulations at risk (P.Correa, personal communication). This is particularly evident for esophageal cancer. Whereas the incidence rate in the capital city of Montevideo, a highly ubanized area, is intermediate (ASR 9.6 per 100,000), the rate present very high figures in northern counties, mainly rural areas (ASR 37.6 per 100,000) [3-4] This high-risk area border with Brazil suggesting the existence of an Uruguayan-Brazilian belt. This area has been called as the *mate*-drinking one [2].

Developed countries (USA and Western Europe) display rather low incidence rates of esophageal cancer. Nevertheless, there are exceptions like the Black male population in US and the population of Calvados, France [3]. In Shanghai, China, esophageal carcinoma has declined more than 50 % in its incidence [5]. The same phenomenon was found in Denmark and other european countries and has been linked with a decreasing consumption of alcohol drinking [6]. In Uruguay, in the period 1953-1991, these was s significant decline of esophageal cancer in men (23 %) and in women (59 %) [7]. According to Blot [8] there are striking differences in the time trends according to histologic type of esophageal cancer. Age adjusted incidence rates of squamous cell carcinoma are now decreasing among black and white men. On the contrary, incidence rates of adenocarcinoma of the esophagus is increasing on the order of 5 % to 10 % per year, a faster pace than virtually any other cancer [8]. These figures applies to USA and to other countries in Western Europe, but not to other areas of the World [9].

Histopathology

According to the World Health Organization Classification of tumours [10], the following tumors ocurr in the esophageal mucosa: squamous cell carcinoma and its variants, adenocarcinoma, salivary gland tumors, carcinoids, mesenchymal tumors, malignant melanoma and non-Hodgkin's lymphomas. By far, squamous cell carcinoma is the more frequent neoplasia, followed by adenocarcinoma. Both types of carcinoma represnts more than 95 % of all malignancies observed in the esophagus. In the following sections we will deal with squamous cell carcinoma and adenocarcinoma of the esophagus. This histologic type has been increasing in frequency in the last years.

Tobacco Smoking

Tobacco smoking is a major risk factor for esophageal carcinoma, specially in Western developed countries like United States and Western Europe. According to the Working Groups of IARC, there are sufficient evidence of carcinogenicity of tobacco in esophageal carcinoma [11-12]. There are increased risks of esophageal cancer associated with smoking intensity and duration. Most studies suggest similar odds ratios for both variables, but Launoy et al. [13] strongly suggested that years of smoking are more important that smoking intensity.

There are evidences that tobacco specific nitrosamines (TSNA), mainly NNN are potent carcinogens for esophageal mucosa [12]. The main source of TSNA is the air-cured tobacco (black tobacco). Epidemiologic studies have shown a consistent effect of black tobacco in esophageal carcinogenesis [14-19].

Another risk factor associated with smoking is the habit of using hand-rolled cigarettes. In fact, hand-rolled cigarettes have been the subject of chemical analyses [12]. The kind of cigarettes contains high yields of benzo[a]pyrene and benzene which are proved carcinogens. The use of hand-rolled cigarettes is particularly prevant in countries like Norway, Australia and Uruguay [11]. Epidemiologic studies on hand-rolled cigarettes suggest that these products are associated with higher risks than manufactured (commercial) cigarettes [14, 18-22].

In Uruguay, men of the low socioeconomic strata have the noxious habit of smoking hand-rolled cigarettes filled with black tobacco. Therefore, this subpopulation is associated with specially high risk of developing squamous cell carcinoma of the esophagus. According to our last estimations the relative risk is about 5.0 for this subset of Uruguayan men. Odds ratios of esophageal cancer for smoking indices are shown in Table 1. The highest risk was associated with smokers of hand-rolled cigarettes filled with black tobacco (OR 4.8, 95 % CI 3.1-7.5). Joint effect of current smokers and smokers of hand-made cigarettes filled with black tobacco was 7.6 (95 % CI 4.8-12.1). Therefore, our studies replicates previous findings, showing that tobacco smoking is an important risk factor in esophageal carcinogenesis. The possible mechanism involved in the effect of tobacco is the high yield of tobacco-specific nitrosamines (TSTA), particularly *N′*-nitrosnornicotine (NNN) which appears to be esophagotropic. In fact, as suggested by Craddock [23], nitrosamines appears to play a central role in esophageal carcinogenesis.

Table 1. Odds ratios of squamous cell carcinoma of the esophagus for tobacco smoking (1)

Smoking variable	Category	Cases/Controls	OR	95 % CI
Smoking status	Never smokers	36/141	1.0	
	Former smokers	227/299	2.1	1.4-3.3
	Current smokers	421/244	5.0	3.2-7.6
	Ever smokers	648/543	3.2	2.2-4.9
Cigarretes/day	Never smokers	36/141	1.0	
	1-9	63/104	2.0	1.2-3.3
	10-19	137/140	2.9	1.8-4.6
	20-29	206/150	3.8	2.5-6.0
	30+	242/149	3.9	2.5-6.1
Years of smoking	Never smokers	36/141	1.0	
	1-24	37/84	1.4	0.8-2.5
	25-34	102/81	3.6	2.2-5.8
	35-44	157/160	2.7	1.7-4.2
	45+	352/218	4.4	2.9-6.8
Years since quit	Current smokers	421/244	1.0	
	1-9	113/92	0.7	0.5-1.0
	10-19	55/105	0.3	0.2-0.4
	20+	59/102	0.3	0.2-0.5
	Never smokers	36/141	0.2	0.1-0.3
Type of tobacco	Blond	299/324	1.0	
	Black	349/219	1.7	1.3-2.1
Type of cigarette	Manufactured	205/228	1.0	
	Hand-rolled	443/315	1.5	1.1-1.9
Smoking pattern	Never smokers	36/141	1.0	
	Commercial+Blond	155/179	2.5	1.6-3.9
	Commercial+Black	50/49	2.7	1.5-4.8
	Hand-rolled+Blond	144/145	2.7	1.7-4.2
	Hand-rolled+Black	299/170	4.8	3.1-7.5

1-Adjusted for age, residence, urban/rural status, education and alcohol drinking.

Alcohol Drinking

According to the Working party of IARC, alcohol drinking is carcinogenic to humans (Group 1) [24]. This assertion applies to esophageal cancer. The pioneer study of Wynder and Bross [25] showed a five-fold increased risk of esophageal cancer associated with heavy drinking of alcohol. Since then prospective and case-control studies replicates this original observation. These studies are reviewed in references 26 and 27. There have been considerable controversy about the type of alcoholic beverage involved in esophageal carcinogenesis. Whereas the majority of authors suggest that ethanol *per se* is responsible of the carcinogenicity of alcohol [24], some researchers have suggested that hard liquor [28] or red wine were responsible of the the alcohol effect [29].

According to our findings, both red wine and hard liquor displayed similar OR's, whereas beer was not associated with risk of esophageal carcinoma. Total alcohol drinking was associated with a significant dose-response and heavy drinkers displayed an increased risk of 3.6 (95 % CI 2.4-5.3) (Table 2). Therefore, we concluded that ethanol *per se* is carcinogenic for esophageal mucosa.

The effect of alcohol appears to be through direct contact, allowing the penetration of carcinogens from tobacco. Thus, alcohol appears to be a promoter in esophageal carcinogenesis. Studies conducted among non-smokers have suggested that alcohol drinking could be also an independent carcinogen [30-31]. This puzzling finding could indicate that alcohol beverages could contain initiators, like nitrosamines and urethane.

Table 3. Odds ratios of squamous cell carcinoma of the esophagus for *mate* drinking (1)

Mate variable	Category	Cases/Controls	OR	95 % CI
Mate status	Never drinkers	10/89	1.0	
	Ever drinkers	150/639	2.7	1.3-5.5
Mate amount	Never drinkers	10/89	1.0	
(liters/day)	0.1-0.9	28/130	1.6	0.9-2.8
	1.0-1.9	87/361	2.9	1.7-4.9
	2.0+	35/148	3.6	2.1-6.4
Yrs of drinking *mate*	Never drinkers	10/89	1.0	
	1-43	47/227	2.7	1.2-6.2
	44-53	38/196	1.9	0.9-4.2
	54+	65/216	3.3	1.5-7.3
Mate temperature	Never drinkers	10/89	1.0	
	Warm	7/19	2.5	0.9-6.9
	Hot	116/594	2.4	1.1-4.9
	Very hot	27/26	8.0	3.2-19.9

1-Adjusted for age, residence, urban/rural status, education, years of smoking and alcohol drinking..

Mate Drinking and Other Hot Beverages and Foods

Since the study of Crespi et al. [32], hot drinks and hot beverages has been suspected as risk factors for esophageal cancer. Hot beverages and soups have been considered as carcinogens through the mechanism of thermal injury. The damaged esophageal mucosa is more susceptible to the penetration of carcinogenes from tobacco, alcohol and diet. In fact, hot drinks leads to cronic esophagitis [33], followed by dysplasia.

In South America, *mate,* an herb derived from the leaves of *Ilex paraguariensis*, is a local tea which is drunk hot or very hot, thus, fulfilling the conditions of other hot beverages drunked in high-risk areas of China and Iran. Since the first study on *mate* drinking and esophageal cancer [34], six studies have been examined the possible role of *mate* as a risk factor for esophageal carcinoma [15,18, 35-38]. Boiled meat from lamb is a major component of stews, a staple food in rural Uruguayan population, a high-risk population. This food item also contains rice, cooked vegetables, potato and red pepper. Stew is rich in fat and is ingested very hot. Both characteristics could play an important role in esophageal carcinogenesis.

The Role of Diet

Diet is an important determinant in esophageal carcinogenesis. In the first studies the role of deficiencies in B vitamins was suspected as a risk factor for squamous cell carcinoma of the esophagus [39-40]. Although both foods and nutrients were studies in realtion with esophageal cancer, the number of studies on foods were far the most frequent reports [41-52]. Two desings were employed: prospective (cohort) studies and retrospective (case-control) studies. The major difference which has been strongly suggested as an advantage of prospective studies is the absence of recall bias [53]. Although the possible role of role of recall bias in case-control studies is a drawback of these reports, a number of well-designed prospective studies essentialy replicates previous findings in case-control studies [54]. As rightly suggested by Potter [55], whereas prospective sudies includes a genetically-susceptible population due to the narrower age-range, case-control studies disminished this likelihood since they usually included elder patients which are less prone to be afflicted by genetic traits. Thus, both kind of studies has drawbacks and advantages.

In Table 5 a summary of the associations between foods and risk of esophageal cancer is shown. Salted meat (*charque*) and barbecued lamb displayed the highest risks (OR for salted meat 4.17, 95 % CI 2.12-8.20 and OR for barbecued lamb 4.60, 95 % CI 2.40-8.51)), followed by high-fat foods (OR 2.80, 95 % CI 1.91-4.00). These findings are extremelly important. According to Craddock [23] exogenous and endogeneous nitrosamines are, perhaps, the most important initiators in the process of esophageal carcinogenesis. Also, heterocyclic amines from barbecued lamb are multiorgan carcinogens and its noxious role have been suggested in the etiology in colorectal, mammary and pulmonary carcinomas [56]. Its role should be further studied in esophageal carcinogenesis. Fat and high-fat foods are, according to Weisburger [57], modifiers of heterocyclic amines, enhancing its effect.

Other foods are strongly protective for esophageal squamous cell carcinoma. High intake of poultry, fish, total white meat, fresh vegetables, green leafy vegetables, citrus fruits, other fruits, total fruits, total vegetables and fruits and legumes are inversely associated with risk of esophageal cancer. Perhaps the most more protective of these food groups are citrus fruits (OR 0.13, 95 % CI 0.06-0.38) and fresh vegetables (OR 0.36, 95 % CI 0.20-0.64). On the reverse, low intake of these items increased the risk of esophageal cancer (OR for total vegetables and fruits 3.2, p-value for trend <0.0001).

Table 5. Odds ratios of squamous cell carcinoma of the esophagus for nutrients (1).

	Tertiles					
	I	II		III		
Nutrient	OR	OR	95 % CI	OR	95 % CI	p-value trend
Protein	1.0	1.60	1.01-2.54	1.79	1.13-2.83	0.01
Carbohydrates	1.0	0.56	0.36-0.88	0.65	0.42-1.02	0.04
Total fat	1.0	0.89	0.55-1.44	1.59	0.99-2.54	0.03
Saturated fat	1.0	0.75	0.46-1.21	1.38	0.87-2.18	0.12
MUFA	1.0	1.01	0.62-1.66	1.79	1.11-2.89	0.01
PUFA	1.0	1.37	0.87-2.18	1.57	0.99-2.48	0.06
Linoleic acid	1.0	1.39	0.88-2.20	1.63	1.03-2.56	0.04
Linolenic acid	1.0	0.91	0.57-1.46	1.21	0.76-1.93	0.41
Cholesterol	1.0	0.97	0.62-1.52	1.23	0.77-1.95	0.38
Vitamin A	1.0	0.68	0.44-1.06	0.45	0.28-0.73	0.001
All carotenoids	1.0	0.75	0.49-1.17	0.52	0.33-0.84	0.007
Beta-carotene	1.0	0.86	0.55-1.33	0.52	0.32-0.84	0.009
Alpha-carotene	1.0	0.73	0.47-1.12	0.62	0.39-0.99	0.04
Lutein	1.0	0.70	0.45-1.09	0.76	0.48-1.20	0.22
Lycopene	1.0	0.76	0.49-1.17	0.69	0.44-1.08	0.09
B-cryptoxanthin	1.0	0.85	0.56-1.29	0.37	0.23-0.61	0.0001
Vitamin C	1.0	0.49	0.31-0.76	0.57	0.36-0.88	0.007
Vitamin E	1.0	0.57	0.37-0.89	0.46	0.29-0.74	0.001
Vitamin B6	1.0	0.74	0.48-1.14	0.55	0.35-0.87	0.01
Vitamin B12	1.0	1.18	0.73-1.90	1.82	1.14-2.88	0.009
Thiamine	1.0	0.96	0.61-1.50	1.13	0.72-1.76	0.59
Riboflavin	1.0	1.23	0.77-1.95	1.62	1.02-2.55	0.04
Folate	1.0	0.49	0.31-0.76	0.41	0.25-0.66	0.0002

1-Adjusted for age, sex, residence, urban/rural status, birthplace, education, body mass index, smoking status, years since quit, number of cigarettes smoked per day, alcohol drinking, *mate* temperature and total energy intake.

In the Table 6 vitamins and antioxidants, which are mainly originated in plant foods, were inversely associated with risk of squamous cell carcinoma of the esophagus. These nutrients incude vitamin A, *beta*-carotene, *alpha*-carotene, *beta*-cryptoxanthin, vitamin C, vitamin C, vitamin E, vitamin B6 and folate. Also, plant sterols and flavonols have been suggested as negatively associated with risk of SCCE. In a previous study, antioxidants presented a weaker protective effect when they were further adjusted for a term for total vegetables and fruits [42]. This finding suggest that, with the exception of *beta*-cryptoxanthin, antioxidants are not fully responsible of the protective effect of plant foods. Therefore, further studies are needed in order to clarify to strong effect of vegetables and fruits in esophageal cancer.

Table 6. Esophageal carcinogenesis. Stages, carcinogens and suspected chemicals.

Stage	Factors	Carcinogens	Chemicals
Chronic esophagitis	Hot drinks	Alcohol	Ethanol
	Hot foods	Mate	*Benzo[*a*]pyrene?*
		Coffee	
		Tea	
Dysplasia	Smoking	Tobacco-specific nitrosamines (TSNA)	NNN
	Poor diet	Low intake of vegetables & fruits	Vitamin C
			Vitamin E
			Vitamin A
			Folate
			B-cryptoxanthin
	Salted meat	Salt	Sodium
		Exogeneous nitrosamines	NDMA
Carcinoma	Tobacco	Promoters	Ethanol
	Alcohol		
	Poor diet		

Among nutrients directly associated with risk of SCCE, protein, monounsaturated fat and vitamin B12 displayed a moderete increased in risk. Perhaps, nitrosodimethylamine (NDMA) is the strongest risk factor for esophageal carcinoma. This is a sort of replication of the effect of salted meat, previously analyzed. Protein and nitrites could act as a nitrosable substances in the gut, generating endogenous N-Nitroso compounds.

Rather recently, factor analysis (principant components) has been used in the field of cancer epidemiology [58]. In fact, the study of food and nutrient patterns by factor analysis have been considered as an useful tool to identify broader eating patterns (Randall et al., 1992; Slattery et al., 1998a; Terry el al., 2001a; Terry et al., 2001b; Palli et al., 2001; Sieri et al., 2004; De Stefani et al., 2004; De Stefani et al., 2005; Fu et al., 2005). Diet represents a complex set of highly correlated exposures. In fact, the traditional reductionist approach [54] accounts only for a small amount of the variance. An unpublished study conducted in Uruguay identified three dietary patterns, labeled as *traditional, healthy* and *dairy*. The *traditional* pattern was characterized by high loadings of barbecue, boiled meat, processed meat (which included salted meat), milk, cooked vegetables, coffee with milk, eine and hard liquor. This dietary pattern was directly associated with risk of SCCE (OR 3.27, 95 % CI 1.69-6.30). The *healthy* pattern showed high loadings for white meat, desserts, fresh vegetables, citrus fruits, other fruits and tea and was inversely associated with esophageal cancer risk. Furthermore, in a model which included age, residence, urban/rural status, smoking, *mate* temperature, total energy intake, alcohol drinking, red meat, dairy foods, total vegetables and fruits and the three dietary patterns forward stepwise regression was carried out. The final model only included three variables: smoking, *treditional* pattern and *healthy* suggesting that broad eating patterns are better than the traditional reductionist approach.

Adenocarcinoma of the Esophagus

In the previous sections, the results are mainly related with squamous cell carcinoma. The other main histologic type is adenocarcinoma of the esophagus. The incidence of this tumor has been increasing for the past 15 years in the United States. According to several studies, tobacco smoking and alcohol drinking are important risk factors for esophageal adenocarcinoma [9]. Furthermore, high body mass index, obesity and the presence of esophagogastric reflux are directly associated with risk of adenocarcinoma of the esophagus [9, 61]. These risk factors are in contrast with SCCE, in which patients are leaner. Since esophageal adenocarcinoma is unfrequent in South America, the role of hot *mate* drinking is unknown. In a recent study of Cheng et al. [61] the authors presented a final model which included the following terms: high body mass index, high intake of fruit, and breastfeeding. The British authors suggested that the etiology of both tumor types could be not identical. They also suggest that aspirin intake could be protective for both histologic types.

Esophageal Carcinogenesis

For squamous cell carcinoma the following steps are recognized: chronic esophagitis, dysplasia, carcinoma *in situ* and invasive carcinoma. In this simple scheme initiators and promoters are well known. In Table 7 these carcinogens and steps are shown. These carcinogens are divided in suspected carcinogens, like benzo[*a*]pyrene in *mate*, urethane and ethanol in alcohol and proved carcinogens like tobacco-specific nitrosamines in tobacco, nitrosodymethilamine in salted meat. Concerning the division in initiators and promoters, nitrosamines are main initiators and alcohol is the main promoter.

Prevention

Since tobacco smoking and alcohol drinking are the major risk factors in developed countries (United States and Western Europe) avoiding or quitting tobacco is the major preventive measure for esophageal cancer. The same applies to alcohol drinking. In a recent study of Brown et al. [62], the attributable risk due to the both factors was close to 92 %. This assertion applies to both whites and blacks. The role of diet also plays a important role. More specifically, diet low in vegetables and fruits is a determinant in esophageal cancer development. Thus, the intake of meals rich in salads and fruits is also a major preventive measure. Both in developed and developing countries hot foods and hot beverages should be avoided. This applies to foods and beverages which usually ingested very hot like tea, coffee, *mate*, soups, stew and other items.

Since a low socioeconomic status, measured by education, income and occupation, is strongly associated with other risk factors and with the disease, general measures which could improve mainly the salaries could result in a decrease in the incidence of esophageal cancer. In fact, squamous cell esophageal cancer is a cancer linked with poverty.

Adenocarcinoma of the esophagus is directly associated with obesity; thus, measures which could disminish the body mass index are extremelly important for this histologic type of esophageal cancer.

Concluding Remarks

According to the current knowledge on the etiology of esophageal cancer there are several points which merit to be emphasizing: 1) In first place, future studies should be focused in each major histologic type of the malignancy, 2) Molecular and genetic epidemiology of esophageal cancer could clarify the etiology of this malignancy, and 3) New approaches leading to establish major dietary patterns could be necessary.

References

[1] Parkin, D.M., Pisani, P. and Ferlay, J.: Estimates of the worldwide incidence of eighteen major cancers in 1985. *Int.J.Cancer* 54, 594-606, 1993.

[2] Muñoz, N. and Day, N.E. (1996) Esophageal cancer. In D. Schottenfeld and J.F. Fraumeni, Jr. (editors): *Cancer epidemiology and prevention.* Second Edition. Oxford University Press, New York-Oxford, 1996. pp. 681-706.

[3] Parkin, D.M., Whelan, S.L., Ferlay, J., Teppo, L. and Thomas, D.B. (eds): *Cancer Incidence in Five Continents.* Vol VIII. IARC Scientific Publications n° 155, Lyon, IARC, 2002.

[4] Vassallo, J.A., Barrios, E., De Stefani, E. and Ronco, A.: Incidencia del cancer en el Uruguay-1991. Montevideo: Comisión Honoraria de Lucha contra el Cáncer, Uruguay, 1996.

[5] Zheng, W., Jin, F., Devesa, S.S., Blot, W.J., Fraumeni, Jr, J.F. and Gao, Y-T.: Declining incidence is greater for esopageal than gastric cancer in Shanghai, People's Republic of China. *Br.J.Cancer* 68, 978-982, 1993.

[6] Moller, H., Boyle, P., Maissoneuve, P. and La Vecchia, C.L.: Changing mortality from esophageal cancer in males in Denmark and other European counties in relation to changing levels of alcohol consumption. *Cancer Causes Control* 1, 181-188, 1990.

[7] De Stefani, E., Fierro, L., Barrios, E. and Ronco, A.: Cancer mortality trends in Uruguay 1953-1991. *Int.J.Cancer* 56, 634-639, 1994.

[8] Blot, W.J.: Esophageal cancer trends and risk factors. Sem.Oncol. 21, 403-410, 1994.

[9] Blot, W.J., Devesa, S.S., Kneller, R:W.: Rising incidence of adenocarcinoma of the adenocarcinoma of the esophagus and gastric cardia. *JAMA* 265, 1287-1289, 1991.

[10] Hamilton, S.R. and Aaltonen, L.A. (editors): Pathology and Genetics. Tumours of the digestive system. World Health Organization Classification of Tumours. IARC*press,* Lyon, 2000.

[11] IARC Monographs on the Evaluation of the Carcinogenic Risk of Chemicals to Humans. Volume 38. *Tobacco Smoking.* IARC, Lyon, France, 1986.

[12] IARC Monographs on the Evaluation of the Carcinogenic Risk of Chemicals to Humans. Volume 83. *Tobacco Smoke and Involuntary Smoking.* IARC, Lyon, France, 2004.

[13] Launoy, G., Milan, C., Faivre, J., Pienkowski, P., Milan, C.J. and Gignoux, M.: Alcohol, tobacco and oesophageal cancer : Effects of duration of consumption, mean intake and current and former consumption. *Br.J.Cancer* 75, 1389-1396, 1997.

[14] Launoy, G., Milan, C., Faivre, J., Pienkowski, P. and Gignoux, M.: Tobacco type and risk of squamous cell cancer of the oesophagus in males: a French milticentre case-control study. *Int.J.Epidemiol.* 29, 36-42, 2000.

[15] De Stefani, E., Muñoz, N., Estève, J., Vasallo, A., Victora, C.G. and Teuchmann, S.: Mate drinking, alcohol, tobacco, diet, and esophageal cancer in Uruguay. *Cancer Res.* 50, 426-431, 1990.

[16] Castellsagué, X., Muñoz, N., De Stefani, E., Victora, C.G., Castelletto, R., Rolón, P.A. and Quintana, M.J.: Independent and joint effects of tobacco smoking and alcohol drinking on the risk of esophageal cancer in men and women. *Int.J.Cancer* 82, 657-664, 1999.

[17] Castellsagué, X., Muñoz, N., De Stefani, E., Victora, C.G., Quintana, M.J., Castelletto, R. and Rolón, P.A.: Smoking and drinking cessation and risk of esophageal cancer (Spain). *Cancer Causes Control* 11, 813-818, 2000.

[18] De Stefani, E., Barrios, E. and Fierro, L.: Black (Air-cured) and Blond (Flue-cured) and Cancer Risk III: Oesophageal Cancer. *Eur.J.Cancer* 29A, 763-766, 1993.

[19] Tuyns, A.J. and Estève, J.: Pipe, commercial and hand-rolled cigarette smoking in oesophageal cancer. *Int.J.Epidemiol.* 12, 110-113, 1983.

[20] Engeland, A., Andersen, A., Halderson, T. and Tretli, S.: Smonking habits and risk of cancers other than lung cancer: 28 years' follow-up of 26,000 Norwegian men and women. *Cancer Causes Control* 5, 497-506, 1996.

[21] Engeland, A., Haldorsen, T., Andersen, A. and Tretli, S.: The impact of smoking habits on lung cancer risk: 28 years' observation of 26,000 Norwegian men and women. *Cancer Causes Control* 7, 366-376, 1996.

[22] De Stefani, E., Fierro, L., Mendilaharsu, M., Larrinaga, M.T., Balbi, J.C. and Lateulade, S.: Factores de riesgo para el cáncer de esófago en Uruguay, *Arch.Med.Int.* 3, 101-108, 1991.

[23] Craddock, V.M.: Aetiology of oesophageal cancer: some operative factors. *Eur.J.Cancer Prev.* 1, 89-103, 1992.

[24] IARC Monographs on the Evaluation of the Carcinogenic Risk of Chemicals to Humans. Volume 44. *Alcohol drinking. IARC*, Lyon, France, 1988.

[25] Wynder, E.L and Bross, I.J.: A study of etiological factors in cancer of the esophagus. *Cancer* 14, 389-413, 1961.

[26] World Cancer Research Fund: Food, nutrition and the prevention of cancer: a global perspective. *American Institute for Cancer Research.* Washington DC, 1997.

[27] Cheng, K.K. and Day, N.E.: Nutrition and esophageal cancer. *Cancer Causes Control* 7, 33-40, 1996.

[28] Victora, C.G., Muñoz, N., Day, N.E., Barcelos, L.B., Peccin, D.A. and Braga, N.M: Hot beverages and oesophageal cancer in southern Brazil: a case-control study. *Int.J.Cancer* 39, 710-716, 1987.

[29] Barra, S., Franceschi, S., Negri, E., Talamini, R. and La Vecchia, C.: Type of alcoholic beverage and cancer of the oral cavity, pharynx and oesophagus in an Italian area with high wine consumption. *Int.J.Cancer* 46, 1017-1020, 1990.

[30] Tavani, A., Negri, E., Franceschi, S. and La Vecchia, C.: Risk factors for esophageal camcer in lilelong nonsmokers. Cancer Epidemiol.Biomarkers Prev.3, 387-392, 1994.

[31] La Vecchia, C. and Negri, E.: The role of alcohol in oesophageal cancer in non-smokers, and of tobacco in non-drinkers. *Int.J.Cancer* 43, 784-785, 1989.

[32] Crespi, M., Muñoz, N., Gras, A., Aramesh, B., Amiri, G. and Mojtabai, A.: Oesophageal lesions in northern Iran: a premalignant condition? *Lancer* II, 217-221, 1979.

[33] Muñoz, N., Victora, C.G., Crespi, M., Saul, C., Braga, N.M. and Correa, P.: Hot maté drinking and precancerous lesions of the oesophagus: an endoscopic survey in southern Brazil. *Int.J.Cancer* 39, 708-709, 1987.

[34] Vassallo, A., Correa, P., De Stefani, E., Cendán, M., Zavala, D., Chen, V., Carzoglio, J. and Deneo-Pellegrini, H.: Esophageal cancer in Uruguay: A case-control study. *J.Natl.Cancer Inst.* 75, 1005-1009, 1985.

[35] Castelletto, R., Castellsagué, X., Muñoz, N., Iscovich, J., Chopita, N. and Jmelnitsky, A.: Alcohol, tobacco, diet, mate drinking, and esophageal cancer in Argentina. *Cancer Epidemiol.Biomarkers Prev.* 3, 557-564, 1994.

[36] Rolón, P.A., Castellsagué, X., Benz, M. and Muñoz, N.: Hot and cold *mate* drinking and esophageal cancer in Paraguay. *Cancer Epidemiol.Biomarkers Prev.* 4, 595-605, 1995.

[37] Castellsagué, X., Muñoz, N., De Stefani, E., Victora, C.G., Castelletto, R. and Rolón, P.A.: Influence of mate drinking, hot beverages and diet on esophageal cancer risk in *South America.* 88, 2000.

[38] Sewram, V., De Stefani, E., Brennan, P. and Boffetta, P.: Maté consumption and the risk of squamous cell esophageal cancer in Uruguay. *Cancer Epidemiol.Biomarkers Prev.* 12, 508-513, 2003.

[39] Blot, W.J., Taylor, P.R., Guo, W., Dawsey, S., Wang, G.Q, Yang, C.S., Zheng, S.F., Gail, M., Li, GY., Yu, , Y., Liu, B., Tangrea, J., Sun, Y., Liu, E., Fraumeni, J.F., Jr., Zhang, Y.H. and Li, B.: Nutrition intervention trials in Linxian, China: supplementation with specific vitamin,/mineral combinantions, cancer incidence, and disease-specific mortality in the general population. *J.natl.Cancer Inst.* 85, 1483-1492, 1993.

[40] Muñoz, N., Wahrendorf, J., Bang, I.J., Crespi, M., Thurnham, D.I., Day, N.E., Ji,, Z.H., Grassi, A. , Yan, L.W., Lin. L.G. et al.: No effect of riboflavin, retinol, and zinc on prevalence of precancerous lesions of oeosphagus. Randomised double-blind intervention study in high-risk population of China. *Lancet* 2, 111-114, 1985.

[41] De Stefani,, E., Deneo-Pellegrini, H., Boffetta, P. and Mendilaharsu, M. Meat intake and risk of squamous cell esophageal cancer: a case-control study in Uruguay. *Int.J.Cancer 1999*; 82: 33-37.

[42] De Stefani, E., Brennan, P., Boffetta, P., Ronco, A.L., Mendilaharsu, M., and Deneo-Pellegrini, H. Vegetables, fruits, related dietary antioxidants and risk of squamous cell carcinoma of the esophagus: A case-control study in Uruguay. *Nutr.Cancer* 2001; 28: 23-29.

[43] De Stefani, E., Deneo-Pellegrini, H., Ronco, A.L., Boffetta, P, Brennan, P., Muñoz, N., Castellsagué, X., Correa, P. and Mendilaharsu, M. Food groups and risk of squamous cell carcinoma of the oesophagus: a case-control study in Uruguay. *Br.J.Cancer* 2003; 89: 1209-1214.

[44] Yu, Y., Taylor, P.R., Li, J-Y., Dawsey, S.M., Wang, , W., Blot, W.J. and Shen, Q. Retrospective cohort study of risk factors for oesophageal cancer in Linxian, People´s Republic of China. *Cancer Causes Control* 1993; 4: 195-202.

[45] Guo, W., Blot, W.J., Li, J-Y., Taylor, P.R., Liu, B.Q., Wang, W., Wu, Y.P, Zheng, W., Dawsey, S.M., Li, B. and Fraumeni, J.F., Jr. A nested case-control study of oesophageal and stomach cancers in the Linxian nutrition intervention trial. *Int.J.Epidemiol.* 1994; 23: 444-450.

[46] Sauvaget, C., Nagano, H.J., Hayashi, M., Spencer, E., Shimizu, Y. and Allen, N. Vegetables and fruit intake and cancer mortality in the Hiroshima/Nagasaki Life span Study. *Br.J.Cancer* 2003; 88: 689-694.

[47] Tran, G.D., Sun, X-D., Abnet, C.C., Fou, J-H., Dawsey, S.M., Dong, Z-W., Mark, S.D., Qiao, X-L. and Taylor, P.R. Prospective study of risk factors for esophageal and gastric cancer in the Linxian General Population Trial cohort in China. *Int.J.Cancer* 115, 456-463, 2005.

[48] Ziegler, R.G., Morris, L.E., Blot, W.J., Pottern, L.M., Hoover, R. and Fraumeni, J.F., Jr. Esophageal cancer among black men in Washington DC. II. Role of nutrition. *J.natl.Cancer Inst.* 1981; 67: 1199-1206.

[49] Brown, L.M., Blot, W.J., Schuman, S.H., Smith, V.M., Ershow, A.G., Marks, R.D. and Fraumeni, J.F., Jr. Environmental factors and high risk of esophageal cancer among men in coastal South Carolina. *J.natl.Cancer Inst.* 1988; 80:1620-1625.

[50] Brown, L.M., Swanson, C.A., Gridley, G., Swanson, G.M., Schoenberg, J.B., Greenberg, R.S., Silverman, D.T., Pottern, L.M., Hayes, R.B. and Schwartz, A.G. Adenocarcinoma of the esophagus: role of obesity and diet. *J.natl.Cancer Inst.* 1995; 87: 140-149.

[51] Brown, L.M., Swanson, C.A., Gridley, G., Swanson, G.M., Silverman, D.T., Greenberg, R.S., Hayes, R.B., Schoenberg, J.B., Pottern, L.M., Schwartz, A.G., Liff, J.M., Hoover, R. and Fraumeni, J.F., Jr.: Dietary factors and the risk of squamous cell esophageal cancer among black and white men in the United States. *Cancer Causes Control* 1998; 9: 467-474.

[52] De Stefani, E., Boffetta, P., Deneo-Pellegrini, H., Ronco, A.L., Correa, P. and Mendilaharsu, M.: The role of vegetable and fruit consumption in the aetiology of squamous cell carcinoma of the oesophagus: A case-control study in Uruguay. *Int.J.Cancer* 116, 130-135, 2005.

[53] González, C.A.: Vegetables, fruit and cereal consumption and gastric cancer risk. In E.Riboli and R.Lambert (editors): *Nutrition and lifestyle: Opportunities for cancer prevention.* IARC Scientific Publications N° 156, IARC*press*, Lyon 2002. pages:79-83.

[54] Balder, H.F., Goldbohm, R.A. and van den Brandt, P.A.: Dietary patterns associated with male lung cancer risk in the Netherlands Cohort Study. *Cancer Epidemiol.Biomarkers Prev.* 14, 483-490, 2005.

[55] Potter, J.D.: Colorectal neoplasia and meat: Epidemiology and mechanisms. In I.T. Johnson and G.R.Fenwick (editors): *Dieraty anticarcinogens and antimutagens. Chemical and Biological Aspects.* Royal Society of Chemistry, 2000.

[56] Sinha, R.: An epidemiologic approach to studying heterocyclic amines. *Mutat.Res.* 2002; 506-507: 197-204.

[57] Weisburger, J.H.: Comments on the history and importance of aromatic and heterocyclic amines in public health. *Mutat.Res.* 2002; 506-507: 9-20.

[58] Randall, E, Marshall, JR, Brasure, J and Graham, S: Dietary patterns and colon cancer in western New York. *Nutr.Cancer* 18, 265-276, 1992.

[59] Palli, D., Russo, A. and Decarli, A. (2001) Dietary patterns, nutrient intake and gastric cancer in a high-risk area of Italy. *Cancer Causes Control* 12, 163-172.

[60] De Stefani, E, Boffetta, P, Ronco, AL, Correa, P, Oreggia, F, Deneo-Pellegrini, H, Mendilaharsu, M and Leiva, J: Dietary patterns and risk of cancer of the oral cavity and pharynx in Uruguay. *Nutr.Cancer* 51, 132-139, 2005.

[61] Cheng, K.K., Sharp, L., McKinney, P.A., Logan, R.F.A., Chilvers, C.E.D,, Cook-Mozaffari, P. and Day, N.E. (2000) The preventability of oesophageal adenocarcinoma in women: a case-control study. *Br.J.Cancer* 83: 127-132.

[62] Brown, L.M., Hoover, R., Silverman, D., Baris, D., Hayes, R., Swanson, G.M, Schoenberg, J., Greenberg, R., Liff, J., Schwartz, A., Dosemeci, M., Pottern, L. and Fraumeni, Jr., J.F.: Excess incidence of squamous cell esophageal cancer among US black men: Role of social class and other risk factors. *Am.J.Epidemiol.* 153, 114-122, 2001.

In: Esophageal Cancer Research Developments
Editor: Henry K. Kinner, pp. 113-126
ISBN: 1-60021-309-X

Chapter III

Global Epidemiology of Esophageal Cancer

Markku Voutilainen[*]
Jyväskylä Central Hospital, Department of Internal Medicine, Jyväskylä, Finland

Abstract

Esophageal cancer is a male-predominant disease which is globally among the ten most common malignant tumours. The two histological main types of esophageal cancer are squamous-cell and adenocarcinoma. Between individual countries and areas, there is great variation in the incidence of esophageal cancer. This phenomenon is explained by different dietary, environmental, and cultural risk factors. The highest incidences of squamous-cell carcinoma have been reported from developing countries, whereas adenocarcinoma incidence has increased rapidly in some western countries. The risk factor profiles for these two histological cancer types are different. Low intake of B vitamins (especially riboflavin and niacin), magnesium, zinc, selenium as well as eating hard and scratchy, spicy or highly salted food, and smoked meat or fish are risk factors for squamous-cell carcinoma. Heavy alcohol use together with smoking is a strong risk factor for squamous-cell carcinoma. Risk factors for esophageal adenocarcinoma are male gender, gastroesophageal reflux disease, Barrett's esophagus, white race, obesity, tobacco smoking, family history, and lower esophageal sphincter pressure lowering medication. Barrett's esophagus is the strongest risk factor for adenocarcinoma. Endoscopic surveillance of Barrett's esophagus to reduce mortality from esophageal adenocarcinoma is recommended by many experts, although definite scientific evidence for this strategy is not available. Effective tobacco and alcohol control would probably be the most efficient way to reduce the global burden of esophageal cancer.

[*] tel. +358-14-2691811 ; Fax +358-14-2692770 ; Email markku.voutilainen@ksshp.fi

Incidence of Esophageal Cancer

In 2002, a total of 462117 cases of esophageal cancer were reported in the whole world (http://www-dep.iarc.fr/, GLOBOCAN 2002 database). Of these cancer patients, 68% (N = 315394) were males. Over 80% (N = 386435) of esophageal cancers were detected in less developed regions of the world. In men, the reported incidence rate was 13.8 (age-standardised world population per 100000, ASR (W)) in less developed and 6.8 in more developed regions of the world, respectively. In women, the respective rates were 6.5 and 1.3. The highest reported incidence was 19.1 (ASR (W)) in males in Eastern Africa. According to the GLOBOCAN database, esophageal cancer is a male-predominant disease in all parts of the globe. Esophageal cancer belongs among the ten most common malignancies in the world [1]. The rates of esophageal cancer vary up to 200-fold from population to population, this variation being wider than in any other cancer [1, 2]. There is also great variation in the incidence rates inside individual countries; i.e. in China there are high incidence foci in central and western parts of the country [3].

Older cancer statistics have revealed even higher incidence rates in certain parts of the world. In northern Gonbad in the Caspian region of Iran, an incidence of 195.3 per 100 000 in women and 165.3 in men has been reported [4].The highest risk area was called Asian esophageal cancer belt, in which Caspian littoral in northern Iran, southern republics of the former USSR, and eastern China were included [4]. During the last thirty years, the incidence of esophageal cancer has decreased dramatically in Iran (nowadays ASR incidence 17.6 in males and 14.4 in females), but still esophageal cancer is the second most common malignancy in both sexes [5] Changes in esophageal cancer incidence in China have an important role in determining the whole esophageal cancer burden at global level, because the Chinese constitute one-fifth of the world population. A recent study estimated that there has been decrease in the incidence of esophageal cancer in China from the year 2000 to 2005: ASR per 100 000 may have decreased from 27.6 to 24.0 in males and from 12.1 to 9.7 in females [3] Still esophageal cancer would be the fourth most common cancer in China with over 230 000 cases in 2005 [3] However, the epidemic of tobacco smoking in China may change this favourable development in the future [3] For comparison, American Cancer Society estimated that in 2005 there would be a total of 14520 new esophageal cancers (incidence 7.7 for males and 2.0 for females) in USA (http://www.seer.cancer.gov/).

Also in Japan esophageal cancer incidence in males reportedly is high (10.0, http://www-dep.iarc.fr/, GLOBOCAN 2002 database). In Europe, highest male incidence rates were reported from United Kingdom (9.6), France (11.0), and Hungary (9.8; GLOBOCAN 2002 database). In South America, the highest incidence rate in males was that of Uruguay (11.1); whereas in Central America the highest rate was in Haiti (15.3, GLOBOCAN 2002 database). In females, high incidence areas are otherwise similar to males, but in Japan, France and Hungary incidence rates are not higher than in their neighbouring countries (Globocan 2002 database). Likewise, there were no high risk areas in South and Central America. Instead, incidence rates were high in females in Yemen (7.5) and India (5.1). Between individual western countries and Japan, there is significant variation in mortality trends from esophageal cancer [6]. However, in all these countries the age-specific mortality rates were highest in males born around 1930 [6].

The histological main types of esophageal cancer are squamous cell cancer and adenocarcinoma, which constitute over 90% of all esophageal cancers [7, 8]. Squamous cell carcinoma has traditionally been significantly more common than adenocarcinoma. In western countries, the incidence of esophageal adenocarcinoma has increased significantly while that of the squamous type has decreased during the last decencies [1, 2, 9-12]. This change has been most marked among white men. A few decades ago, adenocarcinoma comprised about 15% of all esophageal malignancies, but nowadays it accounts up to 75% of esophageal cancers in the USA [2]. However, wide regional variation in incidence rates was detected in the USA [13]. The incidence of esophageal and gastric cardia adenocarcinoma has increased also in many European countries [14]. Significant variation in incidence rates exists also between European countries. Data from the year 1997 showed that in Scotland the incidence rate was 3.9 (per world age-standardized population per 100 000) in men and 1.1 in women, whereas in Finland the respective figures were 0.6 for men and 0.1 for women [10]. In the year 2000, the highest incidence rates of adenocarcinoma were reported from Great Britain [5.0 - 8.7 per 100 000), Australia (4.8), Netherlands (4.4), and USA (3.7) [12]. Also in China the incidence of esophageal adenocarcinoma has been on the increase [15].

The increase in the incidence of esophageal adenocarcinoma in the US is not explained solely by the age affect, but the incidence has increased in all age groups [12]. Recently, age-specific incidence has shifted to younger age groups affecting more persons in the age group from 55 to 65 years [10]. The increase in the prevalence of obesity in western world has been linked to the increase in the incidence of esophageal adenocarcinoma [10, 12, 41]. It has been speculated that the increase in esophageal adenocarcinoma incidence could partly be due to changing classification patterns. However, it has been estimated that only one fourth of the increase in adenocarcinoma incidence could be explained by improvements in classification [12].

Aetiology and Risk Factors of Esophageal Cancers

Squamous-Cell Carcinoma

Risk factor profiles are different for esophageal squamous-cell carcinoma and adenocarcinoma.

The uneven geographical distribution of esophageal cancer is probably due to dietary, environmental and cultural factors. In general, any factor that causes chronic irritation or inflammation of the esophageal mucosa may lead to the development of squamous-cell carcinoma.

In high risk esophageal cancer areas in developing countries, nutritional deficiencies play an important role in the pathogenesis of esophageal squamous cell carcinoma. In areas with high incidence of squamous-cell esophageal carcinoma (Linxian China, Iran and South Africa), low intake of B vitamins (especially riboflavin and niacin), magnesium and zinc have been reported [16]. Also in Europe, low intake of vitamin C and E reportedly was associated with an increased esophageal carcinoma risk [17]. In addition to dietary folate deficiency, also genetic mutations resulting in a decreased activity of methylenetetrahydrofolate and low serum folate concentration seem to increase esophageal cancer risk [18]. Low serum selenium concentration and papilloma virus infection have also been linked to squamous-cell carcinoma [19, 20]. Dietary habits that may increase the risk of esophageal cancer are inadequate chewing of food, eating hard and scratchy food, spicy or highly salted food, and smoked meat or fish [21]. In northern Iran, where esophageal cancer has been more common in women than in men, women mainly eat food containing strong black pepper and sharp crushed pomegranate seeds [21]. The results of a small case-control study from Iran suggested that a riboflavin-deficient diet may also explain a portion of the increased esophageal cancer risk in this area [22]. Vitamin C content is low in plants grown in molybdenum-deficient soil, and vitamin C deficiency may cause esophageal mucosal hyperplasia, which may lead to the development of cancer [2].

Tobacco smoking or chewing and alcohol intake further increase the cancer risk in developing countries [23, 24]. Hand-rolled cigarettes seem to be strongest risk factor for cancers of the lower third of the esophagus, whereas strong cigarettes and dark tobacco increase the risk of upper third cancers [25]. The ingestion of very hot beverages causing caustic injury to the esophagus is a risk factor for squamous cell carcinoma in Russia, Asia, and parts of the Middle East [2]. Likewise, ingestion of hot tea (China), tea-like beverage maté (South America) and calvados (France) are risk factors [26-28]. Low intake of fresh fruits, vegetables and dairy products with low intakes of vitamins A, C, riboflavin and zinc are risk factors for squamous cell carcinoma [1]. Iron deficiency anaemia related to Plummer-Vinson or Paterson-Kelly syndrome harbours also an increased risk of this malignancy. Chronic esophageal stasis due to peptic stricture or achalasia, or previous head and neck cancer may also increase the risk of esophageal squamous cell carcinoma [8, 28]. Hyperkeratosis of palms and soles or tylosis is an autosomal dominant hereditary disease that is related to this tumour also. Consumption of fermented or pickled vegetables containing fungi as well as bread eaten mouldy has been linked to the development of squamous cell carcinoma [2]. Smoked meat and fish contain nitrates and their conversion products, and elevated nitrate levels are detected also in drinking water. Consumption of nitrate containing foods and drinks may increase the risk of squamous-type esophageal carcinoma.

In western world, the role of tobacco smoking and alcohol is more marked than that of dietary deficiencies in the pathogenesis of esophageal squamous cell carcinoma. Compared with non-smokers, smokers have 5- 10-fold risk of acquiring squamous cell carcinoma of the esophagus, and the risk correlates with the duration of tobacco exposure. Also alcohol consumption increases the risk in a dose-dependent way. Tobacco and alcohol together have an additive effect on the cancer risk: heavy tobacco and alcohol consumption may increase the risk 100-fold [2]. A study from Japan reported that in men increased esophageal cancer mortality risk was related to simultaneous consumption of both tobacco and alcohol, but neither of these stimulants alone increased the cancer risk {Sakata K, 2005 #18). Researchers from the US have reported similarities in the epidemiological changes of esophageal and lung squamous and adenocarcinomas suggesting that a carcinogen of tobacco may induce also esophageal adenocarcinoma [29]. High level occupational exposure to nitrosamines, carbon black, sulphuric acid and perchloroethylene reportedly increased the risk of esophageal cancer [30-32].

The tar amount of cigarettes has positive correlation with the esophageal cancer risk [33]. Recent research has elucidated also other the mechanisms by which alcohol and tobacco may act in increasing the risk of upper gastrointestinal tract cancers. Present data suggest that the first metabolite of ethanol metabolizing enzyme alcohol dehydrogenase, asetaldehyde, acts as a local carcinogen in the upper digestive tract [34]. It is formed also by the normal microbe flora of the oral cavity. Tobacco smoke contains also asetaldehyde that dissolves in the saliva during smoking. Nitrosoamines also belong to tobacco carcinogens [8]. Aldehyde dehydrogenase is a mitochondrial enzyme that oxidizes acetaldehyde. Epidemiological studies have revealed that upper gastrointestinal cancers are common among Asians with genetically determined low activity of aldehyde dehydrogenase [34]. The gene coding the alcohol dehydrogenase enzyme is polymorphic and heavy drinking associated with high activity of alcohol dehydrogenase resulting high aldehyde concentration increases significantly the risk of upper aerodigestive cancers [34]. Thus increased risk of esophageal cancer may be related either to decreased ability to eliminate aldehyde or enhanced ability to produce this toxic compound [34].

Other factors predisposing to the development of squamous cell carcinoma are previous (more than ten years) radiotherapy to the mediastinum, i.e. for the treatment of breast cancer or lymphoma [8]. Betel quid chewing with tobacco smoking and alcohol intake reportedly increase the risk of esophageal squamous-cell carcinoma over 40-fold [35]. Mycotoxins called fumonisins may contaminate corn and consumption of fumonisin-contaminated corn has been linked to esophageal cancer in China and southern Africa [36] In high risk areas in China, wheat may contain high concentration of Fusarium mycotoxins [37]. Infection with CagA-positive strains of Helicobacter pylori in association with gastric atrophy may increase the risk of squamous-cell carcinoma [38].

Most cases of esophageal cancers are sporadic, but some genetic risk groups reportedly exist. Patients with palmoplantar keratoderma have up to 40% risk of developing squamous-cell carcinoma of the esophagus [2, 39]. Plummer-Vinson syndrome is characterized by dysphagia, iron deficiency anaemia, and esophageal web(s) and harbours a 10% risk of esophageal cancer [2]. A retrospective cohort study suggested that patients with Fanconi anaemia may also have an increased risk of esophageal cancer [40].

Adenocarcinoma

Risk factors for esophageal adenocarcinoma are different from those of squamous cell carcinoma.

The most often identified risk factors are increasing age, male gender, gastroesophageal reflux disease, Barrett's esophagus, white race, obesity, tobacco smoking, family history, and lower esophageal sphincter pressure lowering medication [10, 12, 41, 42]. Instead, alcohol consumption does not seem to increase the risk of esophageal adenocarcinoma [43]. Compared with squamous cell esophageal carcinoma, tobacco smoking increases the risk of esophageal adenocarcinoma to lesser extend (5-10-fold vs. twofold or less) [2]. Persons born preterm and those with low birth weight may be at increased risk of esophageal adenocarcinoma [44].

The risk of esophageal adenocarcinoma correlates positively with increasing age, but although in general incidence rates have been higher in older age groups (age effect), there is also general trend of higher incidence rates in each 5-year cohorts born more recently (cohort effect) [10]. The increase in the incidence of esophageal adenocarcinoma could be explained by a factor affecting more cohorts born more recently: possible factors could be i.e. GERD, obesity, diet, and drugs [10].

Reflux disease has been linked to the development of esophageal adenocarcinoma. Compared to persons without heartburn and regurgitation, long-standing and severe reflux disease symptoms reportedly increased the risk of this cancer over 43-fold in Sweden [45]. Likewise, studies from USA have revealed a positive correlation between symptomatic reflux disease and esophageal adenocarcinoma [46, 47]. Instead, infection of gastric mucosa by CagA-positive Helicobacter pylori strain may protect from esophageal adenocarcinoma [38, 48, 49]. Also the use of dietary fibre reduces the risk of this malignancy [50]. Obesity has been linked to reflux disease symptoms, erosive esophagitis as well as esophageal adenocarcinoma [51]. Overweight has opposite relationship to the two major histological esophageal cancer types: it increases the risk of esophageal adenocarcinoma but has inverse correlation with squamous-cell carcinoma [52]. A dose-dependent relationship between obesity and esophageal adenocarcinoma has been detected [51, 53].

The only known precursor lesion for esophageal adenocarcinoma is Barrett's esophagus, which is defined as the replacement of normal squamous epithelium by metaplastic mucosa in the distal esophagus. This metaplastic mucosa is histologically defined as specialized columnar epithelium or incomplete intestinal metaplasia [54]. An autopsy study from Olmsted reported that the prevalence of Barrett' esophagus was 0.4% [55]. Very recently published study from Sweden showed that a randomly selected cohort from general population had a prevalence of Barrett's esophagus of 1.6% [56]. The strength of the Swedish study was that the detection of specialized columnar epithelium at the columnar-lined esophagus was used as a diagnostic criterion for Barrett's esophagus. Compared with the rest of the study population, the prevalence of BE was higher in persons with gastroesophageal reflux symptoms and in those with esophagitis, but the difference was not statistically significant [56]. Most esophageal adenocarcinomas are sporadic cases, but familial aggregation of both Barrett's esophagus and esophageal adenocarcinoma have been reported [57, 58], although there are also studies showing no familial susceptibility to esophageal cancer [59].

Approximately 60-80% of patients with esophageal adenocarcinoma have Barrett's esophagus [2]. When symptomatic, esophageal adenocarcinoma has usually sent metastases and the 5-year prognosis of metastasised tumour is dismal: less than 5% are alive after five years from the diagnosis [10]. Instead, about 75% of patients are alive after 5 years if the tumour is detected and treated when it is at the localized phase. In Europe, the 5-year survival rates are from 3% to 15% [60]. Despite the increased use of upper gastrointestinal endoscopy, only about 2% of adenocarcinomas were detected at the localized phase in 1990s in the USA [10]. Compared with 1970s, the results of modern surgical treatment of esophageal cancer are only slightly better [10]. Probably most esophageal adenocarcinomas develop at the Barrett-type mucosa through stepwise progression of dysplasia up to high grade dysplasia and eventually adenocarcinoma [54], and this theoretically provides the opportunity to detect adenocarcinoma before it has sent metastases.

Several study groups have published results of the endoscopic and biopsy surveillance of Barrett's esophagus to detect esophageal adenocarcinoma in its early and possibly curable phase. Some studies have reported that compared with patients with symptom-detected esophageal adenocarcinomas, those with surveillance-detected tumours have a longer survival [61-64]. However, not all studies have observed any value of surveillance [65-67], or the benefit of surveillance may be limited to male patients [68]. Furthermore, the overall mortality of patients with Barrett's esophagus may be similar to that of general population [69].

A recent meta-analysis revealed that published studies may have over-estimated the true cancer risk related to Barrett's esophagus [70]. The results of this study suggested that the actual risk of Barrett's esophagus patients of developing esophageal adenocarcinoma may be 0.5% yearly [70]. However, only approximately 5% of esophageal adenocarcinoma patients have Barrett's esophagus diagnosed before their cancer [71]. Studies reporting the positive impact of Barrett's esophagus surveillance on esophageal cancer mortality are cohort studies with a relatively small numbers of Barrett's patients, and their results may be biased by selection, lead time and length time bias as well as confounding. A randomized prospective trial should be performed to prove the benefit of Barrett's esophagus surveillance in preventing esophageal adenocarcinoma. Such a trial would require thousands of Barrett patients to be followed several years and mortality should be the primary outcome [72, 73]. It is very unlikely that such a study will ever be performed.

Prevention of Esophageal Cancer

World-wide, the vast majority of esophageal cancers are squamous cell carcinomas. Esophageal cancer is a highly lethal disease. However, for the localized esophageal cancer, surgical resection results a 5-year prognosis. It would thus be logical to screen and survey patients at high risk of developing esophageal cancer. Only some rare conditions such as tylosis palmaris and caustic injury of the esophagus harbour an increased risk of squamous-cell carcinoma. Endoscopic screening every 1-3 years is recommended for these small groups [74].

Large-scale primary prevention of squamous cell carcinoma would thus mean reduction of risk factors at population level in developing countries. The main risk factors for esophageal cancer are diet (i.e. low fruit and vegetable intake) and the use of addictive substances (tobacco and alcohol) [75]. There are no controlled studies on the effect of dietary intervention to reduce the incidence of esophageal squamous-cell carcinoma, but epidemiological cohort studies and meta-analysis based on these studies support the opinion that lack of fresh fruits and vegetables is linked to the development of this malignancy [76]. Supplementation of micronutrients may slightly reduce the risk of esophageal cancer in high risk areas. In Linxian with one of the world's highest esophageal cancer incidences, a 5-year supplementation with riboflavin and niacin resulted in a 14% and 10% decline in the incidence and mortality of esophageal cancer [77]. In patients with esophageal dysplasia, vitamin and mineral supplementation did not produce statistically significant protective effect against esophageal cancer [78]. Modern food preserving methods, i.e. refrigerator makes possible to use better diet containing fruit and vegetables with essential vitamins and micronutrients, which might prevent or delay the onset of esophageal cancer [79]. The Mediterranean-type diet with high fruit and vegetables content, and the use of olive oil and other unsaturated fats, but without red meat and refined cardohydrates seems to reduce esophageal cancer risk [80, 81]. Also avoiding high consumption of butter and cholesterol may decrease esophageal cancer risk [82]. It has been estimated that the use of alcohol and tobacco causes up to 80% of esophageal squamous-cell carcinomas, thus effective control of the use of these substances would have a key role in preventing this cancer [28, 34, 83]. Some studies, i.e. that performed in Shanghai [84], have suggested that green tea may protect against upper gastrointestinal cancers, but not all studies have confirmed these findings A small nested case-control study also conducted in Shanghai revealed that urinary tea polyphenols inversely correlated with the risk of esophageal cancer [84]. The 40% reduction of esophageal cancer risk was related to one polyphenol (epigallocatechin). The risk reduction, however, was not statistically significant [84].

Experimental and epidemiologic data suggest that the use on aspirin and non-steroidal anti-anti-inflammatory drugs is associated with a reduced risk of esophageal adenocarcinoma [85]. At present, there are no randomized studies to prove the efficacy of these drugs in practice in preventing mortality from esophageal adenocarcinoma [86]. Neither fundoplication nor high-dose proton pump inhibitor treatment have been shown to prevent esopaheageal adenocarcinoma in patients with Barrrett's esophagus [86].

Appendix

This review is based on a systematic electronic literature search was performed. The following terms were searched from titles and abstracts of articles published in English: esophageal neoplasms/cancer/squamous-cell carcinoma/adenocarcinoma/ neoplasms, and Barrett's esophagus. In the first step, these terms were linked mass/genetic screening and surveillance. In second step, the above-mentioned terms were linked to terms risk factors, protective factors, diet, food, meat, fruit, vegetables, smoking, tobacco, alcohol drinking, beverages, and Helicobacter pylori. The three searches were confined with the following terms: epidemiologic study characteristics, epidemiologic research design, evaluation studies, research design, outcome and process assessment, bias (epidemiology), evidence-based medicine, comparative study. The following terms were used to search scientific evidence. Clinical trial (pt), controlled clinical trial, randomized controlled trial, meta-analysis, guideline, practice guideline, multicenter study, and consensus development conference. Scientific reports were search by following sieve: ("Meta-Analysis"[MeSH] OR "Meta analysis" tw OR "Meta-analysis" tw OR meta analysis[Publication Type]) OR "systematic review"tw OR "systematic reviews"tw OR "systematic overviews"tw OR "systematic overview"tw OR "systematic view"tw OR "Review Literature"[MeSH] exp OR Cochrane ab/ti OR embase ab/ti OR psychlit OR psyclit ab/ti OR psychinfo OR psycinfo ab/ti OR cinahl OR cinhal ab/ti OR "science citation index" ab/ti OR "bids" ab/ti OR cancerlit ab/ti

OR "reference list" ab/ti OR "reference lists" ab/ti OR bibliograph* ab/ti OR "hand search" ab/ti OR "relevant journals" ab/ti OR "manual search" ab/ti OR "manual searches" ab/ti OR "hand searches" ab/ti OR ((review[Publication Type]) AND ("data extraction" OR "selection criteria")), from which the following was excluded: NOT (comment[Publication Type] OR letter[Publication Type] OR editorial[Publication Type]) . Articles on animal studies and therapy were also excluded (NOT ("Animals"[MeSH] NOT (("Humans"[MeSH] AND "Animals"[MeSH])) and NOT ("therapy"[Subheading] OR treatment ti/ab OR Therapy ti/ab.

References

[1] Crew KD, Neugut AI. Epidemiology of upper gastrointestinal malignancies. *Semin in Oncology* 2004;31(4):4540-464.

[2] Lukanich JM. Section I: Epidemiological review. *Semin Thoracic and Cardiovasc Surg* 2003;15(2):158-166.

[3] Yang L, Parkin DM, Ferlay J, Li L, Chen Y. Estimates of cancer incidence in China for 2000 and projections for 2005. *Cancer Epidemiol Biomarkers Prev* 2005;14(1):243-250.

[4] Boyle P. Global burden of cancer. *Lancet* 1997;349((suppl II)):23-26.

[5] Sadjadi A, Nouraie M, Mohagheghi MA, Mousavi-Jrrahi A, Malekezadeh R, Parkin DM. Cancer occurence in Iran in 2002, an international perspective. *Asian Pac J Cancer Prev* 2005;6:359-363.

[6] Qiu D, Kaneko S. Comparison of esophageal cancer mortality in five countries: France, Italy, Japan, UK and USA from the WHO mortality database (1960-2000). *Jpn J Clin Oncol* 2005;35:564-567.

[7] Sarbia M, Becker KF, Höfler H. Pathology of upper gastrointestinal malignancies. *Seminars in Oncology* 2004;31(4):465-475.

[8] Enzinger PC, Mayer RJ. Esophageal cancer. *N Engl J Med* 2003;349(23):2241-2252.

[9] Powell J, McConkey CC, Gillison EW, Spychal RT. Continuing rising trend in oesophageal adenocarcinoma. *Int J Cancer* 2002;102:422-427.

[10] El-Serag HB. The epidemic of esophageal adenocarcinoma. *Gastroenterol Clin N Am* 2002;31:421-440.

[11] Forman D. Review article: oesophagogastric adenocarcinoma- an epidemiological perspective. *Aliment Pharmacol Ther* 2004;20((Suppl 5)):55-60.

[12] Wei JT, Shaheen NJ. The changing epidemiology of esophageal adenocarcinoma. *Semin in Gastrointestinal Disease* 2003;14(3):112-127.

[13] Kubo A, Corley DA. Marked regional variation in adenocarcinomas of the esophagus and the gastric cardia in the United States. *Cancer* 2002;95:2096-2102.

[14] Botterweck AAM, Schouten LJ, Volovics A, Dorant E, van den Brandt PA. Trends in incidence of adenocarcinoma of the oesophagus and gastric cardia in ten European countries. *Int J Epidemiol* 2000;29:645-654.

[15] Zhang H, chen S-H, Li Y-M. Epidemiological investigation of esophageal carcinoma. *World J Gastroenterology* 2004;10:1834-1835.

[16] van Rensburg SJ. Epidemiologic and dietary evidence for a specific nutritional predisposition to esophageal cancer. *J Natl Cancer Inst* 1981;67:243-251.

[17] Bollschweiler E, Wolfgarten E, Nowroth T, Rosendahl U, Monig SP, Holscher AH. Vitamin intake and risk of suntypes of esophageal cancer in Germany. *J Cancer Res Clin Oncol* 2002;128:575-580.

[18] Song C, Xing D, Tan W, Wei Q, D L. Methylenetetrahydrofolate reductase polymorphims increase risk of esophageal squamous cell carcinoma in a Chinese population. *Cancer Research* 2001;61:3272-3275.

[19] Wei W-Q, Abnet CC, Qiao Y-L, Dawsey SM, Dong Z-W, Sun X-D, *et al.* Prospective study of selenium concentrations and esophageal and gastric cardia cancer, heart disease, stroke, and total death. *Am J Clin Nutr* 2004;79:80-85.

[20] Yao P-F, Li G-C, Li J, Xia H-S, Yang X-L, Huang H-Y, *et al.* Evidence if human papilloma virus infection and its epidemiology in esophageal squamous cell carcinoma. *World J Gastroenterology* 2006;12:1352-1355.

[21] Ghadirian P, Ekoe JM, Thouez JP. Food habits and esophageal cancer. *Cancer Detect Prev* 1992;16:163-168.

[22] Siassi F, Ghadirian P. Riboflavin deficiency and esophageal cancer: a case control-household study in the Caspian Littoral of Iran. *Cancer Detect Prev* 2005;29:464-469.

[23] Chen ZM, Xu Z, Collins R, Li WX, Peto R. Early health effects of the merging tobacco epidemic in China. A 16-year propspective study. *JAMA* 1997;278:1500-1504.

[24] Castellsagúe X, Muñoz N, De Stefani E, Victora CG, Casteletto R, Rolón PA, *et al.* Independent and joint efects of tobacco smoking and alcohol drinking on the risk of esophageal cancer i men and women. *Int J Cancer* 1999;82:657-664.

[25] Launoy G, Milan C, Faivre J, Pienkowsi P, M G. Tobacco type and risk os squamous cell cancer of the oesophagus in males: a French multicentre case-control study. *Int J Epidemiol* 2000;29:36-42.

[26] Castellsagúe X, Muñoz N, De Stefani E, Victora CG, Casteletto R, Rolón PA. Influence of mate drinking, hot beverages and diet on esophageal cancer risk in South America. *Int J Cancer* 2000;88:658-664.

[27] Launoy G, Milan C, Day NE, Faivre J, Pienkowsi P, Gignoux M. Oesophageal cancer in France: potential importance of hot alcoholic drinks. *Int J Cancer* 1997;71:917-923.

[28] Rozen P. Cancer of the gastrointestinal tract: early detection or early prevention? *Eur J of Cancer Prevention* 2004;13:71-75.

[29] Cockburn MG, Wu AH, Bernstein L. Etiologic clues from the similarity of histology-specific trends in esophageal and lung cancers. *Cancer Causes Control* 2005;16:1065-1074.

[30] Ruder AM, Ward EM, Brown DP. Mortality in dry-cleaning workers: an update. *Am J Ind Med* 2001;39:121-132.

[31] Parent M-É, Siemiatycki J, Fritschi L. Workplace exposures and oesophageal cancer. *Occup Environ Med* 2000;57:325-334.

[32] Straif K, Weiland SK, Buners M, Holthenrich D, Taeger D, Yi S, *et al.* Exposure to high concentrations of nitrosamines and cancer mortality among a cohort of rubber workers. *Occup Environ Med* 2000;57:180-187.

[33] Gallus S, Altieri A, Boseti C, Franceschi S, Levi F, Negri F, *et al.* Cigarette tar yield and risk of upper digestive tract cancers: case-control studies from Italy and Switzerland. *Annals of Oncology* 2003;14:209-213.

[34] Salaspuro M, Salaspuro V, Seitz HK. Interaction of alcohol and tobacco in upper aerodigestive tract and stomach cancer. In: *Alcohol, tobacco, and cancer.* Editors: Cho CH, Purohit V. Karger AG, Basel 2006.

[35] Lee C-H, Lee J-M, Wu D-C, Hsu H-K, Kao E-L, Huang H-L, *et al.* Independent and combined effects of alcohol intake, tobacco smoking and betel quid chewing on the risk of esophageal cancer in Taiwan. *Int J Cancer* 2005;113:475-482.

[36] Voss KA, Riley RT, Norred WP, Bacon CW, Meredith FI, Howard PC, *et al.* An overview of rodent toxicities: liver and kidney effects of fumonisins nd Fusarium moniliforme. Environmental Health Perspectives 2001;109(S2):259 - 266. (www.ehponline.org).

[37] Luo Y, Yoshizawa T, Katayama T. Comparative study on the natural occurrence of Fusarium mucotoxins (trichothecenes and zearalenone) in corn and whaetbfrom high- and low-risk areas for human esophageal cancer in China. *Applied and Environmental Microbiology* 1990;56:3723-3726.

[38] Ye W, Held M, Lagergren J, Engstrand L, Blot WJ, McLaughlin JK, *et al.* Helicobacter pylori infection and gastric atrophy: risk of adenocarcinoma and squamous-cell carcinoma of the esophagus and adenocarcinoma of the gastric cardia. *J Natl Cancer Inst* 2004;96:388-396.

[39] Ilhan M, Erbaydar T, Akdeniz N, Arslan S. Palmoplantar keratoderma is associated with esophagus squamous cell cancer in Van region of Turkey: a case control study. *BMC Cancer* 2005;5:90.

[40] Rosenberg PS, Greene MH, Alter BP. Cancer incidence in persons with Fanconi anemia. *Blood* 2003;101:822-826.

[41] Chow W-H, Blot WJ, Vaughan TL, Risch HA, Gammon MD, Stanford JL, *et al.* Body mass index and risk of adenocarcinomas of the esophagus and gastric cardia. *J Natl Cancer Inst* 1998;90:150-155.

[42] Lagergren J, Bergström R, Adami H-O, Nyrén O. Association between medications that relax the lower esophageal sphincter and risk for esophageal adenocarcinoma. *Ann Intern Med* 2000;133:165-175.

[43] Lagergren J. Adenocarcinoma of oesophagus: what exactly is the size of the problem and who is at risk? *Gut* 2005;54(Suppl I):i1-i5.

[44] Kaijser M, Akre O, Cnattingius S, Ekbom A. Preterm birth, low birth weight, and risk for esophageal adenocarcinoma. *Gastroenterol* 2005;128:607-609.

[45] Lagergren J, Bergström R, Lindgren A, O N. Symptomatic gastroesophageal reflux is a strong risk factor for esophageal adenocarcinoma. *N Engl J Med* 1999;340:825-831.

[46] Farrow DC, Vaughan TL, Sweeney C, Gammon MD, Chow W-H, Risch HA, *et al.* Gastroesophageal reflux disease, use of H2 receptor antgonists, and risk of esophageal and gastric cancer. *Cancer Causes Control* 2000;11:231-238.

[47] Chow WH, Finkle WD, McLaughlin JK, Frankl H, Ziel HK, FraumeniJr JF. The relation of gastroesophageal reflux disease and its treatment to adenocarcinomas of the esophagus and gastric cardia. *JAMA* 1995;274:474-477.

[48] de Martel C, Llosa A, Farr SM, Friedman GD, Vogelman JH, Orentreich N, *et al.* Helicobacter pylori infection and the risk of developmentb of esophageal adenocarcinoma. *J Infect Diseases* 2005;191:761-767.

[49] Chow W, Blaser MJ, Blot WJ, Gammon MD, Vaughan TL, Risch HA, *et al.* An inverse relation between cagA+ strains of Helicobacter pylori infection and risk of esophageal and gastric cardia adenocarcinoma. *Cancer Res* 1998;58:588-590.

[50] Pera M, Manterola C, Vidal O, Grande L. Epidemilogy of esophageal adenocarcinoma. *J Surg Oncol* 2005;92:151-159.

[51] Hampel H, Abraham NS, El-Serag HB. Meta-analysis: obesity and the risk factors for gastroesophageal reflux disease. *Ann Intern Med* 2005;143:199-211.

[52] Engeland A, Tretli S, Bjorge T. Height and body mass index in realtion to esophageal cancer; 23-year follow-up of two million Norwegian men and women. *Cancer Causes Control* 2004;15:837-843.

[53] Lagergren J, Bergström R, Nyrén O. Assocition between body mass and adenocarcinoma of the esophagus and gastric cardia. *Ann Intern Med* 1999;130:883-890.

[54] Haggitt RC. Barrett's esophagus, dysplasia, and adenocarcinoma. *Hum Pathol* 1994;25:982-993.

[55] Cameron AJ, Zinsmeister AR, Ballard DJ, Carney GA. Prevalence of columnar-lined (Barrett's) esophagus. Comparison of population-based clinical and autopsy findings. *Gastroenterology* 1990;99:918 - 922.

[56] Ronkainen J, Aro P, Storskrubb T, Johansson SE, Lind T, Bolling-Sternevald E, *et al.* Prevalence of Barrett's esophagus in the general population: an endoscopic study. *Gastroenterology* 2005;129:2101-2103.

[57] Drovdlic CM, Goddard KAB, Chak A, Brock W, Chessler L, King JF, *et al.* Demographic and phenotypic features of 70 families segregating Barrett's oesophagus and oesophageal adenocarcinoma. *J Med Genet* 2003;40:651-656.

[58] Chak A, Lee T, Kinnard MF, Brock W, Faulx A, Willis J, *et al.* Familial aggregation of Barrett's oesophagus, oesophageal adenocarcinoma, and oesophagogastric junctional adenocarcinoma in Caucasian adults. *Gut* 2002;51:323-328.

[59] Dhillon PK, Farrow DC, Vaughan TL, Chow W-H, Risch HA, Gammon MD, *et al.* Family history of cancer and risk of esophageal and gastric cancers in the United States. *Int J Cancer* 2001;93:148-152.

[60] Keighley MRB. Gastrointestinal cancers in Europe. *Aliment Pharmacol Ther* 2003;18(Suppl. 3):7-30.

[61] van Sandick JW, Lanschot JJB, Kuiken BW, Tytgat GNJ, Offerhaus GJA, Obertop H. Impact of endoscopic biopsy surveillance of Barrett's oesophagus on pathological stage and clinical outcome of Barrett's carcinoma. *Gut* 1998;43:216-222.

[62] Fountoulakis A, Zafirellis KD, Dolan K, Dexter SPL, Martin IG, Sue-Ling HM. Effect of surveillance of Barrett's oesophagus on the clinical outcome of oesophageal cancer. *Br J Surg* 2004;91:997-1003.

[63] Peters JH, Clark GWB, Ireland AP, Chandrasoma P, Smyrk TC, DeMeester TR. *Outcome of adenocarcinoma arising in Barrett's esophagus in endoscopically surveyed and nonsurveyed patients.* 108 1994(813-822).

[64] Corley DA, Levin TR, Habel LA, Weiss NS, Buffler PA. Surveillance and survival in Barrett's adenocarcinomas: a population-based study. *Gastroenterol* 2002;122:633-640.

[65] Macdonald CE, Wicks AC, Playford RJ. Final results from 10 year cohort of patients undergoing surveillance for Barrett's oesophagus: observational study. *BMJ* 2000;321:1252-1255.

[66] Eckardt VF, Kanzler G, Bernhard G. Life expectancy and cancer risk in patients with Barrett's esophagus: a prospective controlled investigation. *Am J Med* 2001,111:33-37

[67] van der Burgh A, Dees J, Hop WCJ, van Blankenstain M. Oesophageal cancer is an uncommon cause of death in patients with Barrett's oesophagus. *Gut* 1996;39:5-8.

[68] Bani-Hani K, Sue-Ling H, Johnston D, Axon TR, Martin IG. Barrett's oesophagus: results from a 13-year surveillance programme. *Eur Gastroenterol Hepatol* 2000;12:649-654.

[69] Anderson LA, Murray LJ, Murphy SJ, Fitzpatrick DA, Johnston BT, Watson RG, *et al.* Mortality in Barrett's oesophagus: results from a population based study. *Gut* 2003;52:1081-1084.

[70] Shaheen NJ, Crosby MA, Bozymski EM, Sandler RS. Is there publication bias in the reporting of cancer risk in Barrett's esophagus. *Gastroenterol* 2000;119:333 -338.

[71] Dulai GS, Guha S, Kahn KL, Gornbein J, Weinstein WM. Preoperative prevalence of Barrett's esophagus in esophageal adenocarcinoma: a systematic review. *Gastroenterol* 2002;122:26-33.

[72] Kapadia CR. Barrett's esophagus: whom to screen? Still an unanswered question. *Gastroenterology* 1998;114:1101-1102.

[73] Sharma P, Sidorenko EI. Are screening and surveillance for Barrett's oesophagus really worthwhile? *Gut* 2005;54(Suppl I):127-132.

[74] Brown A, Shaheen NJ. Screening for upper gastrointestinal tract malignancies. *Semin in Oncology* 2004;31:487-497.

[75] Ezzati M, Lopez AD, Rodgers A, Vander Hoorn S, Murray CJL, and the Comparative Risk Assessment Collaborating Group. Selected major risk factors and global and regional burden of disease. *Lancet* 2002;360:1347-1360.

[76] Riboli E, Norat T. Epidemiologic evidence of the protective effect of fruit and vegetables on cancer risk. *Am J Clin Nutr* 2003;78(suppl):55S-69S.

[77] Blot WJ, Li J-Y, Taylor PR, Guo W, Dawsey S, Wang G-Q, *et al.* Nutrition intervention trials in Linxian, China: supplementation with specific vitaminImineral combinations, cancer incidence, and disease-specific mortality in general population. *J Natl Cancer Inst* 1993;85:1483-1492.

[78] Li J-Y, Taylor PR, Dawsey S, Wang G-Q, Ershow AG, Guo W, *et al.* Nutrition intervention trials in Linxian, China: multiple vitamin/mineral supplementation, cancer incidence, and disease-specific mortality among adults with esophageal dysplasia. *J Natl Cancer Inst* 1993;85:1492-1498.

[79] Hoffmann K, Boeing H, Volatier J-L, Becker W. Evaluating the potential health gain of the World Health Organization's recommendation concerning vegetable and fruit consumption. *Public Health Nutrition* 2003;6:765-772.

[80] Chen H, Ward MH, Graubard BI, Heineman EF, Markin RM, Potischman NA, *et al.* Dietary patterns and adenocarcinoma of the esophagus and distal stomach. *Am J Clin Nutr* 2002;75:137-144.

[81] La Vecchia C. Mediterranean diet and cancer. *Public Health Nutrition* 2004;7:965-968.

[82] Launoy G, Milan C, Day NE, Pienkowsi MP, Gignoux M, Faivre J. Diet and squamous-cell cancer of the oesophagus: a French multicentre case-control study. *Int J Cancer* 1998;76:7-12.

[83] Key TJ, Allen NE, Spencer EA, RC; T. The effect of diet on risk of cancer. *Lancet* 2002(360):861-868.

[84] Sun C-L, Yuan J-M, Lee M-J, Yang CS, Gao Y-T, Ross RK, *et al.* Urinary tea polyphenols in relation to gastric and esophageal cancers: a prospective study of men in Shanghai, China. *Carcinogenesis* 2002;9:1497-1503.

[85] 85. Sharma P, McQuaid K, Dent J, Fenerty B, Sampliner S, et al. A critical review of the diagnosis and management of Barrett's esophagus: the AGA Chicago workshop. *Gastroenterology* 2004;127: 310-330.

[86] 86. Wang KK, Wongkeesong M, Buttar NS. American Gastroenterological Association technical review on the role of gastroenterologist in the management of esophageal carcinoma. *Gastroenterology* 2005;128: 1471 - 1505.

In: Esophageal Cancer Research Developments
Editor: Henry K. Kinner, pp. 127-147

ISBN: 1-60021-309-X

Chapter IV

Comparative Serum-free Culture of Normal and Malignant Esophageal Epithelial Cells: A Possible Cross Talk between Transforming Growth Factor-β1 and Epidermal Growth Factor Signaling

Masafumi Katayama*[1], *Masaru Shoji*[2] *and Susumu Satomi
Division of Advanced Surgical Science and Technology,
Graduate School of Medicine, Tohoku University, Sendai, Japan
[1]Department of Home Economics, Tohoku Seikatsu Bunka College, Sendai, Japan
[2]Department of Surgery, Koritsu Fukaya Hospital, Mono-gun, Miyagi Japan

Abstract

Although more than 70 cell lines have been established from esophageal cancer, the growth properties of these cell lines have not compared extensively with their normal counterpart. Comparative culture of normal and malignant cells is important for understanding the growth properties of esophageal cancer. We succeeded in establishing a serum-free culture system for normal human esophageal epithelial cells (HEE cells). We also established 15 cell lines from esophageal cancer (TE-cell lines) using serum-containing medium. Among 15 TE-cell lines, 14 cell lines grew well in serum-free medium. The growth properties of HEE cells and 14 TE-cell lines were characterized upon stimulation with various growth-related factors. Among these factors, acidic fibroblast growth factor (aFGF) was the most effective stimulant for both the HEE cells and all the TE-cell lines. Transforming growth factor-β1 (TGF-β1) inhibited the growth of HEE cells and 7 TE-cell lines; however, the other 7 TE-cell lines were resistant to the inhibitory effect of TGF-β1. Interestingly, epidermal growth factor (EGF) had a much greater stimulatory effect on the TGF-β1-resistant cells than the TGF-β1-sensitive cells. Although ethanolamine enhanced the growth-promoting ability of EGF or aFGF in the TGF-β1-sensitive cells, it had no effect on the TGF-β1-resistant cells. These findings suggested a possible cross talk between TGF-β1 and EGF signaling, and an important role of ethanolamine in the signaling pathways of growth factors. We also analyzed the expression of cyclin-dependent kinase inhibitor 2B (CDKN2B) mRNA, which was reported to be related to TGF-β signaling. Induction of CDKN2B mRNA by TGF-β1 treatment was observed in HEE cells using the reverse transcription polymerase chain reaction (RT-PCR). In contrast, 12 TE-cell lines didn't express this mRNA. Although remaining 2 TE-cell lines barely expressed this mRNA, these cells were resistant to the inhibitory effect of TGF-β1. Basic fibroblast growth factor (bFGF) had a much lower growth-promoting activity than did aFGF for HEE cells and all the TE-cell lines with the exception of one TE-cell line (TE11). In TE11 cells, however, the growth promoting activity of bFGF was higher than that of aFGF. Most of TE-cell lines required higher concentration of calcium for their optimal growth, especially two cell lines (TE2 and TE10) required calcium essentially for their growth over 10 holds. Conditioned medium of TE3, TE9 and TE14 stimulated the growth of all TE-cell lines; however had inhibitory effect on HEE cells.

Key words: aFGF; EGF; TGF-β1; ethanolamine; calcium; CDKN2B

Introduction

Esophageal cancer occurs world wide and is characterized by a large variation in incidence (Nishihira et al. 1993; Lewin et al. 1995). Despite efforts aimed at early diagnosis and treatment, the prognosis of esophageal cancer is poor. Studies at the cellular and molecular level are important for advances in the diagnosis and treatment of esophageal cancer. To pursue these studies, we previously established a serum-free culture system for normal human esophageal epithelial (HEE) cells (Katayama et al. 1984; Katayama et al. 1986), and 15 cell lines (TE-cell lines) from human esophageal cancer (Nishihira et al. 1993; Nishihira et al. 1994).

More than 70 cell lines have been established from esophageal cancer and reported in the literature (Nishihira et al. 1993, Nishihira et al. 1994). These cell lines were used for studies of their growth and differentiation properties, chromosomal aberrations, protein kinase C, growth factors and their receptors, oncogenes, and tumor-suppressor genes. Cultured normal human esophageal epithelial cells have been used as the counterpart of cancer cell lines in studies of carcinogenesis, and the growth and differentiation of cancer cells (Nishihira et al. 1993). However, the growth properties of normal and cancer-derived esophageal epithelial cells have not been compared extensively. To perform such studies, we needed defined culture conditions. Serum should be omitted from the medium in experiments involving growth factors, since serum contains unknown factors that affect growth.

We previously reported that crude bovine neural tissue extract was an effective growth stimulant of HEE cells, and serum had an inhibitory effect on the growth of HEE cells

(Katayama et al. 1986). Bovine neural tissue extract has been a widely used source of acidic and basic fibroblast growth factors (aFGF and bFGF, respectively) (Burgess and Maciag 1989). Although we succeeded in establishing cell lines of mouse esophageal epithelial cells using aFGF-supplemented serum-free medium (Katayama and Kan 1991), it was difficult to maintain HEE cells for serial passages in this medium. Recently, we found that ethanolamine was essential for the serial passage of HEE cells (Katatayama et al. 2005). Most of the TE-cell lines also grow well in serum-free medium supplemented with aFGF and ethanolamine. Here we compare the growth properties of HEE cells and the TE-cell lines in defined serum-free conditions. We show that there are several differences in their responses to growth factors, ethanolamine, calcium ions and conditioned medium.

Transforming growth factor-β1 (TGF-β1) has been shown to be a potent inhibitor of the proliferation of most normal cell types in culture as well as *in vivo*, but many tumorigenic cell lines have lost responsiveness to the antiproliferative effects of TGF-β1 (Alexandrow and Moses 1995). The growth inhibitory effect of TGF-β1 was observed in HEE cells and 7 TE cell lines (Katatayama et al. 2005). Other 7 TE cell lines that were resistant to inhibitory effect of TGF-β1 showed different growth properties from HEE cells. Two classes of antiproliferative gene response are involved in TGF-β-mediated growth arrest: gene responses that inhibit cyclin-dependent kinases (CDKs) and those that result in downregulation of c-*myc* (Massagué et al. 2000). The cyclin-dependent kinase inhibitor 2B (CDKN2B) gene lies adjacent to the tumor suppressor gene CDKN2A, in a region that is frequently mutated and deleted in a wide variety of tumors (Lubomierski et al. 2001). The CDKN2B gene encodes a cyclin-dependent kinase inhibitor, which forms a complex with CDK4 or CDK6, and prevents activation of the CDKs; thus the encoded protein functions as a cell growth regulator that controls cell cycle G1 progression (Alexandrow and Moses 1995). Expression of this gene was found to be dramatically increased by TGF-β treatment, which suggests that it plays a role in TGF-β-induced growth inhibition (Hannon and Beach 1994). Therefore, we analyzed the expression of CDKN2B mRNA in HEE cells and TE-cell lines using the reverse transcription polymerase chain reaction (RT-PCR). Induction of CDKN2B mRNA synthesis by TGF-β1 treatment was observed in normal HEE cells. Here we show that expression of the CDKN2B gene is altered in TE cell lines.

Materials and Methods

Materials

aFGF, epidermal growth factor (EGF), transforming growth factor-α (TGF-α), TGF-β1, ethanolamine, L-proline, heparin, bovine serum albumin (BSA), insulin, transferrin, trypsin (type□), soybean trypsin inhibitor (STI, type II-S), dithiothreitol (DTT), and pronase-E were from Sigma Chemical Co. (St. Louis, MO, USA). bFGF was from Becton Dickinson Labware (Bedford, MA, USA). Fetal bovine serum (FBS) was from Cansera International Inc. (Canada). L-Ascorbic acid phosphate, $FeSO_4$□$7H_20$ was from Wako Pure Chemical Industries, Ltd. (Osaka, Japan). Hormone-free, Ca^{++} -free medium RITC 80-7 (Kan and Yamane 1982) and DM-160 (Nishihira et al. 1994) were from Kyokuto Pharmaceutical

Industrial Co. (Tokyo, Japan). Tissue culture flasks (T25, T75: Falcon), 55-□^{2} Petri dishes (Falcon), and 6-well (Ø35mm) tissue culture plates (Corning Costar Corporation, Cambridge, MA, USA) were coated with type □ collagen (Cell Gen: Koken, Tokyo, Japan) (Katayama and Kan 1991). The Taq DNA polymerase, 10X PCR buffer and dNTP mixture were from Takara Syuzo Co. (Kyoto, Japan).

Cell Cultures

Methods for the primary culture of HEE cells were modifications of previously described procedures (Katayama et al. 1984; Katayama et al. 1986; Katayama and Kan 1991). The basal medium (HEE basal medium) consisted of hormone-free, Ca^{++} -free medium RITC 80-7 (Kan and Yamane 1982) containing $CaCl_2$ (30 μM), insulin (1 μg / ml), transferrin (10 μg / ml), $FeSO_4$□$7H_2O$ (1.5 μM), L-proline (175 μg / ml), ethanolamine (10 μg / ml), DTT (50 μM), L-ascorbic acid phosphate (100 μM), and BSA (5 mg / ml). HEE cells were obtained from the non-neoplastic epithelia of esophageal cancer patients. Surgically resected esophagi were cut open longitudinally, and then the epithelium was stained with Lugol's solution to distinguish carcinoma from normal epithelium (Brodmerkel 1971). Normal epithelium that was sufficiently apart from the carcinoma was obtained from the esophagus (Katayama et al. 1984). The epithelium was minced with knives and digested for 90 min. at 37°C in 5ml of Ca^{++}-, Mg^{++}-free phosphate buffered saline (PBS) containing 0.025% (wt/vol) trypsin and 0.02% (wt/vol) EDTA. The suspension was pipetted and centrifuged to pellet the cells, which were then resuspended in 5ml of HEE basal medium supplemented with 10% FBS and an antibiotic antimycotic solution (100X; Sigma Chemical Co., St. Louis, MO, USA), and were inoculated into a collagen coated tissue culture flask. After overnight culture, the flasks were rinsed 5 times with 5 ml of PBS, and this solution was replaced with 5 ml of HEE basal medium containing 10 ng/ml aFGF, 10 μg/ml heparin, and an antibiotic antimycotic solution. Cells were then fed every 3 days with fresh medium without the antibiotic antimycotic solution. After 4-10 days the cultures became subconfluent and were subcultured using published methods as described (Katayama et al. 1986; Katayama and Kan 1991). HEE cells were usually cryopreserved at the first passage in HEE basal medium containing 10 ng/ml aFGF, 10 μg/ml heparin, and 10% dimetyl sulfoxide. Cryopreserved cells were thawed, cultured, and used for growth assays.

TE-cell lines

TE-cell lines (TE1~TE15) had been established in Second Department of Surgery, Tohoku University School of Medicine from 1979 to 1986 (Nishihira et al. 1994). Most of the tumor samples were obtained from resected primary lesions of esophageal cancer. Specimens were minced with knives, digested with collagenase, and inoculated into serum-supplemented medium. Fibroblasts were removed by differential trypsinization or mechanical scraping. Cells were maintained in DM-160 with 10% FBS.

Cell Growth Assays

For the growth assays of the HEE cells and TE-cell lines, we used a low-density culture system and cells that had been serum-starved before the assays to eliminate the conditioning effects of being cultured. Before the growth assays in serum-free medium, cells were maintained in HEE basal medium with 10 ng/ml aFGF and 10 μg/ml of heparin for a minimum of 4 days. The medium was changed to the HEE basal medium without growth factors 1 day before the experiment. The cells were harvested as described (Katayama and Kan 1991), and added at a density of $1x10^4$ cells to 2 ml of HEE basal medium in each well of 6-well collagen-coated tissue culture plates. The medium was changed to the experimental medium 2 hours later. The number of cells in each well was determined by dissociating the culture with 0.0075% (wt/vol) pronase-E and 0.02% EDTA in PBS after 7 days of culture. The dispersed cells were then counted with a Coulter counter (Coulter Electronic, Inc., Hialeah, FL, USA). The number of cells was determined for duplicate wells.

aFGF (or bFGF) solution containing 1000:1 (wt/wt) heparin to aFGF (or bFGF) was used for growth assays as described (Kan et al. 1988). For experiments with ethanolamine or calcium, the experimental factor was omitted from the HEE basal medium (ethanolamine-free HEE basal medium, Ca^{++}-free HEE basal medium). For experiments testing TGF-β1, aFGF was used to stimulate cell growth. Increasing concentrations of TGF-β1 (0, 0.03, 0.1, 0.3, 1, 3 or 10 μg/ml) were added to medium, and the percent inhibition by TGF-β1 was calculated at the concentration that had the greatest effect for each cell line.

Conditioned Medium

For the experiments using conditioned medium, TE3, TE9, or TE14 were grown in HEE basal medium with 10 ng/ml aFGF and 10 μg/ml heparin in T-75 tissue culture flasks. After reaching subconfluence, the cells were washed twice with PBS, and incubated for an additional 2 days in the HEE basal medium free of BSA and insulin. Conditioned medium samples were spun at 500 X g to remove cellular debris and used directly in bioassays. Conditioned medium was added at 3, 10, or 30% to the HEE basal medium, and cell growth was then assayed.

RNA Isolation

HEE cells or TE-cell lines were grown in HEE basal medium supplemented with 10ng/ml aFGF and 10μg/ml heparin in 55-cm^2 petri dishes. After the cells reached subconfluence, the medium was changed to HEE basal medium supplemented with 1ng/ml TGF-β1, 5ng/ml aFGF and 5μg/ml heparin. Total RNA was isolated from 3 dishes of cultures treated with TGF-β1 for 0, 6 or 12 hours (0, 6 or 12 hours for HEE cells; 0 or 12 hours for TE cell lines) using RNAzol (Biotecx Lab. Inc., Houston, TX, USA) according to the manufacturer's instructions.

RT-PCR

Complementary DNA (cDNA) was prepared by random priming from 5μg of total RNA using a First-Strand cDNA Synthesis Kit (Paramecia-LKB, Uppsala, Sweden). Two μl of the cDNA mixture was used for each RT-PCR assay in a 100 μl reaction mixture containing 0.4μg of the primers, 2.5 Units of Taq DNA polymerase, 200μM of dNTP mixture and reaction buffer (Takara Syuzo Co. Kyoto, Japan). The primers 5'-TGCGGAATGCGCGAGGAGAACAA-3' and 5'-ACCACCAGCGTGTCCAGGAA-3' were used to detect CDKN2B/p15INK4b mRNA (a 299 bp band was expected) (Guan et al. 1994, Hannon et al. 1994) and the primers 5'-AACAAGGGCATGCCCAGTGG-3' and 5'-ATCATCATGACCTGGATCGCGCGCCT-3' were subsequently used to amplify the products from the first primer set for cloning and sequencing (a 155 bp band was expected) (Guan et al. 1994, Hannon and Beach 1994). The conditions of RT-PCR were as follows: denaturation at 94° for 5 min followed by 30 cycles of denaturation at 94° for 1 min, annealing at 60° for 1 min, and extension at 72° for 1 min in a thermal cycler (Perkin Elmer Cetus, Norwalk, CT, USA). The RT-PCR products were separated on agarose gel and visualized with ethidium bromide. The human glyceraldehyde 3-phosphate dehydrogenese (G3PDH) control amplimer set (Clonteck Laboratories, Inc., Palo Alto, CA, USA) was utilized as a control in the RT-PCR experiment (5'-TGAAGGTCGGAGTCAACGGATT TGGT-3' and 5'-CATGTGGGCCATGAGGTCCACCAC-3' for G3PDH gene, PCR product: 983bp).

Sequencing of PCR Products

PCR product was isolated and cloned into the pGEM vector (Promega, Madison, WI, USA). The nucleotide sequences were determined by the chain termination method using a dye termination cycle sequencing kit (Applied Biosystems, Foster City, CA, USA) and a 373S DNA sequencer (Applied Biosystems, Foster City, CA, USA).

Statistical Analysis

The statistical significance was calculated using Student's *t* -test. *P* values less than 0.05 were considered to indicate significant differences.

Results

Growth Requirements for Normal HEE Cells

The effects of EGF, TGF-α, aFGF, bFGF, or TGF-β1 on the growth of HEE cells and TE-cell lines were analyzed. Because the effect of TGF-α was very similar to that of EGF, the data for TGF-α are not shown here. The dose-dependent effects of various growth factors are shown in Figure 1. aFGF was the most effective growth stimulant for HEE cells (Figure 1A). EGF also stimulated their growth, but EGF's growth-promoting activity was lower than that of aFGF, and the optimum concentration of EGF for stimulating HEE cell growth was much lower than that of aFGF. Although 30 ng/ml EGF has been used in cultures of human epidermal keratinocytes (Rheinwald and Green 1977), HEE cell growth was inhibited at this concentration; at higher concentrations of EGF, the morphology of the HEE cells became spindle-like (data not shown). bFGF had little growth-promoting activity (Figure 1A). TGF-β1 inhibited the growth of HEE cells in a dose-dependent manner (Figure 1D).

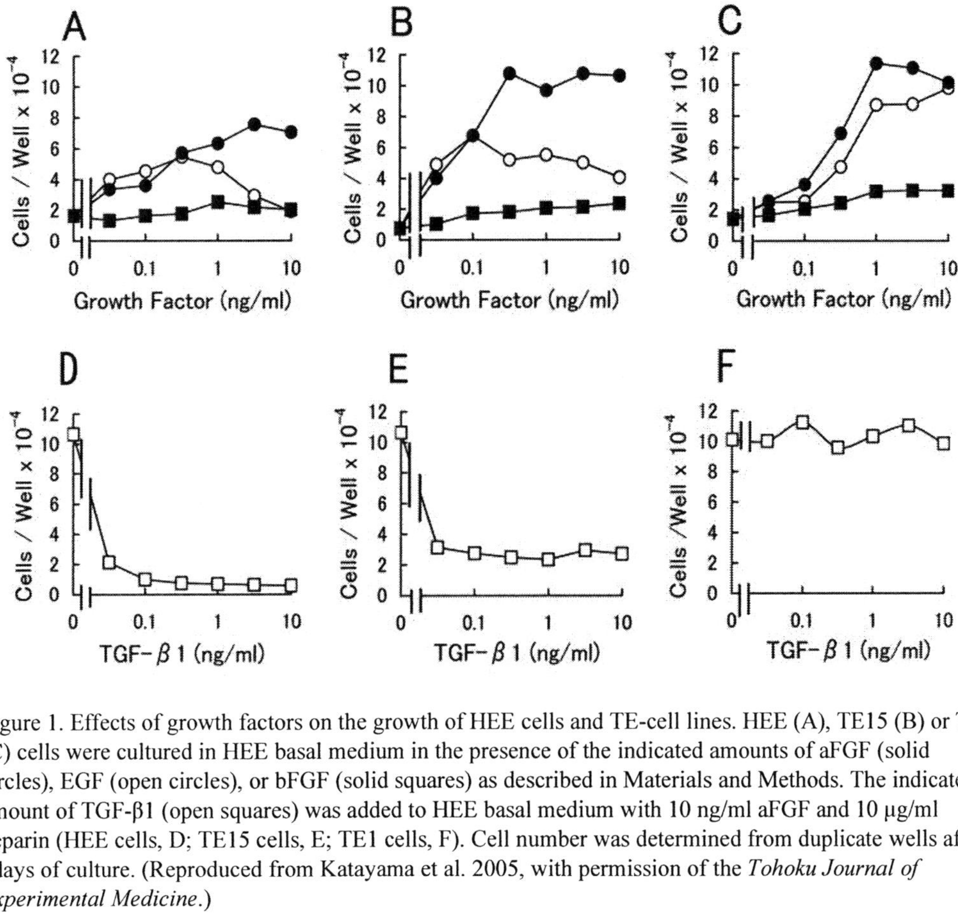

Figure 1. Effects of growth factors on the growth of HEE cells and TE-cell lines. HEE (A), TE15 (B) or TE1 (C) cells were cultured in HEE basal medium in the presence of the indicated amounts of aFGF (solid circles), EGF (open circles), or bFGF (solid squares) as described in Materials and Methods. The indicated amount of TGF-β1 (open squares) was added to HEE basal medium with 10 ng/ml aFGF and 10 μg/ml heparin (HEE cells, D; TE15 cells, E; TE1 cells, F). Cell number was determined from duplicate wells after 7days of culture. (Reproduced from Katayama et al. 2005, with permission of the *Tohoku Journal of Experimental Medicine*.)

Growth Requirements for Malignant TE-cell Lines

The TE-cell lines, except for TE4, could be roughly separated into two groups according to their growth properties. Cell lines in the first group, such as TE15 (Figure 1B, E), grew in a manner similar to that of HEE cells. However, the cell lines in the second group, such as TE1 (Figure 1C, F), had different growth properties. For example, even though aFGF was the most effective growth stimulant for both types of cells, the response of TE1 to EGF or TGF-β1 was different from that of HEE cells: neither EGF nor TGF-β1 inhibited TE1 growth

(Figure 1 C, F), and the optimum concentration of EGF for TE1 was relatively high. Instead, TE1 responded to EGF much as to aFGF. The growth-promoting activity of bFGF was very low for both types of cells (Figure 1).

Of the 15 TE-cell lines, only one, TE4 did not grow in serum-free medium supplemented with a growth factor (Figure 2). These cells required serum or conditioned medium from other TE-cell lines for their attachment and growth (Figure 2.). Because their attachment to dishes seemed very poor in serum-free medium, we analyzed attachment factors such as fibronectin and various types of collagen. However, neither coating the dishes with various attachment factors nor with serum was successful in supporting the attachment and growth of the TE4 cells (data not shown). Therefore, we excluded TE4 cells from the experiments in serum-free medium.

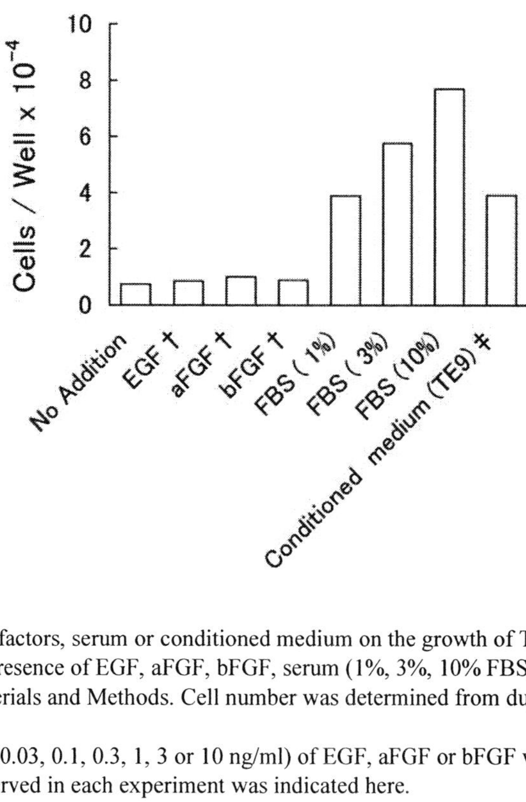

Figure 2. Effects of growth factors, serum or conditioned medium on the growth of TE4. TE4 was cultured in HEE basal medium in the presence of EGF, aFGF, bFGF, serum (1%, 3%, 10% FBS) or conditioned medium of TE9 as described in Materials and Methods. Cell number was determined from duplicate wells after 7days of culture.

† Increasing concentration (0.03, 0.1, 0.3, 1, 3 or 10 ng/ml) of EGF, aFGF or bFGF was added to the culture. Maximum cell number observed in each experiment was indicated here.

‡ Conditioned medium of TE9 was collected as described in Materials and Methods. 30% of conditioned medium was added to the culture.

Different Growth Properties of TGF-β1-sensitive Cells and -resistant Cells

The percent inhibition of HEE and TE cell growth by TGF-β1 is shown in Figure 3. TGF-β1 did not inhibit, but rather stimulated the growth of the TE7, TE10, and TE13 lines. Seven TE-cell lines (TE1, 5, 7, 8, 10, 13, and 14) that showed growth inhibition of less than 10% were designated TGF-β1-resistant cells. The remaining seven lines (TE2, 3, 6, 9, 11, 12, and 15) and HEE cells were designated TGF-β1-sensitive cells. For the HEE cells and the TGF-β1-sensitive TE15 cell line (Figure 1A,B), the optimum concentration and growth-promotion effect of EGF were lower than those of aFGF. However, the TGF-β1-resistant TE1 cells responded similarly to EGF and aFGF (Figure 1C).

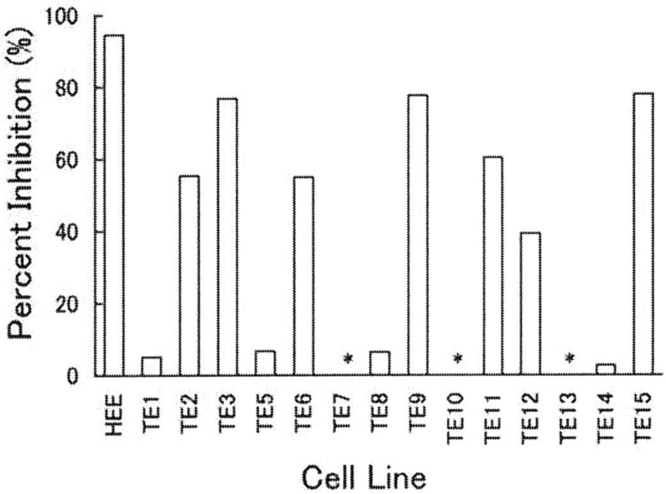

Figure 3. Effect of TGF-β1 on the growth of HEE cells and TE-cell lines. HEE cells and TE-cell lines were cultured in HEE basal medium in the presence of 10 ng/ml aFGF, 10 μg/ml heparin, and increasing concentrations of TGF-β1, as described in Materials and Methods. Cell number was determined from duplicate wells after 7days of culture. Results are expressed as maximum percent inhibition of TGF-β1 for each experiment, for the concentration of TGF-β1 that had the greatest effect.
* TGF-β1 had no inhibitory effect on the growth of TE7, TE10, or TE13 at any concentration. (Reproduced from Katayama et al. 2005, with permission of the *Tohoku Journal of Experimental Medicine*.)

We analyzed quantitatively the differences in the growth properties between the TGF-β1-sensitive and the TGF-β1-resistant cells (Table 1). We separated the HEE cells and TE-cell lines into the two groups described above, according to their TGF-β1 sensitivity (group A, TGF-β1-sensitive cells, n = 8; group B, TGF-β1-resistant cells, n = 7). To examine the optimum concentration for growth factors, the half-maximum effective dose (ED_{50}) was determined from dose-response curves similar to those shown in Figure 1. The mean ED_{50} for EGF in group A (0.09 ng/ml) was much lower than that in group B (0.55 ng/ml). The mean ED_{50} for aFGF in the two groups (group A, 0.32 ng/ml; group B, 0.44 ng/ml) was similar to the ED_{50} for EGF in group B (0.55 ng/ml). To compare the optimum concentration between EGF and aFGF, we calculated the ratio of the ED_{50} of EGF to the ED_{50} of aFGF. This ratio (EGF/aFGF) for group A was significantly lower than for group B ($p < 0.01$). To compare the growth-promoting abilities of EGF and aFGF, we calculated the ratio of maximum cell number for EGF to that for aFGF (EGF /aFGF). This ratio for group A was significantly lower than for group B ($p < 0.05$). In fact, the ED_{50} and maximum cell number ratios in group B were close to one. These results indicated that the TGF-β1-resistant cells responded similarly to EGF and aFGF, but the TGF-β1-sensitive cells responded differently to these two growth factors. The optimum concentration and growth-promotion effect of EGF on TGF-β1-sensitive cells were lower than those of aFGF. Therefore, the responses of the TGF-β1-

resistant cells to EGF and aFGF were significantly different from the responses of the TGF-β1-sensitive cells.

Table 1. Comparison of the growth properties of TGF-β1-sensitive and -resistant cells

	TGF-β1-sensitive cells [a] (% inhibition > 10%) (n = 8)	**TGF-β1-resistant cells [b] (% inhibition < 10%) (n = 7)**
% inhibition by TGF-β1	67.13 ± 17.64*	2.98 ± 3.08*
ED_{50} of EGF (ng/ml) [c]	0.09 ± 0.09	0.55 ± 0.63
ED_{50} of aFGF (ng/ml) [d]	0.32 ± 0.20	0.44 ± 0.42
Ratio of ED_{50} (ED_{50} of EGF / ED_{50} of aFGF) [e]	0.27 ± 0.18*	1.16 ± 0.51*
Ratio of maximum cell number (Cell No. EGF/Cell No. aFGF) [f]	0.73 ± 0.27**	1.25 ± 0.64**

[a,b] The HEE cells and TE-cell lines were separated into two groups according to the percent inhibition by TGF-β1. Cell lines for which the % inhibition by TGF-β1 was more than 10% were designated as TGF-β1-sensitive cells (n = 8), and cell lines that had less than 10% inhibition in response to TGF-β1 were designated as TGF-β1- resistant cells (n = 7). Values are shown as the mean±S.D.

[c,d] Half-maximum effective dose (ED_{50}) of EGF [c)] or aFGF [d)] on cell growth was determined from dose-response curves similar to those shown in Figure 1.

[e] The ratio of ED_{50} of EGF to ED_{50} of aFGF.

[f] The ratio of the maximum cell number in medium with EGF to that in medium with aFGF.

*: $p < 0.01$, **: $p < 0.05$, TGF-β1-sensitive cells compared with TGF-β1-resistant cells by Student's t test.

(Reproduced from Katayama et al. 2005, with permission of the *Tohoku Journal of Experimental Medicine.*)

Different Responses to Ethanolamine between TGF-β1-sensitive and -resistant Cells

Ethanolamine is required for the serial passaging of HEE cells. Because we noticed that ethanolamine modified the inhibitory effects of EGF (1 ng/ml), we analyzed the effect of ethanolamine on the growth of the HEE cells and TE-cell lines (Figure 4). Adding 10 μg/ml ethanolamine to the medium increased the optimum concentration and growth-promoting effect of EGF or aFGF for HEE cells (Figure 4A, B). The growth properties of TE1 cells were not, however, affected by ethanolamine (Figure 4C, D). Interestingly, the different patterns of response to ethanolamine seemed to be correlated with the cells' sensitivity to TGF-β1. We analyzed the differences in response to ethanolamine between the TGF-β1-sensitive (group A) and TGF-β1-resistant (group B) cells quantitatively (Table 2). To compare the optimum concentration of each growth factor, we calculated the ratio of the ED_{50}s obtained with (+) and without (-) ethanolamine. The ED_{50} ratio for group A was significantly higher than that for group B when cell growth was stimulated by EGF ($p < 0.01$) or aFGF ($p < 0.05$). To compare the growth-promoting ability of the growth factors, we calculated the ratio of the maximum cell number with (+) to that without (-) ethanolamine. The maximum cell number ratio for group A was significantly higher than that for group B ($p < 0.05$) when cell growth was stimulated by aFGF. All of the ED_{50} and maximum cell number ratios for group B were close to one. These results indicated that ethanolamine increased the optimum concentration and growth-promoting ability of growth factors in the TGF-β1-sensitive cells, but had no effect on the TGF-β1-resistant cells.

Table 2. Comparison of the effect of ethanolamine on TGF- β 1-sensitiive and -resistant cells

	TGF- β 1-sensitive cells [a] (% inhibition > 10%) ($n = 8$)	TGF- β 1-resistant cells [b] (% inhibition < 10%) ($n = 7$)
Ratio of ED_{50} (+/-) ethanolamine, (EGF)[c]	$4.88 \pm 2.44^{*}$	$1.23 \pm 0.41^{*}$
Ratio of ED_{50} (+/-) ethanolamine, (aFGF)[d]	$2.90 \pm 1.88^{**}$	$1.01 \pm 0.36^{**}$
Ratio of maximum cell number (+/-) ethanolamine, (EGF)[e]	1.34 ± 0.39	1.06 ± 0.10
Ratio of maximum cell number (+/-) ethanolamine, (aFGF)[f]	$1.30 \pm 0.25^{**}$	$1.02 \pm 0.10^{**}$

[a,b] HEE cells and TE-cell lines were separated into two groups as shown in Table 1. Values are shown as the mean ± S.D.

[c,d] The ratio between ED_{50} in the presence or absence (+/-) of ethanolamine when cell growth was stimulated by increasing concentrations of EGF [c)] or aFGF [d)]. ED_{50} was determined from dose-response curves similar to those shown in Figure 3.

[e,f] The ratio between the maximum cell number in the presence or absence (+/-) of ethanolamine, when cell growth was stimulated by increasing concentrations of EGF [e)] or aFGF [f)].

*: $p < 0.01$, **: $p < 0.05$, TGF- β 1-sensitive cells compared with TGF- β 1-resistant cells by Student's *t* test.

(Reproduced from Katayama et al. 2005, with permission of the *Tohoku Journal of Experimental Medicine.*)

Expression of CDKN2B mRNA

To elucidate whether TGF-β1 signal transduction pathways were altered in TE cell lines, we analyzed the expression of CDKNB2B mRNA using RT-PCR. Figure 5 shows the results of the RT-PCR of the HEE cells and 14 TE-cell lines (TE1-TE3, TE5-TE15) that had been treated with TGF-β1 for 12 hours. A 299 bp CDKN2B RT-PCR product was amplified from the HEE cell RNA. In contrast, CDKN2B mRNA was not detected in the TE cell lines, except for TE1 and TE5. Faint 299bp bands were detected in the TE1 and TE5 RT-PCR samples. However, these cell lines were TGF-β1-resistant . Of the 12 TE cell lines that did not express CDKN2B mRNA, 7 were sensitive to the inhibitory effects of TGF-β1. The bands detected in the RT-PCR samples of HEE cells treated with TGF-β1 for 0, 6 and 12 hours were the same when 30 cycles were used (data not shown). Therefore, we reduced the cycle number for the RT-PCR to 27 (Figure 6). Under these conditions, time-dependent induction of CDKN2B mRNA synthesis was observed in HEE cells, and expression of this mRNA in TE1 and TE5 was not recognized (Figure 6).

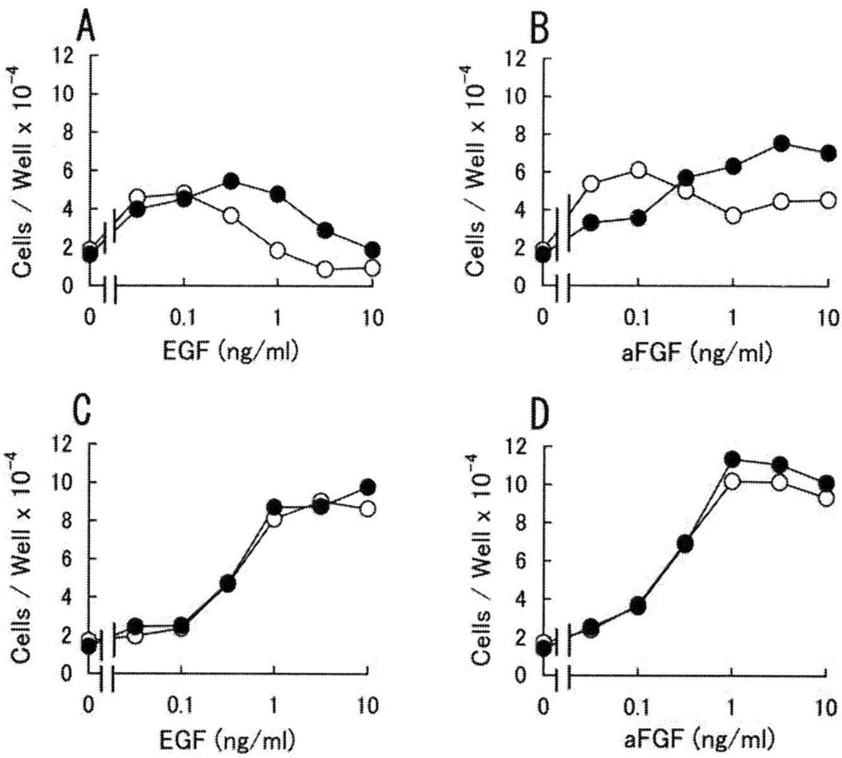

Figure 4. Effect of ethanolamine on the growth of HEE cells and the TE1 cells. HEE (A, B), and TE1 (C, D) cells were cultured in ethanolamine-free HEE basal medium with (closed circles) or without (open circles) 10 μg/ml ethanolamine in the presence of the indicated amounts of EGF (A, C) or aFGF (B, D). The cell number was determined from duplicate wells after 7 days of culture. (Reproduced from Katayama et al. 2005, with permission of the *Tohoku Journal of Experimental Medicine*.)

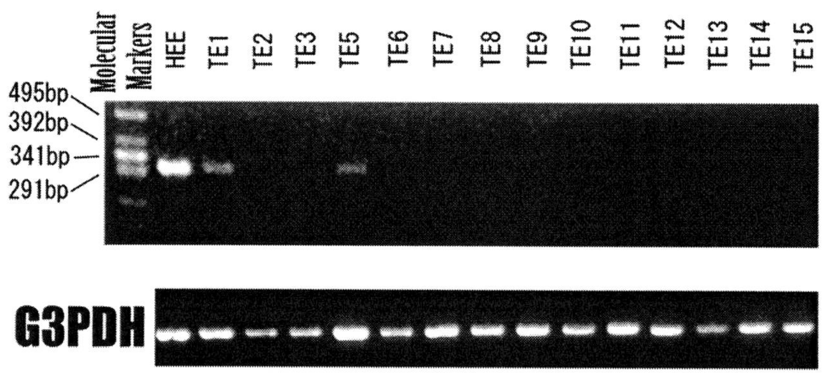

Figure 5. Detection of CDKN2B mRNA in HEE cells and TE cell lines. CDKN2B mRNA was detected by RT-PCR amplification of RNA from cell samples that were obtained 12 hours after treatment with TGF-β1, as described in the Materials and Methods (primers: 5'-TGCGGAATGCGCGAGGAGAACAA-3' and 5'-ACCACCAGCGTGTCCAGGAA-3', a 299 bp band was expected). G3PDH mRNA was used as an internal control.

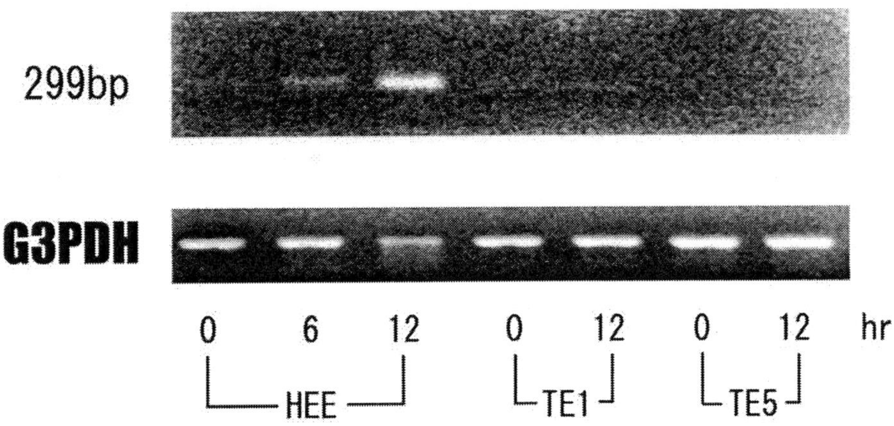

Figure 6. Induction of CDKN2B mRNA synthesis by TGF-β1 treatment. CDKN2B mRNA was detected by RT-PCR amplification of HEE, TE1 and TE5 RNA samples at the indicated times following treatment with TGF-β1, as described in the Materials and Methods. The cycle number for the RT-PCR was reduced from 30 to 27 for this experiment.

Nucleotide Sequence of RT-PCR Product

In order to ascertain that the 299 bp RT-PCR product was really from CDKN2B/p15INK4b mRNA, we analyzed the nucleotide sequence of the RT-PCR product. The 299 bp product was further amplified by RT-PCR using other primers (5'-AACAAGGGCATGCCCAGTGG-3' and 5'-ATCATCATGACCTGGATCGCGCGCCT-3'). This RT-PCR product was cloned and sequenced as described in the Materials and Methods. The nucleotide sequence of the RT-PCR product was slightly different from the sequence of CDKN2B/p15INK4b that was reported earlier (Hannon and Beach 1994). Guan et al. (1994) reported that the CDKN2B/p14INK4b gene nucleotide sequence was slightly different from that of CDKN2B/p15INK4b. Differences at nucleotide positions 379-387 (AGCGCCGCG in p14 and ACGCCG in p15) result in changes of the amino acid sequence from Ser-Ala-Ala in p14 to Thr-Pro in p15, and differences at nucleotide positions 417-423 (AGCTCCT in p14 and ACTCCTG in p15) changes the corresponding amino acid sequence from Asn-Leu-Leu in p14 to His-Ser-Trp in p15 (Guan et al. 1994, Hannon and Beach 1994). The sequence of our RT-PCR product was the same as CDKN2B/p14INK4b. CDKN2B/p14INK4b protein has also been reported to be associated with CDK4 and CDK6 (Guan et al. 1994).

Other Growth Characteristics

bFGF was less effective than aFGF in promoting the growth of HEE cells (Figure 1). In TE11 cells, however, the growth-promoting activity of bFGF was higher than that of aFGF. Therefore, we compared the growth-promoting activity of bFGF with that of aFGF in the HEE cells and all the TE-cell lines (Figure 7). We found bFGF had a much lower growth-promoting activity than did aFGF for HEE cells and all the TE-cell lines except TE11. We also studied the effect of heparin on the growth promoting activity of EGF, aFGF or bFGF in

HEE cells and TE-cell lines (Table 3). Heparin had no effect on EGF, stimulatory effect on aFGF, and inhibitory effect on bFGF (Table 3, Figure 7). It was interesting that heparin had no effect on the growth promoting activity of bFGF in only TE11 among HEE cells and TE cell lines (Figure 7). It should be mentioned that growth promoting activities of aFGF with heparin on HEE cells and TE cell lines except TE11 were higher than that of bFGF without heparin (data not shown). These studies revealed that only TE11 showed different pattern of response to bFGF and heparin from HEE cells.

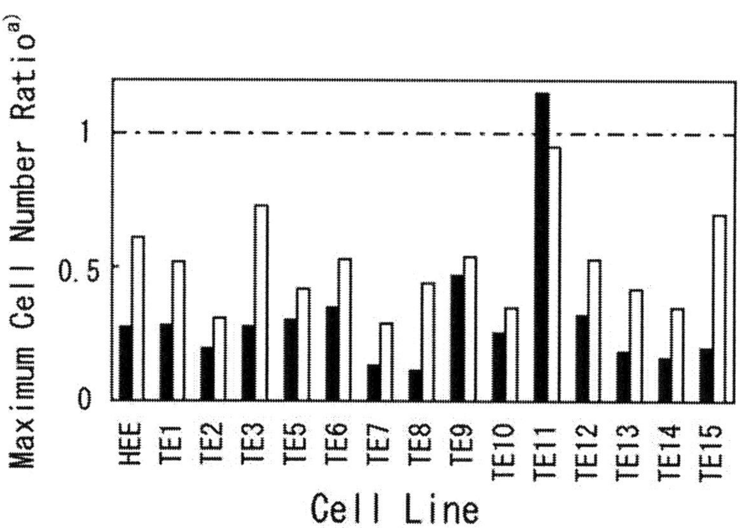

Figure 7. Effects of bFGF and heparin on the growth of HEE cells and TE cell lines. Closed column: Effect of bFGF was compared with the effect of aFGF. Maximum cell number when cell growth was stimulated by increasing concentration of bFGF or aFGF was determined from dose-response curves similar to Figure 1. The ratio[a)] between maximum cell number when cell growth was stimulated by bFGF and aFGF (bFGF / aFGF) was indicated by closed column. Open column: Effect of heparin on the growth-promoting effect of bFGF was studied. The ratio[a)] between maximum cell number under presence and absence (+/-) of heparin when cell growth was stimulated by increasing concentration of bFGF were indicated by open column.

Table 3. Effect of heparin on proliferative response of esophageal cells to EGF, acidic FGF, and basic FGF

Growth factor	Maximum Cell number ratio with heparin / without heparin [a]
EGF	1.01±0.11
aFGF	1.48±0.28
bFGF	0.51±0.18

Esophageal cells (HEE cells, TE1-3, TE5-15) were cultured in HEE basal medium supplemented with increasing concentration (0.03, 0.1, 1, 3 or 10ng/ml) of EGF, aFGF or bFGF with or without heparin as described in Materials and Methods. 1000:1 (wt/wt) heparin to each factor was utilized.

[a] The ratio between maximum cell number when cells were cultured with heparin and without heparin. Values shown as mean±S.D.

The cell growth of the HEE cells and TE5 line was highest in medium without Ca^{++} (Table 4). Because the albumin in the medium contains small amounts of Ca^{++} (Katayama et al. 1986), we suspected that the albumin-supplied Ca^{++} was sufficient for the growth of these cells. However, most of the TE-cell lines required higher concentrations of Ca^{++} to grow (Table 4). In particular, in lines TE2 and TE10, the ratio of the cell number in cultures grown in higher-Ca^{++} concentration medium to that in control medium was more than 10 (Table 4). Therefore, Ca^{++} seemed necessary for the rapid growth of these cell lines. In addition, increasing the concentration of Ca^{++} changed the morphology of the HEE cells and all the TE-cell lines to a stratified epithelial-like shape (data not shown).

Table 4. Effect of calcium concentration on the growth of HEE cells and TE-cell lines

Cell line	Ca^{++} concentration [a] (mM)	Fold stimulation[b]
HEE	-	1 > [c]
TE1	0.3	2.14
TE2	0.3	12.69
TE3	0.3	3.77
TE5	-	1 > [c]
TE6	0.3	1.13
TE7	3	4.23
TE8	0.03	1.17
TE9	0.3	1.76
TE10	3	13.86
TE11	3	2.10
TE12	3	1.50
TE13	0.3	2.00
TE14	0.3	1.22
TE15	0.3	2.36

Cells were cultured in Ca^{++}-free HEE basal medium with 0, 0.03, 0.3, or 3 mM Ca^{++}, 10 ng/ml aFGF and 10 μ g/ml heparin.

[a] Ca^{++} concentration for maximum cell growth.

[b] Fold stimulation was calculated from the cell number of a control culture that contained no Ca^{++} and the culture that yielded the maximum cell number.

[c] In cultures of HEE cells or the TE5 line, the maximum cell number was obtained under Ca^{++}-free conditions.

(Reproduced from Katayama et al. 2005, with permission of the *Tohoku Journal of Experimental Medicine.*)

We compared the effects of conditioned medium from three TE-cell lines on the growth of HEE cells and the TE-cell lines (Figure 8). Interestingly, the conditioned medium from each of the three TE-cell lines inhibited the growth of the HEE cells, although the medium from each TE-cell line stimulated the growth of all the TE-cell lines.

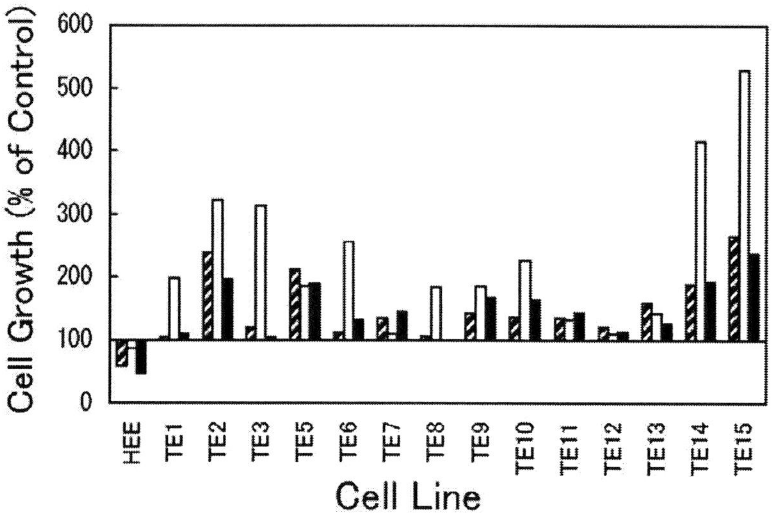

Figure 8. Effect of conditioned medium from TE3, TE9, or TE14 cells on the growth of HEE cells and the TE-cell lines. Conditioned medium from TE3 (hatched column), TE9 (open column), or TE14 (closed column) cells was obtained as described in Materials and Methods. Cultures were grown in 30% conditioned medium. Cell number was determined from duplicate wells after 7 days of culture. Cell growth is expressed as the percent of the control culture (% of control). (Reproduced from Katayama et al. 2005, with permission of the *Tohoku Journal of Experimental Medicine*.)

Discussion

Comparative culture of normal and malignant cells is important for understanding the growth properties of tumors. Although many cell lines have been established from esophageal cancers, the growth properties of normal and cancer-derived esophageal epithelial cells have not been compared extensively (Nishihira et al. 1993, 1994). We succeeded in establishing the assay systems for both normal and cancer-derived esophageal epithelial cells in serum-free, defined medium (Katayama et al. 2005). We used aFGF and ethanolamine as substitutes for bovine neural tissue extracts. We found that it was best to eliminate the conditioning effects of high-density culture by growing cells at low density and starving them of serum or growth factors before using them in experiments to test growth factors. To our knowledge, this is the first reported comparative study of normal and carcinoma cells of the esophagus in serum-free conditions.

The growth-inhibiting effects of EGF on human squamous cell carcinomas, including esophageal carcinoma, grown in serum-containing medium were previously reported (Kamata et al. 1986). In our previous study (Katayama and Kan 1991), we showed that EGF inhibited the growth of TE9, but stimulated the growth of TE11 slightly in serum-free, high-density culture conditions. Here, however, a stimulatory effect of a low concentration of EGF was observed on TE9 cells, using the assay system we had worked out for HEE cells. In serum-free and low-density culture condition, using starved cells, EGF stimulated the growth of HEE cells and all the TE-cell lines. The optimum concentration of EGF for some of these cell lines was very low. The discrepancy between the inhibitory effects of EGF described in the previous paper and its stimulatory effect in this paper could be explained by the presence of small amounts of EGF-like activities in the cultures. Because serum contains variety of growth-promoting activities, and cultured cells produce autocrine factors such as TGF-α and amphiregulin (Cook et al. 1991), it is possible that the growth-promoting effect of EGF at low concentration was masked by such factors. Thus, only the inhibitory effect of EGF was observed in previous studies. Because we utilized serum-free, low-density cultures and starved cells, we were able to observe growth-promoting effects of EGF at low concentrations.

The optimum concentration for EGF to promote HEE cell growth was much lower than that for aFGF. The growth promoting-activity of EGF for HEE cells was also much lower than that of aFGF. For the TE1 line, however, the optimum concentration and growth-promoting activity of EGF was similar to that of aFGF. In fact, there seemed to be two types of cell lines in the series, one that was similar to the HEE cells and one that shared properties with the TE1 line. It was interesting to find that the cell lines similar to TE1 were resistant to the inhibitory effect of TGF-β1. In addition, ethanolamine had no effect on these TGF-β1-resistant cells, although the addition of ethanolamine changed the optimum concentration and growth-promoting effects of EGF and aFGF on TGF-β1-sensitive cells. Thus, the response to TGF-β1 seemed to be correlated with the responses to EGF and to ethanolamine.

Although TGF-β is a potent growth inhibitor in epithelial tissues, it is both a suppressor and a promoter of tumorigenesis (Massagué et al. 2000). On one hand, TGF-βhas a tumor-suppression function that is lost in many tumor-derived cell lines: nearly all pancreatic cancers and colon cancers have mutations disabling a component of the TGF-β-signaling pathway. Some of these mutations occur in the TGF-β receptors, Smad4 or Smad2; others may occur in hitherto untested or unknown components of the signaling pathway (Massagué et al. 2000). On the other hand, in patients with esophageal squamous cell carcinoma, no mutations have been reported in the TGF-β type II receptor, Smad2, Smad3, or Smad4 genes (Osawa et al. 2000). An immunohistochemical study of esophageal carcinoma, however, revealed that Smad4 expression was negative in approximately half the esophageal cancer patients tested (Fukuchi et al. 2002). In our experiments, half the cell lines were resistant to the inhibitory effect of TGF-β1; however, we have not assessed whether these results are related to a loss of Smad4 expression.

Ethanolamine is required for the growth of many types of cultured cells (Arthur et al. 1993). Ethanolamine is thought to be incorporated into phosphatidylethanolamine, which is a major phospholipid of mammalian cell membranes, accounting for 20~30% of the total phospholipid (Arthur et al. 1993). The EGF receptor on cell membranes has been reported to be altered by TGF-β treatment (Takehara et al. 1987): Long-term exposure of rat heart endothelial cells to TGF-βdecreased the number of high-affinity EGF receptors, with almost no change in total receptor number. These results from previous reports suggest that alterations could occur near the cell membrane in TGF-β1-resistant TE-cell lines. However, the reason the response to TGF-β1 was correlated with the responses to EGF and ethanolamine is still unclear.

Many tumorigenic cell lines have lost responsiveness to the negative growth-regulatory effects of TGF-β1, so it is anticipated that an understanding of the molecular mechanisms by which TGF-β1 arrests cell growth will also provide information relevant to molecular events important in neoplastic transformation (Alexandrow and Moses 1995). TGF-β1 inhibits cell proliferation by inducing G1-phase cell cycle arrest. Normal progression through G1 is promoted by the activity of the cyclin-dependent protein kinases CDK4 and CDK6, which are inhibited by CDKN2B (Hannon and Beach 1994). Treatment with TGF-β1 increases CDKN2B/p15INK4b expression by approximately 30-fold in human keratinocytes, suggesting that CDKN2B/p15INK4b may act as an effector of TGF-β-mediated cell cycle arrest (Hannon and Beach 1994). Therefore, we analyzed the expression of CDKN2B/p15INK4b mRNA using RT-PCR. The nucleotide sequence of the PCR product was slightly different from the sequence of CDKN2B/p15INK4b (Hannon and Beach 1994) but was the same as the sequence of the CDKN2B/p14INK4b gene that was reported by Guan et al (1994). Induction of CDKN2B mRNA synthesis by TGF-β1 treatment was observed in HEE cells. In contrast, this was not detected in 12 of the TE cell lines. Low levels of CDKN2B mRNA were detected in 2 TE cell lines, however they did not respond to TGF-β1 treatment. The low levels of CDKN2B expression in those 2 TE cell lines therefore did not correlate with the inhibitory effects of TGF-β1. Seven TGF-β1-sensitive TE cell lines did not express CDKN2B mRNA, so it is possible that alternative TGF-β1 signaling pathways that do not involve CDKN2B remain active in these cell lines.

We utilized heparin to potentiate the growth promoting activities of aFGF (Kan et al. 1988; Burgess and Maciag 1989). We also used heparin for bFGF in order to make the condition same as aFGF in the growth assay, because inhibitory effect of heparin on bFGF was not reported (Burgess and Maciag 1989). However, we found heparin inhibited the growth promoting effect of bFGF on HEE cells and TE cell lines. In the case of TE11, heparin had no effect on the growth promoting activities of bFGF, and the growth promoting activity of bFGF was higher than that of aFGF. FGF activity is regulated by heparan sulfate proteoglycans and by the type of FGF receptor expressed on the target cells (Ornitz and Leder 1992; Kan et al. 1993). Four FGF receptors and their isoforms which result from alternative splicing were reported to have different ligand binding properties (Ornitz and Leder 1992). Therefore, it is possible that only TE11 cells express different type of FGF receptor from HEE cells. The reason why heparin inhibits the growth promoting effect of bFGF is still unclear.

We utilized medium that contained a low concentration of Ca^{++} (0.03 mM), because the cell growth of HEE cells in low-concentration Ca^{++} was similar to that in the standard concentration of Ca^{++}, and terminal differentiation could be avoided in low-Ca^{++} medium (Katayama et al. 1986; Katayama and Kan 1991). Most TE-cell lines required a higher concentration of Ca^{++} for their optimum growth, and in two cell lines in particular, increasing the Ca^{++} concentration increased cell growth more than 10-fold. Optimum growth of human skin keratinocytes has also been reported to occur at a very low level of Ca^{++}, while fibroblasts require higher Ca^{++} levels (Peehl and Ham 1980). The Ca^{++} requirement of the TE-cell lines seems more similar to that of fibroblasts than to that of normal squamous epithelial cells.

Cultured cells have been reported to produce many factors related to cell growth (Katayama and Kan 1991; Hashimoto et al. 1994; Rodeck et al. 1994). We found quite different responses between the HEE cells and the TE-cell lines to conditioned media from three TE-cell lines. It is possible that the conditioned medium from these lines contained growth-promoting factors that are specific to TE-cell lines or growth inhibitory factors that are specific for HEE cells. Further studies will be necessary to clarify what factors have these different growth properties in normal and malignant cells.

This study shows a variety of differences between human esophageal normal and malignant epithelial cells in culture. The identification of the mechanisms affecting the different growth properties of these lines will contribute to our understanding of the process of malignant transformation.

References

Alexandrow, M.G. and Moses, H.L. (1995) Transforming growth factor β and cell cycle regulation. *Cancer Res., 55*, 1452-1457.

Arthur, G. and Lu, X. (1993) The ethanolamine requirement of keratinocytes for growth is not due to defective synthesis of ethanolamine phosphoacylglycerols by the decarboxylation pathway. *Biochem. J., 293*, 125-130.

Brodmerkel, G.J.,Jr. (1971) Shiller's test: An aid in esophagoscopic diagnosis. *Gastroenterology, 60*, 813.

Burgess, W.H. and Maciag, T. (1989) The heparin-binding (fibroblast) growth factor family of proteins. *Annu. Rev. Biochem., 58*, 575-606.

Cook, P.W., Mattox, P.A., Keeble, W.W., Pittelkow, M.R., Plowman, G.D., Shoyab, M., Adelman, J.P. and Shipley, G.D. (1991) A heparin sulfate-regulated human keratinocyte autocrine factor is similar or identical to amphiregulin. *Mol. Cell. Biol., 11*, 2547-2557.

Fukuchi, M., Masuda, N., Miyazaki, T., Nakajima, M., Osawa, H., Kato, H. and Kuwano, H. (2002) Decreased Smad4 expression in the transforming growth factor-β signaling pathway during progression of esophageal squamous cell carcinoma. *Cancer, 95*, 737-743.

Guan, K.-L., Jenkins, C.W., Li, Y., Nichols, M.A., Wu, X., O'Keefe, C.L., Matera, A.G. and Xiong, Y. (1994) Growth suppression by *p*18, a $p16^{INK4/MTS1}$ -and $p14^{INK4B/MTS2}$ -related CDK6 inhibitor, correlates with wild-type pRb function. *Genes and Dev., 8*, 2939-2952.

Hannon, G.J. and Beach, D. (1994) $p15^{INK4B}$ is a potential effector of TGF-β- induced cell cycle arrest. *Nature, 371*, 257-261.

Hashimoto, K., Higashiyama, S., Asada, H., Hashimura, E., Kobayashi, T., Sudo, K., Nakagawa, T., Damm, D., Yoshikawa, K. and Taniguchi, N. (1994) Heparin-binding epidermal growth factor-like growth factor is an autocrine growth factor for human keratinocytes. *J. Biol. Chem., 269*, 20060-20066.

Kamata, N., Chida, K., Rikimaru, K., Horikoshi, M., Enomoto, S. and Kuroki, T. (1986) Growth-inhibitory effects of epidermal growth factor and overexpression of its receptors on human squamous cell carcinomas in culture. *Cancer Res., 46*, 1648-1653.

Kan, M., DiSorbo, D., Hou, J., Hoshi, H., Mansson, P-E. and McKeehan, W.L. (1988) High and low affinity binding of heparin-binding growth factor to a130-kDa receptor correlates with stimulation and inhibition of growth of a differentiated human hepatoma cell. *J. Biol. Chem., 263*, 11306-11313.

Kan, M., Wang, F., Xu, J., Crabb, J.W., Hou, J., and McKeehan, W.L. (1993) An essential heparin-binding domain in the fibroblast growth factor receptor kinase. *Science*, *259*, 1918-1921.

Kan, M. and Yamane, I. (1982) In vitro proliferation and lifespan of human diploid fibroblasts in serum-free BSA-containing medium. *J. Cell Physiol.*, *111*, 155-162.

Katayama, M., Akaishi, T., Nishihira, T., Kasai, M., Kan, M. and Yamane, I. (1984) Primary culture of human esophageal epithelial cells. *Tohoku J. exp. Med.*, *143*, 129-140.

Katayama, M., Akaishi, T., Nishihira, T., Kasai, M., Kan, M. and Yamane, I. (1986) Primary cultures and serial passages of normal human esophageal epithelial cells in a serum-free medium. In M. Kasai (Eds.), Esophageal Cancer: Proceedings of the International Symposium on Cancer of the Esophagus (pp. 31-34). Princeton, NJ: Excerpta Medica.

Katayama, M. and Kan, M. (1991) Heparin-binding (fibroblast) growth factors are potential autocrine regulators of esophageal epithelial cell proliferation. *In Vitro Cell. Dev. Biol.*, *27A*, 533-541.

Katayama, M., Shoji, M. and Satomi, S. (2005) Differential growth properties of normal and malignant esophageal epithelial cells: A possible cross talk between transforming growth factor-β1 and epidermal growth factor signaling. *Tohoku J. Exp. Med.*, *206*, 61-71.

Lewin, K.J., Dawsey, S.M. and Wang, G.Q. (1995) Squamous carcinoma of the esophagus in China and the West: are they different disorders? *Diseases of the Esophagus*, *8*, 44-47.

Lubomierski, N., Kersting, M., Bert, T., Muench, K., Wulbrand, U., Schuermann, M., Bartsch, D. and Simon, B. (2001) Tumor suppressor genes in the *9p21* gene cluster are selective targets of inactivation in neuroendocrine gastroenteropancreatic tumors. *Cancer Res.*, *61*, 5905-5910.

Massagué, J., Blain, S.W. and Lo, R.S. (2000) TGFβ signaling in growth control, cancer, and heritable disorders. *Cell*, *103*, 295-309.

Nishihira, T., Hashimoto, Y., Katayama, M., Mori, S. and Kuroki, T. (1993) Molecular and cellular features of esophageal cancer cells. *J. Cancer Res. Clin. Oncol.*, *119*, 441-449.

Nishihira, T., Katayama, M., Hashimoto, Y. and Akaishi, T. (1994). Cell lines from esophageal tumors. In J-G. Park, and A. Gazdar (Eds.), Atlas of Human Tumor Cell Lines (pp. 269-285). San Diego, CA: Academic Press.

Ornitz, D.M. and Leder, P. (1992) Ligand specificity and heparin dependence of fibroblast growth factor receptors 1 and 3. *J. Biol. Chem.*, *267*, 16305-16311.

Osawa, H., Shitara, Y., Shoji, H., Mogi, A., Kuwano, H., Hagiwara, K., and Takenoshita, S. (2000) Mutation analysis of transforming growth factor β type II receptor, Smad2, Smad3, and Smad4 in esophageal squamous cell carcinoma. *Int. J. Oncol.*, *17*, 723-728.

Peehl, D.M. and Ham, R.G. (1980) Clonal growth of human keratinocytes with small amounts of dialyzed serum. *In Vitro*, *16*, 526-538.

Rheinwald, J. G. and Green, H. (1977) Epidermal growth factor and the multiplication of cultured human epidermal keratinocytes. *Nature*, *265*, 421-424.

Rodeck, U., Bossler, A., Graeven, U., Fox, F.E., Nowell, P.C., Knabbe, C. and Kari, C. (1994) Transforming growth factorβproduction and responsiveness in normal human melanocytes and melanoma cells. *Cancer Res.*, *54*, 575-581.

Takehara, K., LeRoy, E.C. and Grotendorst, G.R. (1987) TGF-βinhibition of endothelial cell proliferation: alteration of EGF binding and EGF-induced growth-regulatory (competence) gene expression. *Cell*, *49*, 415-422.

In: Esophageal Cancer Research Developments
Editor: Henry K. Kinner, pp. 149-170 ISBN: 1-60021-309-X

Chapter V

Atypical Regenerative Hyperplasia of the Esophagus: A Lesion that Could Simulate Esophagic Carcinoma

Julián Arista-Nasr*, Braulio Martínez-Benitez and Héctor Maldonado Martínez
Department of Pathology, Instituto Nacional de Ciencias Médicas y Nutrición, S. Z. (INCMNSZ), México City, México

Abstract

Atypical epithelial regeneration (AER) is one of the most common changes in surgical pathology. Among locations where not infrequently can AER be seen, are gastric mucosa, pancreas, urinary bladder, prostate gland, gallbladder, mesothelial lining of pleura and peritoneum, skin, squamous epithelium of the cervix and eventually the esophagus [1-7] (table 1). Only slight or moderate cellular atypia is usually evident in the majority of cases with epithelial regeneration, and therefore the condition can then be easily recognized as benign. However, in some other cases atypical regeneration may closely resemble carcinoma or mesothelioma in both cytological samples and biopsies.

In the present chapter we deal with atypical regeneration from several of the locations cited above and in particular squamous epithelium of the esophagus. We also briefly comment on the difficulties of the interpretation of atypical reglands with generative changes in biopsies of the gastro-esophageal junction, which can result in false positive results for dysplasia or malignant neoplasia.

* Correspondence: Julian Arista-Nasr, MD, Department of Pathology, Instituto Nacional de Ciencias Médicas y Nutrición S.Z., Vasco de Quiroga No.15, Tlalpan, 14000, México D.F, Mexico. Fax number: (+52-5) 5485-3489; Telephone number: (+52-5) 5487-0900, exts- 2186 y 21; E mail: pipa5@hotmail.com (J. Arista-Nasr, MD) and brauliomb77@yahoo.com.mx (Braulio Martinez-Benitez MD).

Table 1. Organs in which atypical regenerative changes may simulate carcinoma

- Stomach
- Esophagus
- Pancreas
- Gallbladder
- Prostate
- Bladder
- Skin
- Cervix
- Mesothelial linings from pleura and peritoneum

Atypical Regeneration in Diverse Organs

Although slightly different histological data are seen in different locations with AER, in any of these cases the common worry is confusion of these changes with carcinomas or mesotheliomas. As we will see different strategies are used in different locations to deal with this diagnostic problem.

The presence of atypical regeneration on gastric mucosa may simulate high grade dysplasia or early carcinoma. Glands with an irregular growth pattern, enlarged nuclei, prominent nucleoli, nuclear stratification, mitosis and absence or reduction in cytoplasmic mucus, are some of the data that may lead to misinterpretation. In some cases regenerative changes cannot be convincingly separated from dysplasia. In these cases the diagnosis should be "indefinite for dysplasia", and after adequate treatment, periodic biopsies obtained to determine whether atypical changes disappear or otherwise persist or even become more apparent [2].

Atypical regeneration was found in 13.5% in 200 consecutive cholecystectomy specimens excised for cholelithiasis or cholecystitis, while 3.5% showed carcinoma in situ. The distinction between both lesions is essential, since the treatment of carcinoma in situ could be aggressive [1].

In pancreas, atypical regeneration may also resemble, well differentiated or moderately differentiated adenocarcinoma, in particular in frozen sections. Enlarged nuclei, evident nucleoli and on occasion pseudoinfiltrative pattern account for the confusion. In case of doubt it is recommended to evaluate definitive H and E sections to distinguish regenerative changes seen in chronic pancreatitis from adenocarcinoma.

Recognition of AER of the urinary bladder in urinary cytology represents an special challenge. Not uncommonly, it is necessary to obtain one or several endoscopic biopsies, after conservative treatment has been given, in order to be able to separate regenerative from neoplastic changes [3].

Cases of acute or chronic prostatitis may show glands with atypical changes comprising evident nucleoli and irregular growth pattern that may be interpreted as prostatic carcinoma. However in inflammatory conditions basal cells are preserved and can be observed in conventional H and E slides. When elusive, immunohistochemistry reactions for high molecular weight cytokeratin (903 or 5/6 keratins) may be used to document their presence [4].

Pseudoepitheliomatous hyperplasia is the name given to regenerative atypical hyperplasia of the skin [5]. This condition is associated with different clinical situations including burns, wounds, stasis ulcers, chromomycosis and pigmented lesions. The histological picture is that of an acanthotic hyperplasic epithelium from which thin and markedly elongated proliferating strands of epithelium grow into the dermis. If tissue sections are not properly oriented, the prongs may have a pseudoinfiltrative pattern as they appear as epithelial islands surrounded by stroma. Proliferating keratinocytes in these islands show slight to moderate atypia that make them similar to the infiltrative strands of neoplastic epithelium in epidermoid carcinoma. The presence of inflammatory reaction, acute or chronic, within the epithelial islands accompanying the atypical cells points towards its benign nature.

Controversy on atypical changes in squamous epithelium of the cervix has been a major issue. As one can learn from the many consensus conferences held and reviews published on atypical squamous changes of the cervix (ASCUS) over the last decades, its biological meaning as well as its terminology have been subject to constant review [7,8]. As a result different classifications of ASCUS have been in use; however no precise guidelines existed for its management until recently. The coordination of the 2001 Bethesda and ASCCP consensus meetings resulted in new subcategories of ASCUS along with well-defined management strategies for these interpretations. This new and clinically relevant terminology should lead to a reduction in difficulties at the clinical level and a more uniform management of patients. Standardization of reports and clinical management will also allow more reliable evaluation of patient outcomes and cost analysis [7].

Attempts have been made to use p53 and other molecular markers to make the distinction between reactive squamous proliferation and well differentiated epidermoid carcinoma. However, even though less p53 expression is seen in atypical hyperplasia than in carcinoma, results are not always consistent and the use of such markers is still limited. The role of these and other molecular techniques in this field is extensively discussed later.

The histological interpretation of atypical regeneration in different organs is relevant to clinical practice since management of patients may vary according to that interpretation. Attending physician and pathologist must establish close communication to avoid false positive results that may result in serious psychological and legal issues.

One example of an organ where these errors can have very important consequences is the esophagus. In this organ, atypical regenerative changes misdiagnosed as a malignant lesion may result in aggressive surgical treatment with concomitant morbidity and mortality [9,10]. We will describe as a first step the normal reaction to acute and chronic injury of the squamous epithelium of the esophagus, and then move on to the atypical regenerative response to that injury and the factors that cause confusion with squamous carcinoma. Additionally we will also discuss atypical regenerative changes of gastric mucosa in gastro-esophageal junction biopsies which may resemble well or moderate differentiated adenocarcinoma.

Acute and Chronic Esophagitis and Regeneration of Esophageal Mucosa

Gastroesophageal reflux (GERD) disease is a chronic disease affecting up to 40% of people in the Western world [11]. Risk factors associated with GERD include age and lifestyle habits. However the actual contribution of these factors to disease development and its clinical presentation is unclear. In GERD, refluxed gastric acid into the esophagus lowers its pH, and generally when it falls below 4, mucosal damage overcomes. Presenting symptoms vary from patient to patient, but most commonly heartburn and regurgitation are the main complaints, these two are also the most indicative symptoms of GERD. Less common GERD-associated symptoms include chest pain; a range of ear, nose and throat conditions, and asthma. An important fact is that symptoms severity is frequently unrelated to disease severity, which makes clinical diagnosis of GERD problematic. While endoscopic monitoring may be used to assess the presence and severity of GERD, a lack of visible damage does not necessarily indicate an absence of GERD. Esophageal complications associated with GERD include erosions, ulcers, peptic strictures, and Barrett's esophagus. Because of this, in contrast to perception of the disease as 'merely' heartburn, the impact on the patients' quality of life can be profound. Increasing awareness of GERD by health care professionals has led to improved diagnosis and a greater appreciation of the need for maintenance therapy. Proton pump inhibitors are effective in the treatment of GERD since they reduce hydrogen ion concentration in the gastric juice, elevate esophageal pH in these patients and hence reduce mucosal damage to this organ.

The presence of adaptive responses of squamous esophageal lining to different acute and chronic stimuli as gastro-esophageal reflux, oral drug administration, radiation or chemical irritants, has long been recognized. Mucosal esophageal damage is characterized by hyperemia, balloon degeneration, dyskeratosis, and in later stages erosion, necrosis and ulcers. Acute damage is associated with the presence of polymorphnuclear leucocytes, while in chronic damage inflammatory infiltrate is composed mainly of lymphocytes and eosinophils.

Hyperplasia of basal cells usually seen in squamous epithelial regeneration is generally present in chronic or acute damage. It is the consequence of increases in cell turnover, and results in turn in expansion of the basal proliferative zone and elongation of lamina propria papillae. Proliferating basal cells can be distinguished by its intense basophilic stain. They constitute, as their name indicates, the most basal layer of the epithelium, however, in situations of accelerated turnover they may well comprise the entire lower half of the epithelium. Though a slight to moderate degree of atypia may be found in proliferating basal cells, they can be recognized as benign by their regular disposition, even in situations of quite active regeneration.

In the case of GERD intestinal metaplasia of the squamous epithelium (Barrett's esophagus) is implicated in the development of esophageal adenocarcinoma. Therefore proper recognition of atypical epithelial regenerative changes in these patients may also be of great importance for an adequate diagnosis and treatment.

Recent data add a new histological feature to the morphological spectrum of esophageal mucosal damage, the presence of multinucleated squamous epithelial giant cells simulating viral cytopathic effect and/or dysplasia. In a study of Surendra et al. [12] the authors found in 14 cases of esophagitis multiple multinucleated squamous epithelial cells confined to the basal zone or involving the basal and superficial epithelium. Several multinucleated giant cells were observed (range 2 to 11 cells per biopsy) with a mean of three nuclei per cell,. Nuclei contained a single or multiple eosinophilic nucleoli with a perinucleolar halo, but no inclusions, hyperchromaticity, or atypical mitoses. Immunohistochemical study for cytokeratin AE1/AE3 and S-100 was performed to demonstrate the squamous nature of the multinucleated giant cells. The multinucleated giant cells, in all cases, were strongly positive for cytokeratin AE1/AE3 and negative for S-100. The patients were from both genders, different ethnic backgrounds and comprised a wide range of ages (range 23-87 years, mean age 59 years). Presenting symptoms, etiology and macroscopic findings were also varied. Patients presented with dysphagia or odynophagia, upper gastrointestinal bleeding, heartburn, or abdominal pain. Esophagitis was attributed to gastro-esophageal reflux in ten cases, radiotherapy in one, Candida infection in one, drug intake (alendronate) in one and to an unknown cause in another one. Endoscopically, seven patients had an ulcer or erosion, four only erythema, two stricture formations, and one white mucosal plaques.

This diversity of etiology, clinical and macroscopic findings supports the idea that while this histological feature is not attributable to the genesis of the inflammatory process is present along with mucosal damage. In support of this view, all cases showed associated nonspecific features of active esophagitis such as ulceration, neutrophilic and eosinophilic inflammation, basal cell hyperplasia, and elongation of the lamina propria papillae. After treatment, all patients demonstrated clinical improvement, and in three patients in whom follow-up biopsies were performed no evidence of esophagitis, epithelial cell multinucleation, or dysplasia was found. These follow up data suggest multinucleation to be a direct consequence of mucosal damage and do not seem to involve dysplastic or neoplastic change. Further analysis of etiology was conducted to exclude possible etiologies with similar findings. Immunostaining for herpes simplex virus 1 and 2 (HSV), cytomegalovirus (CMV), as well as DNA in situ hybridization for human papilloma virus (HPV-ISH) were performed in all cases, and electron microscopic evaluation for viral particles was performed in three cases. The multinucleated giant cells, in all cases, were negative for HSV I and II, CMV, and HPV-ISH. Electron microscopy failed to show viral particles in three of three cases. The authors concluded that multinucleated epithelial giant cell changes may rarely be seen in patients with esophagitis of varying etiology and probably represent a regenerative response to injury. It is important to distinguish this feature from either viral cytopathic effect or dysplasia.

We have occasionally found multinucleated giant cells similar to those described in this review. In most of our cases multinucleated giant cells have been associated with keratinocytes easily recognized as benign but showing regenerative changes and slight to moderate atypia. Taking into account the changes on these surrounding cells, we consider a more significant problem distinction of these multinucleated giant cells from viral cytopathic effects than from dysplastic or neoplastic cells. Immunohistochemistry should be done in case of doubt to rule out the possibility of viral infection. Similar multinucleated giant cells have been described in vulva or skin as a regenerative change and should not be mistaken for viral cytopathic changes either.

Atypical Regenerative Changes of Esophageal Squamous Epithelium: A Mimicker of Squamous Cell Carcinoma

In spite of a lesser frequency of regenerative atypical changes in esophageal mucosa in comparison to other sites, its interpretation is usually more difficult. This fact warrants the effort to make available endoscopic and radiological information at the time of the histological evaluation in order to reduce false positive results. Close communication among pathologist and attending physician is of particular importance to correlate the endoscopic findings with those obtained during histological examination. An issue worth of discussion is the amount of the tissue fragments needed for evaluation. We have found at our institution that errors in the interpretation of some esophageal endoscopic biopsies with atypical regenerative changes are due to the limited sample size. Although further studies are needed in order to determine the optimal number of fragments required to distinguish atypical regenerative changes from carcinoma, once the diagnosis of atypical regeneration has been established in a first biopsy, a greater number of fragments than normal should be obtained for a reliable interpretation in subsequent biopsies.

Squamous esophageal epithelium may show considerable atypia that might closely resemble carcinoma, especially when speaking of endoscopic biopsies. However, there is very scarce available literature on this subject. Only a few specialized textbooks mention briefly some of the differences between atypical regeneration and epidermoid carcinoma, and to our knowledge, there are no previous reports that analyze the histological differences between AER and esophageal squamous cell carcinoma (SCC). As mention, the distinction between atypical regenerative changes and SCC is extremely important because the treatment of squamous carcinoma is associated with high morbidity and significant mortality [9,10]

Lewin and Appelman [13], in the Atlas of Tumor Pathology of the Armed Forces Institute of Pathology, mention that the most significant differential diagnosis with SCC in endoscopic biopsies is the squamous epithelial proliferation that occurs at the edges and bases of chronic reflux–induced ulcers. These appear as uniformly thick, parallel prongs of primitive squamous cells that penetrate the granulation tissue of the ulcer bed. Early in the healing process, cells in these prongs are cuboidal with little cytoplasm, and have large, often misshapen, hypercromatic nuclei, and a very high nuclear-cytoplasmatic ratio. Mitotic figures are often prominent. Hence in bias-cut samples where the prongs are often sectioned across, they appear separated from the superficial epithelium and trapped within the stroma. In this situation they can closely resemble nests of invasive carcinoma. If these foci are properly oriented and sectioned, diagnosis is not difficult because the parallel arrangement of the prongs is obvious. Chandrasoma [14] has also pointed out that these ulcer-related changes can mimic malignancy. Careful attention to both architectural and cytological changes is thus necessary to avoid false-positive diagnoses.

Recently, we have described several useful histological features to distinguish atypical regenerative hyperplasia from esophageal carcinoma in endoscopic biopsies [15]. We reviewed 600 endoscopic biopsies (555 with chronic esophagitis and 45 with carcinomas of the esophagus), in order to study the frequency and histological appearance of atypical regenerative hyperplasia, simulating carcinoma. We selected those cases in which the differential diagnosis comprised regenerative atypical hyperplasia versus esophageal carcinoma, and cases of atypical regenerative hyperplasia that were mistaken for carcinoma. For comparative purposes, we studied 10 cases of endoscopic biopsies with esophageal carcinoma that was confirmed by esophagectomy. Among the cases with chronic esophagitis, we found 10 biopsies (1.8%) in which atypical regenerative hyperplasia mimicked carcinoma. In 7 cases there had been 4 to 12 years of follow-up, and no patient had developed an esophageal neoplasm. The remaining 3 patients had been submitted to esophagectomy. None of these patients had shown carcinoma or dysplasia in the esophageal resection (false-positive biopsies). Similarities and differences found between atypical regenerative changes and epidermoid carcinoma in endoscopic biopsies are commented below.

Clinical Features

The main clinical and endoscopic findings in both groups are summarized in table 2. The mean age at which AER and SCC developed was similar. The age range for atypical hyperplasia went from 49 to 80 years and from 50 to 81 years in the epidermoid carcinoma group. Patients in both groups had had clinical manifestations as dysphagia, nausea and vomit. However dysphagia had increased with time in the carcinoma group while it had followed an erratic course in those with atypical hyperplasia. In some of the latter patients this symptom had disappeared with conservative treatment alone. Most of the patients with either condition had presented nausea and vomit at some time during their illness. In contrast bleeding associated to vomit had been more frequent in the patients with carcinoma.

Table 2. Main clinical and endoscopic features in atypical regenerative atypia and squamous cell carcinoma

Clinical features:		
Age	49-80	50-81
Symptoms	Substernal pain	Substernal Pain
	Dysphagia	Increasing dysphagia
	Vomiting	Vomiting/bleeding
	6 months - 10 years	8 months - one year
Endoscopic features:		
Superficial ulcers		Fungating masses
Esphageal epithelium erosion		Ulcers, irregular borders
Erythematous and granular mucosa		

(Modified from ref. 15).

The duration of symptoms had also been different between both groups. Patients with atypical regeneration had referred symptoms elapsing from months to years. One of the patients had referred remissions and relapses of the symptoms for a period of ten years. Patients with epidermoid carcinoma on the contrary had shown a fast progression of symptoms once they had appeared. The symptoms had initiated no more than a year before the diagnosis in all of these patients. Besides, patients with carcinoma had presented systemic symptoms not shown by atypical hyperplasia patients, like asthenia, anorexia and weight loss.

Endoscopic Features

Another important set of differences between both groups became evident when considering their endoscopic features (Table 2). Of ten patients with atypical regenerative hyperplasia, four had shown shallow ulcers, three only erosion of squamous epithelium and three velvety erythema mucosa. These changes were similar to those observed on most of the esophagitis biopsies without atypical regenerative changes, and accordingly the possibility of malignant neoplasia hadn't been suggested in any of these ten cases. On the other hand, six of the patients with epidermoid carcinoma had presented with an exophytic mass on the endoscopic study and the remaining four, as ulcers of irregular borders. The fungating lesions occluded totally or partly the lumen had irregular borders and areas of necrosis. This endoscopic image had been regarded most of the times as consistent with malignancy. Ulcers had been interpreted as either probably malignant (two cases) or suggestive of malignancy (the other two).

As shown in this review, endoscopic data are closely related to histological findings and one should be cautious when malignancy is suspected in the absence of any macroscopic suggestive lesion. Just as shown with endoscopic data, the absence of radiological data suggestive of malignancy should alert clinicians that the diagnosis might be atypical regenerative hyperplasia rather than carcinoma.

Carcinoma in situ constitutes the exception to this rule since early lesions limited to esophagic epithelium are difficult to recognize. However as we further comment below, histological findings and tissue reaction of atypical regenerative hyperplasia is distinct form that of intramucosal carcinoma. Most in situ esophagic carcinomas show epithelial disarrangement with progressive loss of cellular orientation, increase in pleomorphism and presence of atypical keratinocytes with enlarged nuclei, hypercromatism and increase of mitotic activity. When these changes are only partially present in the mucosa the lesion should be diagnosed as dysplastic. Esophagic carcinoma in situ and its precursors are not associated with stromal changes, and most of them show no inflammatory or only a slight inflammatory reaction.

Histological Features

Architectural as well as cytological criteria must be assess when evaluating biopsies with lesions that pose the differential diagnosis of atypical regenerative hyperplasia vs. squamous esophageal carcinoma (Table 3). The sum of both sets of criteria will allow the correct classification of most of these lesions. We consider the following differences as the most helpful in making this distinction:

I) Growth pattern: Proliferating epithelium in epidermoid carcinoma arranges in nests of atypical keratinocytes or irregular projections of epithelium that clearly invade the subjacent squamous epithelium. The degree of squamous differentiation is variable, but most tumors are well or moderately differentiated. Squamous neoplastic cells characteristically form palisades at the periphery of their nest (Figure 1). Some of them may show glandular differentiation and sometimes the neoplasm may show only small and basophilic hyperchromatic cells (small poorly differentiated carcinoma or oat cell carcinoma). Desmoplastic reaction may or may not develop once carcinoma invades the submucosa. Besides, careful analysis of nerves and vessels may reveal infiltration of these structures. Metastatic involvement of other areas of the esophagus wall may be apparent in esophagectomy specimens.

In contrast, atypical regeneration is, constituted by irregular islands of epithelium detached for the surface mucosa (Figure 2). These tissue fragments show no desmoplasia, palisading or nesting of neoplastic keratinocytes. Atypical regenerative squamous cells are closely associated and intermixed to a variable degree of inflammatory infiltrate composed of lymphocytes, plasma cells, eosinophils and neutrophils. It should be mentioned however that areas of hemorrhage or inflammation may also be observed in biopsies of squamous carcinoma.

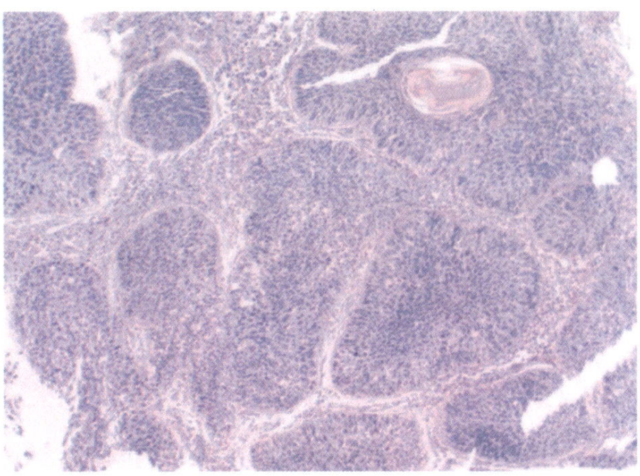

Figure 1. Well differentiated infiltrating squamous carcinoma. Most esophagic SCC are well or moderately differentiated. In this case, nest of malignant keratinocytes palisading and a pearl corn up, right) are shown. (hematoxylin-eosin, 100x)

II) Cytological features: Neoplastic keratinocytes show a variable degree of atypia, though most commonly they show well to moderate differentiation. Squamous cells vary in shape and size, show different degrees of nuclear enlargement, prominent nuclei and frequently show atypical mitoses. Individual keratinization and keratin pearls, are not consistent findings but the last is always absent in atypical regeneration. In some cases of SCC, foci of carcinoma in situ may be found in the same tissue fragment or in other tissue fragments obtained in the biopsy.

Small squamous cell carcinoma (oat cell carcinoma) is rather more infrequent than its moderately or well differentiated counterparts. This variety of carcinoma shows intensely basophilic cells with similarly hypercromatic nuclei and less striking atypia. Atypical regenerative changes may bear greater resemblance to small cell carcinoma given its monotonous appearance (Figure 2, right). Actually in small tissue fragments composed entirely of this kind of carcinoma it may be really difficult or impossible to differentiate it from atypical regeneration. In case of doubt, Immunohistochemical studies, additional levels or additional several fragment biopsies may be required to establish the diagnosis.

Table 3. Histologic changes in squamous carcinoma versus atypical regenerative hyperplasia

Squamous carcinoma	Atypical Regenerative hyperplasia
-Stromal infiltration by keratinocytes	-Pseudoinfiltration by thick prongs of malignant. immature keratinocytes in granulated or inflamed tissues -Detached nests fragments without stromal infiltration.
-Occasional desmoplastic stroma.	-No desmoplastic stroma.
-Occasional palisades	-No palisades
-Pleomorphic cells	-Medium sized keratinocytes
-Common individual keratinization	-Occasional individual keratinization
-Atypical mitosis	-Non atypical mitosis
-Mild acute and chronic inflammation	-Marked acute and chronic inflammation
-Occasional dysplasia, or in situ carcinoma in adjacent epithelium	-No dysplasia or in situ carcinoma
-No lymphoid hyperplasia	-Occasional lymphoid hyperplasia

(Modified from ref.15).

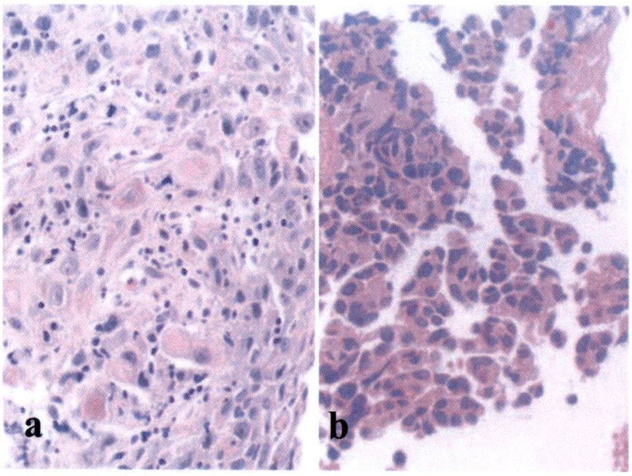

Figure 2 A, (left). Atypical regeneration is, constituted by irregular islands of epithelium detached for the surface mucosa. Keratinocytes show hyperchromatism and individual queratinization. Few lymphocytes and eosinophils can be seen intermixed with atypical keratinocytes (hematoxylin-eosin, original magnification ×575). Figure 2-B, (right). Atypical regenerative hyperplasia. Noncohesive, basophilic reactive small keratinocytes resemble small poorly differentiated carcinoma. (hematoxylin-eosin, 575×)

Cytological data shown by regenerative atypia are in most cases homogeneous medium sized atypical keratinocytes with hypercromatic nucleus and a monotonous and immature appearance (Figure 3). Compared with most SCC, the pleomorphism is less apparent in AER, mitosis are present but atypical mitosis are not, and dysplastic changes in the epithelium adjacent to the lesion are not seen either. In cases where the biopsy shows small or medium relatively monotonous atypical keratinocytes, the association with inflammation or granulation tissue, as well as with the other criteria summarized in Table 3, the diagnosis of atypical regeneration must be favored.

Pseudoinfiltrative pattern may be on occasion found in biopsies otherwise characterized only by regenerative changes. In these cases atypical nests of keratinocytes surrounded by stroma are the result of tangential sectioning and do not represent real neoplastic infiltration (Figure 4). The presence of moderately atypical medium sized squamous cells with hypercromatic nuclei and a monotonous appearance, forming epithelial stripes at the base of an ulcer or surrounded by inflammation or granulation tissue, should arise suspicion over the absence of real infiltration. Similarity of cytological findings between pseudoinfiltrative epithelium and other areas of squamous cells in the regenerative lesion are also of help in recognizing the benign nature of the process and avoiding an erroneous diagnosis of carcinoma.

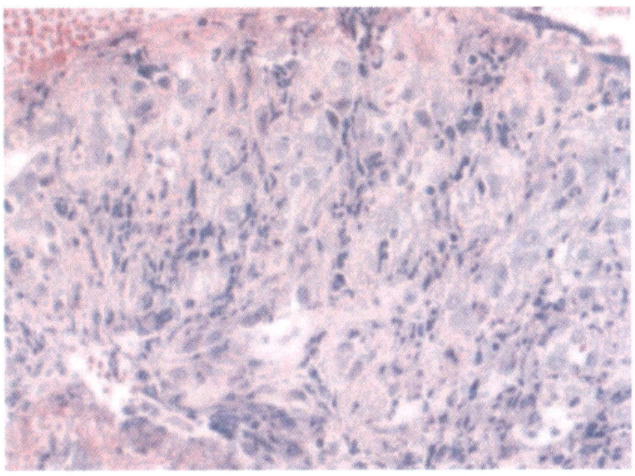

Figure 3. Atypical epithelial regeneration. Esophageal epithelial fragments are arranged in nests and are composed of monotonous medium-sized keratinocytes. Recent hemorrhage and fibrin are shown can be seen (upper right). Other tissue fragments of the same biopsy only showed granulated tissue and hemorrhage. (hematoxylin-eosin, 400x).

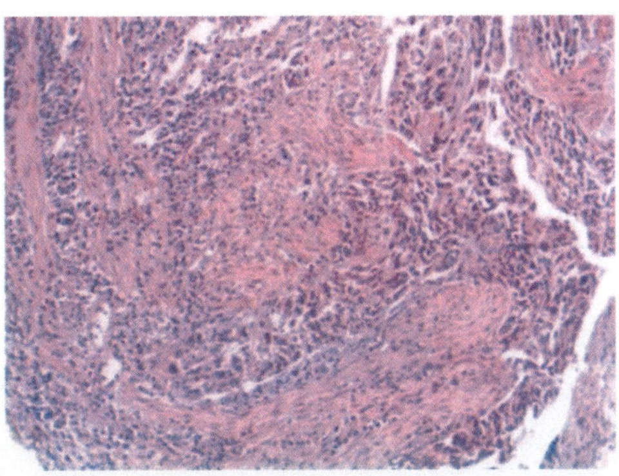

Figure 4. Pseudoinfiltration in atypical regenerative hyperplasia. A thick, round, long prong is surrounded by stroma with abundant inflammatory infiltrate in tangential section. This change should not be confused with infiltrating carcinoma Keratinocytes are monotonous with moderate atypia. Marked stromal inflammation is seen around atypical cells, and its presence may point to a benign process (hematoxylin-eosin, 200x).

Granulation tissue is more frequently identified in biopsies with regenerative changes. Most of the time it is possible to identify readily patent empty spaces lined by evident endothelial cells, interspersed with new formed smaller vessels and inflammatory infiltrates. However in some cases, granulation tissue shows collapsed or obliterated vessels, lined by reactive atypical basophilic endothelial cells with little cytoplasm, that resemble nests of neoplastic epithelial cell rather than vessels. In this instance confusion with a poorly differentiated malignant neoplasm may occur. Even more, the combination of histological findings of atypical regenerative changes with the presence of granulation tissue with such vessels can be confusing and result in a diagnosis of carcinoma. The use of Immunohistochemical studies will reveal positivity for CD34 or Factor VIII, and negativity for cytokeratin and other epithelial markers in the vessels in these cases.

In conclusion, the settings where atypical regenerative hyperplasia can be confused with squamous cell carcinoma include mainly atypical keratinocytes intermixed with inflammatory cells, monotonous keratinocytes simulating small cell esophagic carcinoma, tangential sectioning, dense inflammatory infiltrate obscuring epithelial/stromal infiltrate, and prominent endothelial cells simulating neoplastic keratinocytes [15].

Another histological change that the surgical pathologist must recognize to avoid misclassification of reactive lesions as malignant neoplasms is the presence of hyperplasic lymphoid tissue [16-19]. This lymphoid tissue may arise as a consequence of the chronic damage of the tissue that elicits a chronic inflammatory response. When abundant, inflammatory infiltrate may be confused with lymphoma. Although lymphoma of the gastro-esophageal junction is infrequent, lymphomas may localize to this area or be the result of extension of a gastric lymphoma. Distinction of lymphoma from carcinoma of the gastro-esophageal junction is aided by the inflammatory and reactive response observed in esophagitis, including previously commented data such as epithelial erosion, presence of granulation tissue, fragments of atypical epithelial tissue detached from surface epithelium, etc.

Atypical Regenerative Changes of the Gastric Glands in the Cardias and Gastro-esophageal Junction

Gastro-esophageal junction shows glands in continuity with esophageal squamous epithelium. This glandular epithelium may also show atypical regenerative changes in response to etiologic agents similar to those of atypical regenerative hyperplasia of the squamous epithelium of the esophagus. Additionally, radiation to the gastro-esophageal junction produces striking atypia in both the esophageal and gastric mucosa, and it may be the cause of false positive results when such information is not available at the time of diagnosis.

In order to avoid confusion between reactive conditions the surgical pathologist must distinguish the main histological features of low and high grade dysplasia. This distinction is based on the severity of both architectural a cytolgic findings that may suggest neoplastic transformation of the columnar epithelium. Although either architectural or cytologic criteria may predominate, the main histological changes of gastric dysplasia include: budding, branching, crowded, or irregularly shaped glands; papillary extensions into gland lumina, and villiform configuration of the mucosal surface. Nuclear changes may include marked variations in size and shape, nuclear and/or nucleolar enlargement, increased cytoplasmatic ratio, hypercromatism, and increased number of atypical mitosis (Figure 5). Diagnostic features as cytoplasmic basophilic with loss of mucus and excessive nuclear stratification often extending from the epithelial basement to the luminal surface are recognizable at low power.

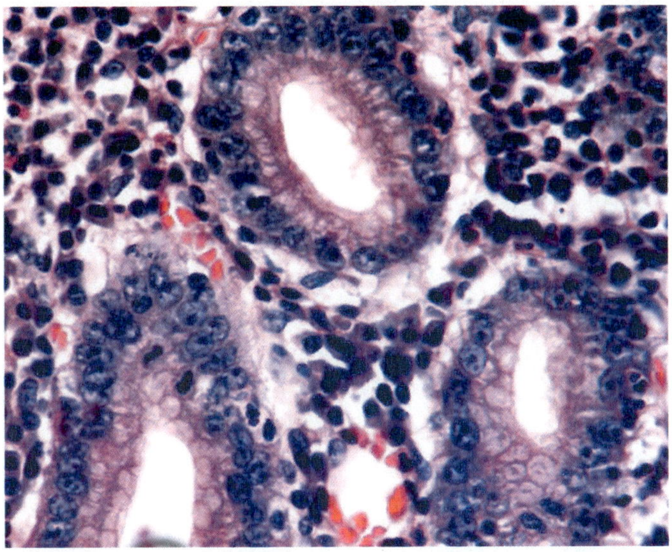

Figure 5. Gastric Dysplasia. Nuclear and nucleoli enlargement are present in most cells. There is increased cytoplasmtic ratio, hyperchromatism, and an atypical mitosis (left). (Hematoxylin and eosin 575x)

Atypical regenerative changes are characterized by glands that may show an irregular distribution or in irregular nests and may also show irregular shaped glands (Figure z. In contrast with true dysplasia they don't simulate colonic adenomas at low power; however, they are lined by cells with scarce cytoplasm in which loss of mucus is evident and which have hypercromatic nuclei that may be in some areas arranged in multiple strata or in intraglandular papillary fronds. These changes simulate gastric dysplasia and sometimes well differentiated adenocarcinoma. It must be emphasized that in most cases the changes previously mentioned in atypical glands are associated with inflammatory infiltrate in which neutrophils; eosinophils, lymphocytes and plasma cells can be recognized. Additionally, the atypical glands may be found intermixed with necrosis and a hemorrhagic background with fibrin intermixed with inflammatory cells (Figure 6). In contrast to dysplasia, atypical regeneration in most cases does not affect surface epithelium and this could be a useful criterion in the differential diagnosis. Another set of histological criteria that can be useful to favor atypical regeneration includes superficial ulceration and the presence of hyperplasic foveolar glands.

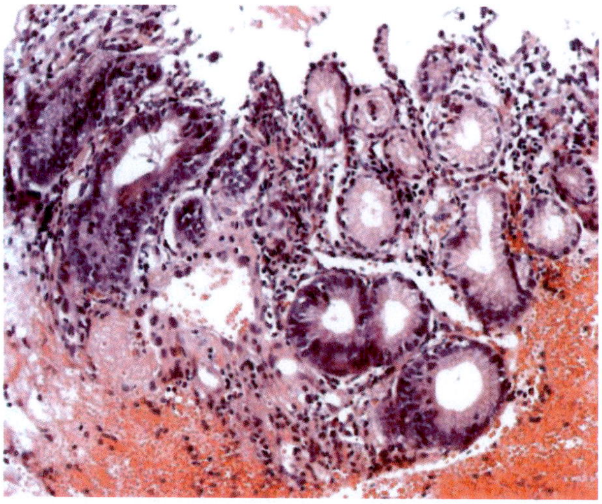

Figure 6. Atypical regenerative changes in gastric mucosa. Glands show an irregular nest of hyperchromatic glands. The presence of ulceration, fibrin and hemorrhage and inflammation are common findings in reactive conditions, however they occasionally may be found in dysplastic lesions (hematoxylin-eosin 100x).

It must be remember that an overlap of findings exists in biopsies with atypical regenerative changes and those with dysplasia. Therefore, some biopsies with dysplasia may show only isolated architectural or cytological changes associated with mild inflammation and, on the other hand some biopsies with regenerative changes with marked atypia may be associated with mild inflammation and show no superficial ulceration nor hemorrhagic background or other changes that could be useful to recognize the lesion as reactive. In this setting, differential diagnosis may again be very difficult. Literature on this topic recommends that lesions, in which the distinction between epithelial regenerative changes and dysplasia cannot be established, should be diagnosed as indefinite for dysplasia [16, 20]. Additional new biopsies should also be obtained after treatment to determine whether the changes show involution or they rather persist or progress indicating the presence of true dysplasia. In most cases, atypical regenerative changes in squamous and glandular epithelium will disappear with treatment in a few weeks. High grade dysplasia on the contrary, would persist or progress to adenocarcinoma in an undefined length of time [21]. When atypia arises in metaplastic Barrett's esophagus glands this latter approach is also adequate as follow-up for potentially malignant lesions.

The distinction between atypical regenerative changes vs. adenocarcinoma is particularly important now, since during the last few decades there has been an unusually increment rise in the incidence of adenocarcinomas originating at the esophagogastric junction. The tumors of the esophago-gastric junction can be categorized in two types of cancer divided according to their anatomical origin: distal esophageal adenocarcinoma and adenocarcinoma of the gastric cardia. However, due to their location, in the transitional zone of the esophagus and stomach, there is constant debate about the proper classification, staging, and management of these tumors. The etiology of distal esophageal adenocarcinoma is clearly related to gastroesophageal reflux disease and the development of Barrett's esophagus. The etiology of adenocarcinoma of the gastric cardia is less well understood [22]. Recent evidence links obesity with the rising incidence of esophageal adenocarcinoma; in one study 82% of patients who developed adenocarcinoma of the esophagus either had overweight or were obese compared with 59% of the healthy control population ($P<0.001$), [23]. However others authors feel that the association between body mass, reflux, and esophageal adenocarcinoma might seem easily interpreted, but a more thorough assessment of the published work shows that several factors are missing. [24]. While tobacco and alcohol are well established risk factors for esophageal squamous-cell carcinoma, their roles in the etiology of the increasingly common esophageal adenocarcinoma remains uncertain. In one study it is concluded that tobacco smoking, a strong risk factor for esophageal squamous-cell carcinoma and cardia adenocarcinoma, does not play an important role in the etiology of esophageal adenocarcinoma [25]

Barrett's esophagus is a complication of chronic gastroesophageal reflux disease. The diagnosis of this complication rests on the presence of specialized columnar epithelium of intestinal type, characterized by the presence of acid mucin-containing goblet cells. Much of the controversy surrounding Barrett's esophagus in the literature relates to the complex anatomy of the esophagogastric junction and the difficulty in precisely identifying this landmark during endoscopic studies. By definition, in Barrett's esophagus, the squamocolumnar junction is proximal to the esophagogastric junction. Although fundic-type or cardial-type (junctional) columnar epithelium may be present in Barrett's esophagus, it is only the presence of specialized columnar epithelium that is diagnostic of this condition. Patients with Barrett's esophagus are at risk of progressing to esophageal dysplasia and adenocarcinoma. As mentioned before, there are several problems in using dysplasia as a marker for esophageal adenocarcinoma risk in these patients. These problems include sampling errors and intra- and interobserver variation in the recognition of dysplasia, and the fact that may be difficult to distinguish regenerative epithelial changes from dysplasia [26].

No consensus exists in literature with respect to the number of biopsies required for the differential diagnosis of these two conditions. However a greater number of biopsies than in normal conditions are needed in biopsies with regenerative atypia to be able to classify adequately the lesion. In any case, it should be kept in mind that this is not always a definitive solution, because the absence of a well-defined lesion leaves open the possibility that the region of concern will not be accurately sampled on a second biopsy.

In conclusion, we believe that there are sufficient architectural and cytological criteria to establish a distinction between esophageal carcinoma and ARH in most endoscopic biopsies. If the histological changes are not absolutely convincing or the surgical pathologist does not feel comfortable establishing a definite diagnosis, it is better to recommend medical treatment and repeat the biopsy, particularly if there are no endoscopic or radiological data to support the presence of a neoplasm. In those cases in which the differential diagnosis includes dysplastic glands vs. glands with atypical regenerative changes, should be diagnosed as indefinite for dysplasia. The patient should be then reevaluated to resolve the case in subsequent biopsies.

Molecular Biology of Esophageal Epidermoid Carcinoma

As we have been commenting over the chapter there are certain histological criteria that allow distinction of invasive epidermoid carcinoma from atypical regenerative hyperplasia (ARH). Although a number of molecular and immunohistochemical markers have been studied, none of them have demonstrated to improve sensibility over histological criteria. Nonetheless some recent data on a set of these markers prove useful in separating both conditions. In the following paragraphs we make a review of the molecular alterations in esophageal squamous carcinoma and emphasize a group of them that may be of greater help for daily surgical pathology practice. We also comment on molecular alterations present in adenocarcinoma of the esophagus.

Different molecular, genetic and epigenetic alterations have been identified in esophageal squamous carcinoma, including cytogenetic abnormalities, protooncogen amplifications, mutations in tumor suppressor or other genes, and aberrant or diminished expression of adhesion molecules, extracelullar matrix proteins or metalloproteinases.

Among these alterations, examples have been identified of genes involved in cell cycling, particularly of those involved in the G1-S checkpoint. Of these, mutations in TP53 (17p13) are an early event, sometimes detectable in intraepithelial neoplasia [27]. Frequency and site of ocurrence for this mutation vary from one geographic area to another, which suggests that mutations in TP53 may result from specific risk factors present in each region.

Amplification of different protooncogenes have been implicated in esophageal malignancy. Some of them are HST-1, HST-2, EGFR and MYC. Another set of genes which promote growth because they directly regulate the cell cycle are cyclins. Of these cyclin D1 gene (11q13) is also amplified in 20% to 40% of esophageal epidermoid carcinomas. [27] Besides, it is more frequently encountered in malignant lesions with preserved RB protein expression. Accordingly involvement of both genes in the same signaling cascade has been proposed.

Loss of heterocigocity (LOH) in different locus has been reported. Analyses of alterations in 3p21.3 have shown LOH of this locus in esophageal carcinoma affecting the DLC1 (deleted in lung and esophageal cancer 1) gene. The function of this gene is not known; however its detection using RT-PCR show negative results in 33% of primary lung and esophageal malignancies. In some cases there is absence of transcripts, in some others there are raised levels of non functional RNAm [28]. LOH has also been found in multiple suppressor genes in 5q15 in most esophageal carcinomas.

Methylation of CpG5' islands leading to transcriptional inactivation of the FHIT gene is an example of epigenetic mechanisms of regulation associated with carcinoma of this organ [28].

Deletion of the TOC gene in chromosome 17q25 is another alteration observed in this form of cancer. This gene was identified from familial studies for esophageal carcinoma. The familial presentation of esophageal carcinomas has received little attention except for its association with palmoplantar epidermolythic keratoderma (Tylosis) [29]. This dermatologic condition affects palms and soles, is inherited in a dominant autosomal fashion, and segregates along with esophageal carcinoma. The gene was named TOC after Tylosis

Oesophageal Cancer. The illness is characterized at the molecular level by altered proteins important for the physical structure of stratified squamous epithelia, whereby loss of function of the gene may alter epithelial integrity thus making it more susceptible to enviromental mutagens.

Further studies are needed to clarify how all these alterations relate among themselves and to the phenotypical changes of esophageal epidermoid carcinoma. The sequence of events that lead to the appearance of squamous esophageal carcinoma are also on the wait for the results of such studies. A proposed sequence of steps based on our current knowledge is presented in figure 7.

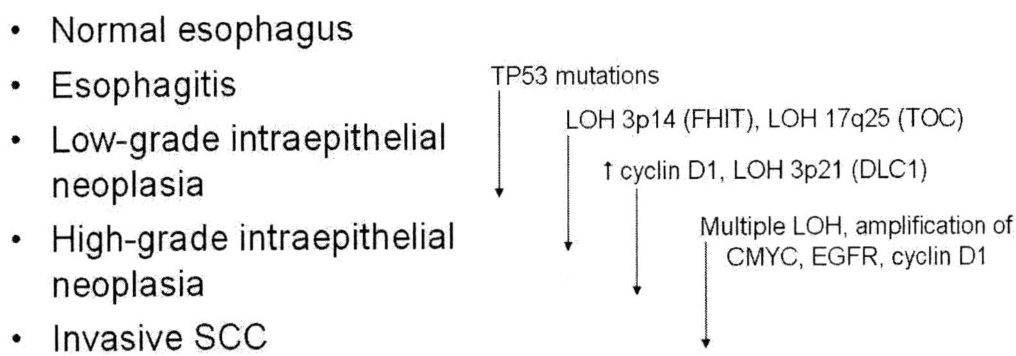

Figure 7. Proposed sequence of events in esophageal carcinogenesis.

Recently various molecular alterations have been identified which may be of some help in distinguishing esophageal squamous carcinoma from regenerative hyperplasia. In a study from Zarovnaya and Black [30] the immunohystochemical expression of p53, matrix metalloproteinase-1 (MMP-1), E-cadherina and collagen IV was employed to differentiate between these two entities.

The p53 tumor suppressor gene is thought to be the gene most commonly mutated in human cancer. Mutation of the p53 gene has been found in mucosa from patients exposed to some of the putative esophageal carcinoma risk factors, i.e. alcohol and tobacco. Even more, loss of chromosome 17p, affecting this gene, has been observed in 50% of SCC. It may be involved in the hyperplasia-dysplasia step, or it may be involved in later steps of tumor progression as in tumor invasion. In accordance to this hypothesis, mutations have been found to increase with tumor progression and are associated with distant metastasis and poor prognosis [27]. Diffuse tumor nuclear positivity in the cases of invasive SCC, in opposition to AER biopsies, supports the theory that mutation of the p53 gene is present in cancer but not AER.

Matrix metalloproteinase 1 is a member of the metalloproteinase family of enzymes. It cleaves the α chains of type I, II, and III collagen. Low or absent levels of this enzyme have been demonstrated in dysplastic, hyperplastic, and benign mucosal tissues. It has therefore been concluded that these metalloproteinases are important for tumor invasion through collagenous stroma. In invasive SCC cases the tumor cells showed cytoplasmic staining, and the surrounding stroma showed a moderate blush [30, 31]. In the cases of HEA, there was no stromal blush. This fact supports the theory of an increased collagenase activity in the invasive tumor cells.

E-cadherin is a calcium dependent cellular adhesion molecule whose expression has been reported to be diminished in aggressive carcinomas of the breast, esophagus, colon, and stomach [30, 32]. The posible underlying mechanism may be promoter methylation of CpG islands, which correlates with negative immunohistochemical staining for the protein. E-cadherin expression has been found to be associated with invasion, metastasis, and poor prognosis in SCC. This findings confirm the observation of decreased E-cadherin staining proportional to tumor cell discohesiveness.

Immunofluorescence and immunoperoxidase methods have shown collagen IV, a ubiquitous component of basement membrane, to be intact and stain in linear fashion in HEA in contrast to invasive carcinomas. While well-differentiated SCC may show small foci of extracellular staining, basement membrane staining is absent in most invasive tumors. However, collagen IV reactivity may be present in invasive tumors with well-defined tumor host borders and cohesive patterns of stromal invasion. In contrast, invasive carcinomas with irregular cords or single tumor cells distributed through the host stroma lack basement membrane at the tumor-stromal interface [30]. On the other hand, prominent and continuous basement membrane staining using collagen IV characterizes benign and reactive hyperplastic squamous mucosa. In the case of epithelial dysplasia the basement membrane varies greatly. It is usually prominent and continuous in mild to moderate dysplasias. In severe dysplasia or carcinoma in situ, the basement membrane is often thinned and occasionally discontinuous.

Alterations in the distribution of type IV collagen in squamous premalignant and malignant lesions indicate that the loss of continuity of the subepithelial basement membrane parallels the progression of the neoplastic transformation process in squamous epithelium.

We have found it efficient in the practice to cut tissue sections for these stains at the same time as additional routine sections are cut to optimize the tissue and reduce trimming. Although a properly oriented hematoxylin-eosin–stained section remains the ideal for diagnostic purposes, use of a p53, MMP-1, and E-cadherin immunoperoxidase panel is helpful as an adjunct in difficult cases, especially those with suboptimal orientation or dense inflammation

Molecular Biology of Esophageal Adenocarcinoma

A variety of molecular disturbances have been found to correlate with the progression of lesions through the metaplasia-dysplasia-carcinoma sequence, when studying patients with Barrett's esophagus. Alterations in TP53 and CDKN2A have been found in early stages of the sequence in tissue form endoscopic biopsies [33].

TP53

High grade intraepithelial neoplasia shows even 60% prevalence of TP53 mutations. This figure is similar to that of adenocarcinoma. It is frequent to find mutation of an allele associated with loss of the other one (17p13.1). Mutations occur in diploid cells, before aneuploidy presents. Eventhough TP53 mutations are also present in squamous carcinoma, the pattern of changes in both types of carcinomas is rather different.

CDKN2A

This gene codes for the cyclin-dependent kinase inhibitor $p16^{ink4a}$ involved in cell cycle control. This protein prevents the formation of a functional kinase capable of phosphorylating the retinoblastoma protein and thereby inhibits cell cycle progression from the G1 to the S phase.

FHIT

Transcript alterations of FHIT are among the early changes observed in metaplasia stages. FHIT is a probably a tumor suppressor gene that extends over a common fragil region, FRA3B.

LOH and Gene Amplification

Altered loci in advanced stages of the sequence have been reported in more than 50% of the cases in chromosome 4 (4q) and 5 (multiple loci including APC). ERBB2 shows amplification.

Among phenotypic changes in Barrett's esophagus is Ki67 overexpression. Besides, this marker correlates with intraepithelial neoplasia grade.

References

[1] Albores-Saavedra J, Henson DE, Klimstra DS. Tumors of the Gallbladder, Extrahepatic Bile Ducts, and Ampulla of Vater. Washington, DC: Armed Forces Institute of Pathology; 1998:208–209. *Atlas of Tumor Pathology,* 3rd series, fascicle 27.

[2] Martinez-Madrigal F, Ortiz-Hidalgo C, Torres-Vega C. et al. Atypical regenerative changes, dysplasia, and carcinoma in situ in chronic gastritis associated with. Helicobacter pylori. *Rev Gastroenterol Mex* 2000;65:11-17.

[3] Epstein JI, Amin MB, Reuter VE. Flat urothelial lesions. In: *Bladder Biopsy Interpretation. Philadelphia* Pa: Lippincott Williams and Wilkins; 2003:30.

[4] Epstein JI, Yang XJ. *Prostate Biopsy Interpretation.* 3rd ed. Philadelphia, Pa: Lippincott Williams and Wilkins 2002:180.

[5] Civatte J. Pseudo-carcinomatous hyperplasia. J Cutan Pathol 1985;12:214–223. McCaughey WTE, Al-Jabi M. Differentiation of serosal hyperplasia and neoplasia in biopsies. *Pathol Annu* 186; 21:271-293.

[6] Rosai J, Dehner LP. Nodular mesothelial hyperplasia in hernia sac: a benign reactive condition simulating a neoplastic process. *Cancer* 1975;35:165-175.

[7] 7- Nayar R. Atypical squamous cells: update on current concepts. *Clin Lab Med.* 2003 Sep;23(3):605-32. Review.

[8] 8- Cox JT, American Society for Colposcopy and Cervical pathology. The clinician's view: role of human papillomavirus testing in the American Society for Colposcopy and Cervical Pathology Guidelines for the management of abnormal cervical cytology and cervical cancer. *Arch Pathol Lab Med.* 2003 Aug;127:950-8.

[9] Iizuka AN, Ide H, Ishida K. et al. Surgery plus chemotherapy compared with surgery alone for localized squamous cell carcinoma of the thoracic esophagus: a Japan Clinical Oncology Group study-JCOG9204. *J Clin Oncol* 2003;21:4592-96.

[10] Aroori S, Parshad R, Kapoor A, Gupta SD, Kumar A. Neoadjuvant chemotherapy in squamous cell carcinoma of the esophagus using low dose continuous infusion 5-fluoracil and cisplatin: results of a prospective study. *Indian J Cancer* 2004;4:3-7.

[11] Pettit M. Gastroesophageal reflux disease: clinical features. *Pharm World Sci* 200; 27:432-435.

[12] Surendra S, Odze R. Multinucleated epithelial giant cell changes in esophagitis: A clinicopathologic study of 14 cases. *Am J Surg Pathol* 1998;22:93-99.

[13] Lewin KJ, Appelman HD. Tumors of the Esophagus and Stomach. Washington, DC: Armed Forces Institute of Pathology; 1995:74–75. Atlas of Tumor Pathology; 3rd series, fascicle18.

[14] Chandrasoma P. *Gastrointestinal Pathology.* Stanford, Com: Appleton and Lange; 1999:50–52.

[15] Arista- Nasr J, Rivera I, Martínez Benitez B, Bornstein L, Orozco H, Lugo Guevara Y. Atypical regenerative hyperplasia of the esophagus in endoscopic biopsy: a mimicker of squamous esophagic carcinoma. *Arch Pathol Lab Med.* 2005;129:899-904.

[16] Rosai J. Rosai and Ackerman's Surgical Pathology. 9th ed. St Louis,Mo:The CV Mosby Co;2004:619–20.

[17] Sheahan DG, West AB. Focal lymphoid hyperplasia (pseudolymphoma) of the esophagus. *Am J Surg Pathol* 1985;9:141-47

[18] Hosaka S, Nakamura N, Akamatsu T. et al. A case of primary low grade mucosa associated lymphoid tissue (MALT) lymphoma of the esophagus. *Gut* 2002; 51: 281-4

[19] Wotherspoon AC. Extragastric MALT lymphoma. *Gut* 2002;51:148–149.

[20] Misdraji J Lauwers JY. Gastric epithelial dysplasia. *Semin Diagn Pathol* 2002; 19:20-30.

[21] Ming SC. The importance of follow-up studies in gastric dysplasia: the pathologist's view. *Endoscopy.* 1993 May;25:294-5.

[22] Marsman WA, Tytgat GN, ten Kate FJ, van Lanschot JJ. Differences and similarities of adenocarcinomas of the esophagus and esophagogastric junction. *J Surg Oncol* 2005;92:160-8.

[23] Ryan AM. Adenocarcinoma of the oesophagus and gastric cardia: Male preponderance in association with obesity. *Eur J Cancer.* 2006 Apr 18; [Epub ahead of print]

[24] Lagergren J. Controversies surrounding body mass, reflux, and risk of oesophageal adenocarcinoma. *Lancet Oncol* 2006;7:347-49.

[25] Lagergren J, Bergstrom R. The role of tobacco, snuff and alcohol use in the aetiology of cancer of the oesophagus and gastric cardia. Int J Cancer 2000 Feb 1;85:340-6.

[26] Goldblum JR Barrett's esophagus and Barrett's-related dysplasia. *Mod Pathol* 2003;16:316-24.

[27] Kawakubo H, Ozawa S, Ando N, Kitagawa Y, Mukai M, Ueda M, Kitajima M. Alterations of p53, cyclin D1 and pRB expression in the carcinogenesis of esophageal squamous cell carcinoma. *Oncol Rep* 2005;14:1453-9.

[28] Yamabuki T, Daigo Y, Kato T, Hayama S, Tsunoda T, Miyamoto M, Ito T, Fujita M, Hosokawa M, Kondo S, Nakamura Y. Genome-wide gene expression profile analysis of esophageal squamous cell carcinomas. *Int J Oncol* 2006;28:1375-84.

[29] von Brevern M, Hollstein MC, Risk JM, Garde J, Bennett WP, Harris CC, Muehlbauer KR, Field JK. Loss of heterozygosity in sporadic oesophageal tumors in the tylosis oesophageal cancer (TOC) gene region of chromosome 17q. *Oncogene* 1998;17:2101-5.

[30] Zarovnaya E, Black C. Distinguishing pseudoepitheliomatous hyperplasia from squamous cell carcinoma in mucosal biopsy specimens from the head and neck. *Arch Pathol Lab Med* 2005;129:1032-6.

[31] Shima I, Sasaguri Y, Kusukawa J, Yamana H, Fujita H, Kakegawa T, Morimatsu M. Production of matrix metalloproteinase-2 and metalloproteinase-3 related to malignant behavior of esophageal carcinoma. A clinicopathologic study. *Cancer.* 1992;70:2747-53.

[32] Nair KS, Naidoo R, Chetty R. Expression of cell adhesion molecules in oesophageal carcinoma and its prognostic value. *J Clin Pathol* 2005 ;58:343-51.

[33] Koppert LB, Wijnhoven BP, van Dekken H, Tilanus HW, Dinjens WN.The molecular biology of esophageal adenocarcinoma. *J Surg Oncol* 2005;92:169-90.

In: Esophageal Cancer Research Developments
Editor: Henry K. Kinner, pp. 171-206
ISBN: 1-60021-309-X

Chapter VI

Genomic Instability of Esophageal Cancer

Kazuhiro Ishikawa and Hideshi Ishii**
Center for Molecular Medicine, Jichi Medical University, Tochigi, Japan

Abstract

Esophageal cancer is the major common cancer of the digestive tract, and recent evidence shows that its incidence is increasing. Esophageal cancers are highly lethal neoplasms, with an overall 5-year survival rate of less than 10%. Recent studies have revealed that alterations of specific chromosome loci, leading to gains of oncogenes, or losses of tumor suppressor genes, including p53 and p16, as well as epigenetic alterations, are early events in the carcinogenesis. Those alterations are associated with loss of cell cycle checkpoints, and evade apoptosis by down- or up-regulation of pro-apoptotic or anti-apoptotic proteins. Ongoing genomic instability leads to cumulative genetic errors and thereby the generation of multiple transformed clones, which are generally refractory to conventional multidisciplinary interventions. In this article, we have noted recent studies of the molecular biology of esophageal cancer, which may characterize its genomic instability in the carcinogenesis scenario.

Key words: esophageal cancer, genomic instability, tumor suppressor genes, oncogenes

* Correspondence to Hideshi Ishii M.D.; Center for Molecular Medicine, Jichi Medical University, Tochigi 329-0498, Japan. FAX: 81-285-44-7501; TEL: 81-285-58-7400; E-mail: hishii@ms.jichi.ac.jp

(1) Introduction

According to the National Cancer Institute (NCI) (http://www.cancer.gov/cancer_information/cancer_type/esophageal), esophageal cancer is the third most common cancer of the digestive tract and the seventh leading cause of cancer-related deaths worldwide. Incidence of esophageal cancer varies considerably according to geographic location. It is more common in northern China, northern Iran, and the southern republics of the former Soviet Union, and is relatively less common in Japan, Great Britain, Europe and Canada.

The two most common types of esophageal cancer are named for the type of cells that become malignant [1]: squamous cell carcinoma and adenocarcinoma. Squamous cell carcinoma begins in squamous cells, which are the thin, flat cells lining the esophagus. Approximately 60% of squamous cell carcinomas develop in the middle third of the esophagus, 30% occur in the lower third, and 10% occur in the upper third, but this can occur anywhere along the esophagus. Squamous cell carcinoma is associated with cigarette smoking and high intake of alcohol. In contrast, adenocarcinoma begins in glandular secretory cells, which produce and release mucus in the lining of the esophagus and is associated with a condition called Barrett's esophagus. Adenocarcinomas usually form in the lower third of the esophagus, near the stomach. According to the NCI, incidence of adenocarcinoma of the esophagus, which is associated with Barrett's esophagus, is rising in the United States.

The esophageal organ is exposed to environmental carcinogens, which act as potential genotoxins. Esophageal cancer is not an exception in the notion that the tumor progression is generally characterized by the accumulation of multiple genetic alterations including loss of tumor suppressor gene function [1, 2]. Identification of the genomic alterations involved in initiation and progression of premalignant and malignant conditions will help address many questions concerning the mechanisms of tumor progression and facilitate the discovery of diagnostic and prognostic markers and potential therapeutic targets [1-3].

(2) Genomic Alterations

In two decades, considerable evidence has accumulated suggesting that cancer has a genetic origin, based on the development of genomic alterations, such as chromosome deletions, translocations, DNA mutations, and/or methylation of DNA or chromatin in critical genes or genomic regions for homeostasis of cellular functions, including cell cycle control, cell survival, and DNA replication. Genomic alterations that inactivate suppressor genes or activate oncogenes could relieve cells from growth constraints, allowing tumor growth [4-6].

(2-1) Loss of Genomic Regions

By measuring the proportion of one allele relative to the other in cells from individuals that are constitutionally heterozygous at a given locus, the balance of copy numbers of specific loci of chromosome or genomic regions can be assessed. Allelic imbalance can arise from the complete loss of an allele or from an increase in copy number of one allele relative to the other. Allelic imbalance includes loss of heterozygosity (LOH). It has been shown that the classic mechanism of tumor suppressor gene inactivation is described by the two-hit model in which one allele is mutated and the other allele is lost through a number of possible mechanisms, resulting in loss of heterozygosity (LOH) at multiple loci [6-9].

A variety of genetic mechanisms are involved in the occurrence of LOH in carcinogenesis or tumor development, including physical deletion, chromosome nondisjunction, and mitotic nondisjunction; those processes can be followed by reduplication of the remaining chromosome, mitotic recombination and gene conversion. Although LOH involves complete loss of one of the two alleles at a locus, it has been shown that normal cell contamination can confound the diagnosis of true LOH and other mechanisms of allelic imbalance; however, studies using flow-cytometrically purified or micro-dissected samples have shown that complete LOH can be clearly detected in tissue samples, including esophageal cancer [3, 10-13]. Previous studies have shown that tumor progression is often associated with the accumulation of multiple genetic changes as the tumor progresses to advanced stages [3, 14-23]. In this article, esophageal cancer is discussed below.

(2-1-1) Squamous cell carcinoma

Previous reports have shown that alterations of various regions of chromosomes, which can lead to inactivation of tumor suppressor genes, activation of oncogenes and development of genomic instability. The following alterations are reported.

Chromosome 1

The alterations of regions on 1q21 [24]; a common polymorphism of p73, a novel p53 homologue, associated with a reduced incidence of esophageal carcinoma, which genetically maps to chromosome 1p36, a region commonly deleted in esophageal cancers [25].

Chromosome 2

A high frequency of LOH at D2S123 on 2p16 near MSH2 gene [26]; high frequency LOH on 2q [27].

Chromosome 3

The ribosomal protein L14 gene (RPL14) on chromosome 3p21.3 (43% LOH and no mutations) [28] ; allele loss and promoter hypermethylation of von Hippel-Lindau disease (VHL), retinoic acid receptor beta (RAR-beta), RAS association domain family 1A (RASSF1A), and fragile histidine triad (FHIT) genes [29]; a high frequency of LOH at D3S1611 on 3p22 near MLH1 gene [26]; LOH (80%) at the locus of the tumor suppressor gene, the fragile histidine triad (FHIT) gene, which higher than LOH in esophageal adenocarcinomas (44%) [30]; high frequency of LOH at 3p25 locus [31]; LOH on 3p21.3 and 9q31as early genetic events during esophageal carcinogenesis; LOH on 9p22 and 17p13, which probably plays roles in progression [32].

Chromosome 4

High frequency LOH on 4p and 4q [27].

Chromosome 5

LOH at approximately 272 kb regions located on 5q23 from D5S1384 to D5S1505 [33]; LOH at APC and MCC loci around D5S82 in human papillomavirus (HPV)-positive samples [34]; LOH at 5p [35]; LOH on 5q (around adenomatous polyposis coli (APC) locus, on 5q21) [36]; LOH at 5q [37] (the study suggested that a gene other than APC is involved in esophageal tumorigenesis, although allelic loss at the APC locus is frequent in squamous cell carcinomas of the esophagus).

Chromosome 6

LOH at p21/Waf-1and HLA loci on D6S497 in Human papillomavirus (HPV)-positive samples [34]; high frequency of LOH nearby p21WAF1 locus [38]; HLA class I gene expression, frequently down-regulated at both the protein and mRNA levels presumably through the mechanism involving hypermethylation of the promoter regions of the HLA-A, -B and -C genes [38];high frequency LOH on 6q [27].

Chromosome 7

LOH (22%) at 7q31-q35 (excluding suppressor of tumorigenicity 7 (ST7) as the target gene of this somatic event) [39].

Chromosome 8

High frequency LOH on 8p [27]; high frequency LOH on 8p22 [40].

Chromosome 9

Previous studies have shown that loss of heterozygosity (LOH) is common on chromosome arm 9p (especially 9p21-p22) in esophageal squamous cell carcinoma and that genetic alterations in CDKN2A and CDKN2B on 9p are also common.

The alterations include: allelic loss of chromosome 9 [41]; absent or low expression of a 9q32 candidate tumor suppressor gene (TSG), DEC1 (deleted in esophageal cancer 1) on a 2.4 Mb 9q33-q34 [42]; CDKN2A mutations including point mutations, deletions, insertions were observed in 25% [43]; intragenic allelic loss in CDKN2A [43]; LOH on 9p23-p22, 9q13-q22.3, and 9q34 (40-80%) [44]; p16 silencing and homozygous deletion (HD) as a later event [45]; a relatively low rate of p16 mutation compared with the frequency of LOH on chromosome 9 suggesting the possible involvement of another tumour suppressor gene located on chromosome 9 in esophageal carcinogenesis [46]; an association of microsatellite instability (MSI) such as DNA replication errors (RER+) with concomitant loss of p16 and pRb (p16-/pRb- phenotype) [47]; the high frequency of LOH and the reduction of the expression (~50%) but no mutations detected in DEC1 (deleted in esophageal cancer 1) gene on 9q32 [48]; aberrant methylation of p16INK4a and deletion of p15INK4b as frequent events, showing that the alteration mode at 9p21 was not uniform, and the two genes were inactivated by distinct mechanisms [49]; LOH at 9p in adenocarcinoma and squamous cell carcinoma of esophagus [50]; LOH on 3p21.3 and 9q31as early genetic events, and LOH on 9p22 and 17p13 probably play roles in progression [32].

Chromosome 10

Allelic loss on 10p14-p15 [51].

Chromosome 11

High frequencies (51.9-61.1%) of allelic loss observed at D11S1338, D11S2000, D11S1990, and D11S1647 loci, in two chromosomal regions, 11p15.5 and 11q22.3 [52].

Chromosome 13

The 6.36 Mb region on 13q33.1-q34 (>50%) [53]; the deletions on chromosome 13 (related to family history) [54]; high frequencies of allelic loss [41]; LOH at BRCA2 locus around D13S260in Human papillomavirus (HPV)-positive samples [34]; the highest frequency of LOH (95%) on 13q a familial esophageal cancer [55]; allelic loss (73%) on chromosome 13q14 and mutations (5%) in deleted in cancer 1 (DICE1) gene in esophageal squamous cell carcinoma [56]; the LOH on13q12.3 and 13q14.1-q14.3 [57]; the High frequency of LOH at 13q [58]; the allelic loss on chromosome 13q12-13 associated with lymph node metastasis (however, inactivation of the RB1 gene might not be responsible for lymph node metastasis), suggesting a possible involvement of unidentified tumor suppressor gene(s) in this region [59]; LOH on the Rb locus correlated with pRb protein expression, which associates with p53 alteration suggesting that defects in the Rb and p53 pathways and their potential synergistic effect are major mechanisms for esophageal carcinogenesis [60]; LOH on 13q (around RB) [36].

Chromosome 14

High frequency LOH on 14q [27].

Chromosome 15

High frequency LOH on 15q [27].

Chromosome 16

LOH and internal tandem duplication mutations of the CBP gene on the chromosome 16p13.3, encoding cyclic AMP response element binding protein binding protein, a nuclear transcriptional coactivator protein, which is an important component of the cAMP signal transduction pathway [61]; LOH (39%) at the WWOX (WW domain containing oxidoreductase) locus , a tumor suppressor gene at 16q23.3-24.1 [62].

chromosome 17

Allelic loss at the TP53 gene locus [41]; the Envoplakin (EVPL) on chromosome 17q25 (LOH and a mutation) [63]; high frequency of allelic imbalance and LOH (83.8%), as well as p53 gene mutation (28.9%) [26]; the high frequency of LOH (>60%) in sporadic esophageal cancer in the region at 17q, in which tylosis-associated cancer susceptibility gene (TOC gene, tylosis oesophageal cancer gene) is mapped (tylosis is the inherited cancer predisposition syndrome with focal nonepidermolytic palmoplantar keratoderma associated with a high risk of esophageal cancer development) [64]; LOH (>70%) in sporadic esophageal squamous cancer at TOC locus on 17q25, suggesting that tumor-suppressor genes found in inherited cancer predisposition syndromes are also responsible for sporadic cancers of this type [65]; LOH on 17p13 [32]; LOH on 17p (around p53) [36].

Chromosome 18

LOH (54%) on 18p11.2 and LOH (42%) on 18q12.2 [66]; the frequent LOH (72%) at D18S978 on 18q12.2, which coincides with a known fragile site FRA18A, the region is 4 cM proximal to known tumor suppressor genes suggesting the possible existence of a yet undiscovered tumor suppressor gene [67]; LOH on 18q around DCC (detected in colon cancer tumor suppressor gene) locus on 18q21.1 [36].

Chromosome 21

Frequent loss (~60%) of copy number on the long arm of chromosome 21 [68].

(2-1-2) Adenocarcinoma

The study of adenocarcinoma of the esophagus and gastroesophageal junction incicated that loss of heterozygosity can occur in histologically normal epithelium in the esophagus and stomach [69]. Previous report also have shown that alterations of various regions of chromosomes, which can lead to inactivation of tumor suppressor genes, activation of oncogenes and developemment of genomic instability, similarly to squamous cell carcinoma. The following alterations are reported.

Chromosome 1

A common polymorphism of p73, a novel p53 homologue, associated with a reduced incidence of esophageal carcinoma, which genetically maps to chromosome 1p36, a region commonly deleted in oesophageal cancers, suggesting Clinical implications in aiding identification of high-risk Barrett's esophagus patients [25].

Chromosome 3

High frequency of LOH 3p24.2-26 loss in (~65%) (no mutations in two candidate tumor-suppressor genes von Hippel-Lindau (VHL) at 3p25-26and peroxisome proliferator-activated receptor gamma (PPAR) gamma at 3p24.2-25excluding possible involvements) in adenocarcinomas of the gastroesophageal junction [70]; LOH at 3p (64% of tumours) [71]; frequent deletions of FHIT and FRA3B in Barrett's metaplasia and esophageal adenocarcinomas [72].

Chromosome 4

The high rate of LOH for 4q in esophageal adenocarcinoma [73].

Chromosome 5

LOH at 5q (45%) [71].

Chromosome 7

LOH (32%) at 7q31-q35 (excluding suppressor of tumorigenicity 7 (ST7) as the target gene of this somatic event) [39].

Chromosome 9

The INK4a-ARF (CDKN2A)- locus on chromosome 9p21 encodes for two tumour suppressor proteins, p16INK4a and p14ARF, which act as upstream regulators of the Rb-Cdk4 and p53 pathways.

The alterations include: 9p21 LOH, p16INK4a methylation and p16INK4a mutations in Barrett's adenocarcinomas in 5 of 15 (33%), 8 of 15 (53%) and 1 of 15 (7%) patients, respectively [74]; p14ARF methylated in 3 of 15 (20%) adenocarcinomas [74]; LOH at 9p (52%) [71]; hypermethylation of p16INK4a promoter in esophageal adenocarcinomas [75]; homozygous deletion of the p16 gene [76]; LOH at 9p in adenocarcinoma and squamous cell carcinoma of esophagus [50].

Chromosome 11

LOH at 11p (61%) [71]; LOH at 11p15 (>50%) [77].

Chromosome 13

LOH at 13q [71].

Chromosome 17

LOH at 17p (96%) and 17q (55%) [71]; LOH (>50%) at 17q [78].

Chromosome 18

LOH at 18q (70%) [71].

(2-1-3) Barrett's Esophagus and Cancer Development

The investigation of the evolutionary relationships of non-random LOH, TP53 and CDKN2A mutations, CDKN2A CpG-island methylation and ploidy during neoplastic progression to study evolution of neoplastic cell lineages in Barrett's esophagus indicated that diploid cell progenitors with somatic genetic or epigenetic abnormalities in TP53 and CDKN2A were capable of clonal expansion, spreading to large regions of esophageal mucosa. The study suggests that clonal evolution is more complex than predicted by linear models [79].

Another study indicated that the tumour suppressor genes TP53, p16INK4a, and APC are often deleted in adenocarcinomas derived from Barrett's esophagus; mutations at these genes are also found in the adenocarcinomas, including the homozygous deletion of the p16 gene; however, the absence of genetic alterations in the Barrett's oesophagus and the low grade dysplastic epithelia, suggesting that mutations at these genes develop later in the progression from Barrett's oesophagus to adenocarcinoma [76]. Numerous genomic alterations in Barrett's esophagus are reported, including:

Chromosome 3

Allelic deletions on 3p21 as potential markers of individuals with a high risk of developing adenocarcinoma in Barrett's epithelium without dysplasia [80]; frequent deletions of FHIT and FRA3B in Barrett's metaplasia and esophageal adenocarcinomas, suggesting that the fragility and recombination-prone nature of FRA3B is related to tumor-specific chromosomal instability affecting the FHIT gene in esophageal adenocarcinoma development [72].

Chromosome 5

Previous report have shown that alterations of the APC gene on 5q (allelic loss in 27%), which are after ablative therapy [81]; allelic deletions at APC locus on 5q21 as potential markers of individuals with a high risk of developing adenocarcinoma in Barrett's epithelium without dysplasia [80]; LOH at the sites of the DCC (detected in colon cancer tumor suppressor gene on 18q21.1), APC (adenomatous polyposis coli on 5q21), and TP53 (on 17p13.1), which occur before the development of adenocarcinoma in Barrett's esophagus [82].

Chromosome 9

p16INK4a on 9p (allelic loss in 56%), which are after ablative therapy [81]; p16INK4a alterations in 12 of 20 (60%) high-grade and in 4 of 20 (20%) low-grade intraepithelial neoplasms of Barrett's esophagus [74]; methylation of p16INK4a (7%), and intact p14ARF in Barrett's mucosa without intraepithelial neoplasia [74]; allelic deletions at p16INK4a locus on 9p21 as potential markers of individuals with a high risk of developing adenocarcinoma in Barrett's epithelium without dysplasia [80]; LOH at 9p [83]; p16INK4a inactivation by methylation of the CDKN2A promoter, which occurs early during neoplastic progression in Barrett's esophagus [84]; LOH at 9p and 17p chromosomes as highly prevalent somatic genetic lesions in premalignant Barrett's tissue, in which LOH on 9p is more common than LOH on 17p in diploid samples, suggesting that LOH on 9p and 17p may be useful biomarkers to stratify patients' risk of progression to esophageal cancer [85].

Chromosome 17

Allelic loss in P53 gene locus on 17p (31%), which are after ablative therapy [81];a certain implication of 17p (p53) LOH as a molecular biomarkers (although authors do not recommend its routine clinical use at the present time) [86]; allelic deletions at TP53 on 17p13.1, as potential markers of individuals with a high risk of developing adenocarcinoma in Barrett's epithelium without dysplasia [80]; LOH at 17p [83];17p (p53) LOH as a possible predictor of progression in Barrett's esophagus that can be combined with a panel of other validated biomarkers [87]; LOH at the sites of the DCC (detected in colon cancer tumor suppressor gene on 18q21.1), APC (adenomatous polyposis coli on 5q21), and TP53 (on 17p13.1), which occur before the development of adenocarcinoma in Barrett's esophagus [82]; LOH on 17p in Barrett's adenocarcinoma [88]; frequent mutation of p53 in Barrett's metaplasia of the intestinal type [89]; frequent loss of heterozygosity on 17q11.2-q12 in Barrett's adenocarcinoma [90].

Chromosome 18

Previous report have shown that alterations of DCC and SMAD4 on 18q (allelic loss in 6%), which are after ablative therapy [81]; LOH at the sites of the DCC (on 18q21.1), APC (5q21), and TP53 (17p13.1), which occur before the development of adenocarcinoma in Barrett's esophagus [82].

(2-1-4) dysplasia or Precancerous Lesions of Squamous Cell Carcinoma

Study of dysplasia samples suggests that LOH may be associated with early stages of tumor initiation and progression and an involvement of considerable genetic heterogeneity in addition to the association between disease progression and these genetic changes [91]. The alterations include:

Chromosome 3

Alterations of the ribosomal protein L14 gene (RPL14) on chromosome 3p21.3 (>40% allelic loss or microsatellite instability) [28]; LOH at marker D3S1766 (3p) [91].

Chromosome 4

LOH at marker D4S2632 (4p) [91].

Chromosome 9

Allelic loss [41]; LOH at marker D9S910 (9q) [91].

Chromosome 10

Allelic loss on 10p14-p15 in esophageal dysplasia [51].

Chromosome 13

Allelic loss [41]; LOH at marker D13S1493 (13q) [91].

Chromosome 17

Allelic loss at TP53 gene locus [41].

(2-2) Gains or Amplification of Genomic Regions

(2-2-1) Squamous Cell Carcinoma

Gene amplification in esophageal squamous cell carcinoma was reported in EGFR [92-95], CCND1 [92, 93, 96-104], ZASC1 (encoding a Kruppel-like zinc finger protein) [105], FGF3/FGF4 [93, 97], EMS1 [93, 97], SAS [97], ERBB2 [93, 97], PDGFRA [97], MYC [93, 97, 106], BCL2 [97], TP63 (amplified in early stage of esophageal squamous cell carcinoma but down-regulated in advanced stage of disease [107], MYEOV (coamplified with CCND1, but epigenetically inactivated [98]), MDM2 [93, 108, 109], FGFR1 [93], cIAP1 [110], SNO [111], ATF3 [112], CENPF [112], GASC1 [113], and DPI gene [114].

(2-2-2) Adenocarcinoma

The previous studied have shown that the relatively high frequency of gene amplification in esophageal adenocarcinomas and high-grade dysplasia indicates the important role of these events in esophageal adenocarcinoma development [115].

Gene amplification in esophageal adenocarcinomas and Barrett's with high-grade dysplasia was reported in the ERBB2 gene [115-120], HER-2/neu [121], GRB7 [118],

CCNE1 [115], GATA4 [115], KRA [115], EGFR [115, 117, 120], CCND1 [115], CCNE [122], HNF3alpha [115, 123], PIK3CA [115], C-MYC [115], DYRK2 [115], DYRK2 (dual-specificity tyrosine-(Y)-phosphorylation regulated kinase 2) [124], CTSB (cathepsin B) [125]. AIB1, AKT1, and IGF1R, amplified in 0 of 87 (0%) of the tumors [115]; CCND1 amplification, correlated negatively with survival ($P < 0.05$) [115]; the ERBB2 amplicon, positively correlated ($P < 0.05$) with GATA4 amplification [115].

(3) Epigenetic Alterations; Hypermethylation of Gene Promoter

(3-1) p16 (MTS-1/CDKN2/INK4A)

Numerous studies have shown that hypermethylation of the p16 (MTS-1/CDKN2/INK4A) tumor suppressor gene is a common event in both esophageal adenocarcinoma and in preinvasive Barrett's esophagus. For examples, p16INK4a was methylated in 0/10 (0%) normal tissues, 4/12 (33%) Barrett's metaplasias, 3/11 (27%) low-grade dysplasias, 3/10 (30%) high-grade dysplasias, and 18/22 (82%) adenocarcinomas [84]; methylated in 27/41 (66%) Barrett's metaplasias, 21/45 (47%) indefinite or low-grade dysplasias, and 17/21 (81%) high-grade dysplasias [126]; methylated in 2/67 (3%) normal-appearing tissues from patients with Barrett's metaplasia, 3/5 (60%) tissues indefinite for dysplasia, 10/18 (56%) low-grade dysplasias, and 3/4 (75%) high-grade dysplasias; and methylated in 9/10 (90%) patients with dysplasia at some time during surveillance and 0/42 (0%) patients without dysplasia during surveillance [127].

Methylation is detected in metaplasias and is similar in all grades of dysplasia or even in normal-appearing tissues from patients with Barrett's metaplasia, suggesting that this epigenetic alteration occurs very early in tumor progression [75, 84, 126, 127]. A recent study using bioinformatic algorithms to identify genetic and/or epigenetic lesions that provide a selective advantage shows that hypermethylation of the p16INK4a promoter provided a clear and strong advantageous effect on cells early in the progression of Barrett's esophagus [128, 129]. This study showed that p16 loss of heterozygosity, promoter methylation, and sequence mutations have strong, independent, advantageous effects on Barrett's cells early in progression; second lesions in p16 and p53 are associated with later selective sweeps [128].

(3-2) APC

Previous studies of both major histologic subtypes of esophageal cancer, adenocarcinoma and squamous cell carcinoma have shown high rates of LOH at chromosome 5q21-22, the locus containing the APC gene [10, 79, 130-133]. The inactivating point mutations, however, were rarely detected within the APC gene, suggesting that the APC gene unlikely follow the classic tumor suppressor gene paradigm in the esophagus [76, 133-135], and that the APC gene might be inactivated by an alternative mechanism or a neighboring gene on chromosome 5q was the target of LOH in esophageal cancers [134, 135]. In this context, studies reported that methylation of the promoter region of the APC gene is strongly associated with silencing of its expression in gastric [136] and esophageal cancer [137-140].

An intensive study report that APC methylated in 0/20 (0%) normal tissues, 17/43 (40%) Barrett's metaplasias, 48/52 (92%) adenocarcinomas, and 16/32 (50%) squamous cell carcinomas [137]. Interestingly, the examination of indicates that monoallelic methylation of the APC promoter is altered in normal gastric mucosa associated with neoplastic lesions [139]; the APC methylation pattern is frequently altered in normal-appearing gastric mucosa of gastric or esophageal adenocarcinoma patients and such alterations also occur in normal gastric mucosa adjacent to intestinal metaplasia [139].

(3-3) Other Genes in the Genome

An intensive study of the methylation status of 20 genes in different stages of Barrett's esophagus and/or associated adenocarcinoma identified distinct "epigenetic fingerprints" in different histologic stages of the disease [141]. Some genes were minimally informative because the frequency of hypermethylation was less than 5% (p14ARF, CDH1, p15INK4b, GSTP1, hMLH1, COX2, and THBS1), completely absent (CTNNB1, RB1, TGFBR2, and TYMS1), or ubiquitous (HIC1 and MTHFR), regardless of tissue type [129, 141]. Genes with an intermediate frequency of hypermethylation were more informative, with methylation ranging from 15% (p16INK4a) to 60% (MGMT) of samples [129, 141].

Previous studies have shown genomic alterations at the tumor suppressor *FHIT* locus such as deletions and translocations and loss of Fhit protein expression in a number of premalignant and malignant lesions of lung, cervix, esophagus, breast, and other organs [142-146], representing progress toward a better understanding of the mechanism of gene inactivation. While point mutations within the FHIT gene are rare [147-151], deletions are common [152, 153], and to a lesser extent, silencing by methylation is involved in esophageal cancer [154].

Aberrant methylation can occur at numerous loci in the same tissue of esophagus [129]. This suggests a generalized deregulation of methylation control, although a discrete group of tumors with the CpG island methylator phenotype was not identified [129]. Nevertheless it is worth while to note that normal and metaplastic tissues as an early stage from patients with associated dysplasia or cancer had a higher incidence of hypermethylation than similar tissues from patients without further evidence of progression, suggesting that hypermethylation may identify esophageal lesions at risk for progression [129], and that hypermethylation might be involved in an early stage of carcinogenesis.

(4) Checkpoints Alterations

(4-1) Checkpoints

Exogenous genotoxin or endogenous DNA damaging stress mainly causes G1/S arrest or reduces S phase progression to prevent replication of damaged DNA or G2/M arrest to prevent segregation of damaged chromosomes during mitosis or S-phase checkpoints activated by genotoxic insults [155-157]. Atm (ataxia-telangiectasia mutated) and Atr (Atm- and Rad3-related) play an important role in an initiation of the signaling [156].

(4-1-1) G1/S Checkpoint

The checkpoint response to genotoxic insults traversing through G1 depends on the Atm(Atr)/Chk2(Chk1)/p53/Mdm2/p21Waf1 pathway, which is capable of inducing sustained, and sometimes even permanent G1 arrest [156-158]. Atm/Atr directly phosphorylate the p53 protein within its amino-terminal transactivation domain, particularly on serine 15; threonin 18, serine 20 and probably some additional p53 sequence(s) could be targeted by Chk1/Chk2 [157]. The ubiquitin ligase Mdm2 is involved in rapid p53 turnover, which is targeted after DNA damage by Atm/Atr, as well as by Chk2/Chk1 [157]. The modifications of p53 and Mdm2 play a role in regulation of the stabilization and accumulation of the p53 protein, and then its transcription activity for target genes, including p21Waf1, an inhibitor of cyclin-dependent kinases [159], which is involved in silencing the G1/S-promoting cyclin/Cdk complexes (cyclin E/Cdk2 kinase) and G1 arrest. The signal inhibits the Rb/E2f pathway and prevents G1/S promotion. At G1/S checkpoint, Atm(Atr)/Chk2(Chk1) plays a critical role in regulation of p53/Mdm2/p21Waf1-dependent checkpoint response. The G1 checkpoint response targets two critical tumour suppressor pathways, p53 and pRb, which are governed commonly deregulated in human cancer [157, 159, 160].

The Atr/Chk1, but not Atm/Chk2, maintains an appropriate abundance of Cdc25A through its ubiquitin-dependent, proteasome-mediated turnover during unperturbed proliferation through moderate constitutive phosphorylation of Cdc25A on its several serine residues [157, 158, 161]. The basal turnover of Cdc25A is estimated as T1/2 ~20-30 min [161]. In response to genotoxic stress, increased activity of Chk1 and Chk2, leads to downregulation of Cdc25A and consequently to the inhibition of cyclin E(A)/Cdk2 complexes [157, 158, 161]. The accelerated turnover of Cdc25A is T1/2 ~10 min [161]. Notably, the Chk1/Chk2/Cdc25A checkpoint is implemented rapidly, independent of p53, which allows immediate G1/S arrest or delay, without the sustained p53-dependent G1 arrest [157].

One common feature of multiple functions of the tumor suppressors, p53 and pRb, is their ability to target the activity of the E2F family of transcription factors [162]. E2F-1 has unique and somewhat paradoxical activities. Forced E2F-1 expression induces quiescent cells to enter S phase, whose activity correlates with the ability of E2F-1 to activate transcription [163], and the deregulated E2F-1 expression can lead to oncogenic consequences [164]. It was shown, however, that the extopic expression of E2F-1 induces apoptosis in several cell types [165, 166], and the studies of E2F-1 deficient mice suggest that E2F-1 functions as a tumor suppressor [167, 168], indicating that E2F-1 plays roles not only in promoting cell cycle progression, but also in stimulating programmed cell death. Like the other members of the E2F family, the activity of E2F-1 is executed by regulating gene expression. The implication of E2F-1 binding has been shown in gene promoters of several regulators of cell cycle, such as Cdc2 [169], Cyclin E1 [170], Cyclin E2 [171], Cyclin A [172], and E2F-1 itself [173], but also of proapoptotic molecules, including Afaf-1 [174], Caspases [175] and

Fhit [176]. Recent report showed that increased E2F-1 expression via tumour cell proliferation and decreased apoptosis are correlated with adverse prognosis in patients with squamous cell carcinoma of the esophagus [177], suggesting that, although E2F-1 remains a controversial prognostic factor, its expression was closely associated with tumor cell proliferation and might influence clinical outcome, mainly via cell cycle progression [177].

(4-1-2) S-phase Checkpoints

In response to genotoxic stress in S phase of cell cycle, which can be caused by environmental or endogenous genotoxic insults such as ionizing or ultraviolet radiartion, various chemicals and drugs, and reactive cellular metabolites, S-phase checkpoints can be activated, which arrests or reduces cell cycle progression so as to allow for repair and prevention of the transmission of damaged or incompletely replicated chromosomes [156]. The genotoxic stress can result not only from DNA-damaging insults, but also from difficulties with the replication process it self. The creation of the duplicated sister chromatids offers more opportunities for error-free repair by homologous recombination, whereas the double strand breaks can be treated via non-homologous end-joining repair [161]. The vast of amounts of data that have been obtained from yeast models now clearly indicated that the S-phase checkpoints are considerably more significant for preventing genomic instability than the G1 or G2 checkpoint or the mitotic-spindle checkpoint [161].

The processes of S phase checkpoints are classified into three types: (1) the replication-dependent intra-S-phase checkpoint (commonly referred as to the replication checkpoint); (2) the replication-independent intra-S-phase checkpoint (generally referred as to the intra-S-phase checkpoint), which can be induced by double strand breaks-induced; (3) the S-M checkpoint, which, like replication checkpoint, depends on DNA replication [161]. They are mutually closely coordinated and share some components. The molecular effectors remain to be more fully defined. Importantly, none of these three S-phase checkpoints requires p53, which is the key target of the G1/S checkpoint [161].

(4-1-3) G2/M Checkpoint

The G2/M checkpoint prevents cells from initiating mitosis when some DNA damages are unrepaired [157]. The G2/M arrest is largely dependent on Chk1-mediated signaling pathway leading to inhibition of the Cyclin B1/Cdc2 activity. Chk1 is activated by phosphorylation on at least two residues (Ser345 and Ser317) located in a Ser/Thr-Gln-rich domain by the DNA damage-activated Atr and Atm kinases [155-158, 161]. When the G2/M checkpoint is activated, Atm/Atr, Chk1/Chk2 and or p38-kinase lead to subcellular sequestration, degradation or inhibition of the Cdc25 family of phosphatases, which normally activate Cdk1 at the G2/M boundary [155, 157].

(4-2) Alterations of Checkpoints in Esophageal Cancer

Generally, in normal cells, the p21 cyclin-dependent kinase inhibitor arrests the cell cycle following DNA damage at the G1-S checkpoint. Recent studies suggest a role of p21 in G2 arrest after DNA damage in esophageal cancer cells [178, 179]. Studies with esophageal squamous cell carcinoma lines showed that antisense blockade of p21/Waf1 decreases radiation-induced G2 arrest in esophageal squamous cell carcinoma, suggesting a role for p21 in mediating G2 arrest in these cells. [178]. The notion is further supported by the examination showing that hypophosphorylated retinoblastoma protein is associated with G2 arrest in esophageal squamous cell carcinoma, suggesting a role for p21 in mediating G2 arrest in esophageal cancer cells [179]. Those observations suggest that normal checkpoint function at G1/S is altered in esophageal cancer cells. As described above, the G1 checkpoint response targets, p53 and pRb, are governed commonly deregulated in human cancer [157, 159, 160]. Since the G1/S checkpoint is frequently inactivated in tumor, including esophageal cancer cells, due to mutation of p53, it is suggested the G2/M checkpoint may be significant in response to cancerous cells to DNA-damaging chemotherapeutic agents [155, 156, 158].

The immunohistochemical investigation using radio-sensitive and radio-resistant esophageal cancer showed that p53 expression was negative in 71% (10 of 14) of radio-sensitive cases and positive in 91% (30 of 33) of radio-resistant cases [180]. The association was strong between high radiation sensitivity and negative p53 expression ($P < 0.0001$) [180]. Cdc25B, which is not expressed in normal epithelium but is in the cytoplasm of esophageal cancers, was strongly expressed in 46% (6 of 14) of radio-sensitive cases and in 6% (2 of 23) of radio-resistant cases, suggesting that the sensitivity for radiation therapy was significantly correlated with Cdc25B overexpression [180]. Nevertheless, p53 and Cdc25B expressions showed no significant associations, and multivariate analysis revealed that both p53 and Cdc25B are significant independent markers for predicting radiation sensitivity [180]. The study suggests that Cdc25B is a novel predictor of radiation sensitivity in esophageal cancers [180]. The importance of a p53-independent G2-M checkpoint in radiation therapy is suggested.

Similarly to Cdc25B, the usefulness of Cdc25A is suggested as a new prognostic factor for squamous cell carcinoma of the esophagus [181]. Multivariate analysis revealed the Cdc25A status to be an independent significant prognostic factor among these biological and clinico-pathological factors [181]. It is suggested that the status of the G1 checkpoint in the cell cycle may vary in esophageal cancer, and the assessment of the function of G1 checkpoint might be usefulness for the efficient therapeutic strategy. In the cases with deregulated G1 checkpoint, the targeting G2 checkpoint might be beneficial.

Recent studies also indicated the prognostic value of elevated levels of Cyclin B1 expression indicating a poor prognosis for patients with esophageal squamous cell carcinoma [182]; HSP27 expression was reduced in metaplasia and then significantly increased with progression, and HSP27 expression was higher in poorly-differentiated adenocarcinomas than in well-differentiated forms, suggesting that HSP27 transcriptional patterns may present potential as a prognostic tool to predict the aggressiveness of esophageal adenocarcinomas [183].

(5) Molecular Diagnosis at Early Stages

The elucidation of molecular mechanism at early stages of carcinogenesis is important for development of novel diagnostic and therapeutic approaches. Barrett's oesophagus is now clearly recognised as a pre-neoplasic condition, which is diagnosed endoscopically with an histological confirmation, showing in most cases incomplete intestinal metaplasia [184]. A diagnosis of dysplasia in Barrett's oesophagus has major clinical and therapeutic consequences, although numerous studies have demonstrated that it is not perfectly reproducible [184]. In addition and parallel to the morphological sequence of events leading from metaplasia to carcinoma of esophagus, chromosomal changes and accompanying genetic alterations occur, with ensuing abnormalities in gene expression and cell cycle regulation. Although the frequency and timing of these alterations are not as well established as in colorectal carcinogenesis [184], some authors have proposed a molecular cancer progression scheme of Barrett's esophagus as early stages of carcinogenesis [185-188]. Some of these changes may be used as criteria for characterizing cancerous condition of esophagus.

(5-1) DNA Content Showing Genomic Instability

Over 90% of high grade dysplasia and adenocarcinomas developed in Barrett's esophagus are DNA aneuploid [184], and there is a significant relation between the presence of DNA aneuploid population and the progression from non-dysplastic Barrett's intestinal mucosa to dysplasia and adenocarcinoma [184], suggesting that changes in DNA content or DNA aneuploidy is involved in development of cancer. Here, cells that contain any other formulation of chromosomes than 2N (diploid) and 4N (tetraploid) are said to be aneuploid. These cells can be detected by flow cytometry. The technique is based on staining single cells with quantitative fluorescent DNA dye in a suspension, which was analyzed by a detection system for assessment of an amount of fluorescence proportional to the amount of DNA. Importantly, aneuploidy does not correlate with any single mutation of the genes such as those of p53, Rb, p16, APC, Cyclin D, Cyclin E, EGF, EGFR, c-erbB2, E-cadherin, beta-catenin, and COX-2, but reflects large DNA changes due to genomic instability [184]. Although some studies have confirmed that DNA aneuploidy is a prognostic factor for malignant transformation in Barrett's esophagus as early stage disorders [189],other groups have reported frequent discordance between histology and DNA ploidy [190], which may be due to technical issues [184].

(5-2) High Rate of Proliferation

Numerous studies have shown that have two markers of cell proliferation, proliferating cell nuclear antigen (PCNA) and Ki67 (usually stained with the monoclonal antibody MIB1), are useful for assessment of cell proliferation of the esophageal regions, although most genetic changes that occur during the carcinogenesis affect genes involved in the regulation of cell cycle, with an ensuing increased proliferation, as mentioned above. Both PCNA and Ki67 can be evaluated by immunohistochemistry on routinely processed esophageal biopsies [184, 191] . PCNA is an indicator of cell cycle progression at the G1/S transition, and Ki67 is expressed in proliferating cells (G1, S, G2, and M phases). Numerous studies have shown an increased proportion of cells stained by both antibodies parallel to the progression of histological changes from metaplasia to increasing grades of dysplasia [184].

(5-3) Alterations of Tumor Related Genes

(5-3-1) p53

It has been shown that mutations of the TP53 gene are the most common event in human cancer [192]. TP53 gene mutations are occasionally found in metaplastic non-dysplastic mucosa and in low grade dysplasia, and that the frequency of mutations increases dramatically in high grade dysplasia and adenocarcinoma, reaching 80% of cases in some series, with an even higher frequency of loss of heterozygosity at the TP53 locus [184]. The allelic losses at 17p usually occur before the loss of 5q (APC locus), suggesting that suggests that TP53 mutation is a relatively early carcinogenesis event in Barrett' esophagus [184]. Recent study indicates that the evaluation of p53 over expression by immunohistochemistry, which reflects the prolonged half life of the mutant p53 protein, is useful for assessment of molecular events of early stage carcinogenesis of esophagus retrospectively [184], but also the study of p53 expression by immunohistochemistry is of interest in the surveillance of patients, especially in those patients with a mucosa indefinite for dysplasia or with low grade dysplasia [184, 193-195].

(5-3-2) HPV

An etiological role of Human papillomavirus (HPV) infections has been implicated in precancer lesions (intraepithelial neoplasia) and squamous cell carcinoma of both the genital tract and other sites (oral cavity, larynx, skin, esophagus, nasal cavity, bronchus) since more than two decades ago [196]. The study of a series of 80 esophageal cytologic specimens derived from the same number of patients with previously diagnosed squamous cell dysplasia of the esophagus in China indicates that the presence of HPV infection (HPV-DNA-positive) in a total of 53 cases (66.3%) by filter in situ hybridization (FISH), using a mixed DNA probe containing HPV types 11, 16, and 18 [197]. Also the study of 51 biopsies from esophageal lesions by in situ DNA hybridization indicated that HPV 6, 11, 16 or 18 DNA sequences were detected in 22/51 (43.1%) of the esophageal specimens [198]. Of the lesions with morphological HPV changes, 64% (16/22) were shown to contain HPV DNA [198]. High-risk types HPV 16 and/or 18 DNA sequences were found in 16 of the 22 HPV DNA-positive cases (72.7%) [198]. They support the hypothesis that HPV is a possible etiologic agent in esophageal carcinogenesis, most probably acting synergistically with physical, chemical, and/or nutritional factors that have previously been related with this malignancy in the high-risk areas of China [197, 198].

(5-3-3) FHIT and Others

A significant correlation between heavy smoking and alcohol use and FHIT inactivation has been shown in esophageal cancer [144]. The study that Fhit expression was lost in 32 of 46 (70%) esophageal carcinoma cases tested, in 8 of 12 (67%) with carcinoma in situ, and 9 of 21 (43%) with dysplasia [144]. The study demonstrated that tumors from heavy users of both tobacco and alcohol show a significantly higher frequency of loss of Fhit expression than those of light users [144], suggesting that loss of Fhit expression is an early event in the development of esophageal carcinoma and predisposing genetic changes have occurred even in normal-appearing epithelium in cases heavily exposed to environmental carcinogens [144, 199]. The other studies suggest that not only smoking but also dietary factors could be involved in esophageal carcinogenesis [200]; smoking may contribute to carcinogenesis in the digestive tract as well as respiratory organs [153]; loss of fragile histidine triad gene expression is associated with progression of esophageal squamous cell carcinoma, but not with the patient's prognosis and smoking history [201]. Studies of large panels of cancers will be needed to reconcile and clarify these varying observations.

It would be reasonable that the combination of several markers as mentioned above may be effective as a diagnostic tool at the early stages of esophageal cancer. Recent studies indicated that increased DNA content and heteroploid rate, accumulation of p53 protein, and over-expression of p21, telomerase and Cyclin D1 proteins were early molecular events during the development of esophageal cancer [202].

(6) Therapeutic Applications of Molecular Basis

(6-1) Gene Therapy

Current studies of the molecular basis of cancer provide a strong rationale to consider gene therapy approaches to cancer as a complement to other cancer therapies [203, 204]. Restoration of the function of a single or several pivotal suppressor gene products appears sufficient to mediate antitumor effects that could be clinically significant [203, 204]. Tumor suppressor genes, such as TP53 or p16 (MTS-1/CDKN2/INK4A), which encode proteins regarded as key effectors of cellular senescence and control exit from the G1 phase of the cell cycle, are inactivated in human cancers [192, 205, 206]; another tumor suppressor FHIT, which frequently inactivated in numerous tumors including esophageal and lung cancer [153, 199, 207, 208]. *In vitro* and *in vivo* studies have shown that restoring tumor suppressor function in cancer cells leads to suppression of cancer cell growth through cell cycle arrest or apoptosis [192, 205, 206].

Preclinical studies or clinical trials has been considered for esophageal cancer by way of the introduction of tumor suppressors or their related genes, including those of p53 [209], p16 [210, 211], interferon-beta [212], manganese superoxide dismutase transgene with irradiation [213-217] and Fhit [199, 218]. The gene delivery systems include adenovirus [209, 219], recombinant adenovirus containing the tetracycline (Tet)-regulated gene expression system in esophageal smooth muscle cells [220], liposome-mediated gene transfer [212-215, 217, 221], EpCAM-targeted adenoviral vectors [222], HIV-1 env DNA vaccine by intranasal administration [223], M9 shuttle peptide increasing transfection efficiency in esophageal mucosal cells [224]. An esophageal explant system was reported to test the transduction of adenoviral vectors ex vivo [225]. Recently the gene therapy research targeting stem cells in esophageal epithelium to protect irradiation-induced damage is reported [226].

The field of gene therapy of cancer including that of esophagus has evolved tremendously, with promising developments especially in the field of targeting and novel therapeutic genes [227]. However, substantial research is still needed before gene therapy will turn out to play a significant role in the management and useful of esophageal cancer [227]. Although it is unlikely that cancer gene therapy can replace the conventional methods of treatment at the present moment, selective and efficient vectors may be used for patients with a certain type of esophageal tumor or as adjuvant therapy for patients in the future [227].

(6-2) Targeted Agents

Erlotinib (Tarcevatrade mark), an inhibitor of epidermal growth factor receptor tyrosine kinase (EGFR-TK), potently suppresses the growth of various tumors [228, 229]. The study using esophageal cancer cells showed that Erlotinib induced cell cycle arrest at the G1/S checkpoint [230]. In esophageal cancer cells, the erlotinib-mediated signaling involved the inactivation of EGFR-TK and ERK1/2, the upregulation of the Cyclin-dependent kinase inhibitors p21 (Waf1) and p27 (Kip1), and the downregulation of the cell cycle promoter Cyclin D1 [230]. However, erlotinib did not induce immediate cytotoxicity or apoptosis in esophageal cancer cells [230], although it is noted that the toxicity profile of erlotinib includes diarrhoea and rash, with no haematological side effects in the clinical trials [229]. The observation indicated that the inhibition of EGFR-TK by erlotinib appears to be a promising novel approach for innovative treatment strategies of esophageal cancer, as it powerfully induced growth inhibition and cell cycle arrest in human esophageal cancer cells and enhanced the antineoplastic effects of other targeted agents [230].

The EGF (epidermal growth factor) receptor-tyrosine kinase inhibitor ZD1839 (Gefitinib, 'Iressa') blocks the cell signaling pathways involved in cell proliferation, survival, and angiogenesis in various cancer cells [231]. The recent study indicated that ZD1839 has anti-cancer properties against human esophageal cancer cells [232]. ZD1839 augmented the anti-cancer activity of TRAIL, even in TRAIL-resistant tumors [232]; TNF-related death apoptosis inducing ligand (TRAIL) acts as an anticancer agent. The study suggests that treatment with ZD1839 and TRAIL may have potential in the treatment of ESCC patients [232].

References

[1] Castell, Donald O., and Richter, Joel E., eds. The Esophagus. 3d ed. Philadelphia, Lippincott Williams and Wilkins, 1999.

[2] McCabe, M.L. and Dlamini, Z. (2005). The molecular mechanisms of oesophageal cancer. *Int Immunopharmacol., 5*,1113-30.

[3] Mei, R., Galipeau, P.C., Prass, C., Berno, A., Ghandour, G., Patil, N., Wolff, R.K., Chee, M.S., Reid, B.J., and Lockhart, D.J. (2000). Genome-wide Detection of Allelic Imbalance Using Human SNPs and High-density DNA Arrays. *Cancer Res.,* 10,1126-1137.

[4] Nowell, P.C. (1993). Chromosomes and cancer: the evolution of an idea. *Adv. Cancer Res., 62*,1-17.

[5] Nowell, P.C., and Croce, C.M. (1986). Chromosomes, genes, and cancer. *Am. J. Pathol., 125*,8-15.

[6] Knudson,, A.G. (1993). Antioncogenes and human cancer. *Proc. Natl. Acad. Sci. U.S.A., 90*,10914-21.

[7] Knudson, A.G. (1985). Hereditary cancer, oncogenes, and anti-oncogenes. *Cancer Res., 45*,1437-43.

[8] Hansen, M.F.a.C., W.B. (1987). Genetics of cancer predisposition. *Cancer Res., 47*,5518-27.

[9] Brown, M.A. (1997). Tumor suppressor genes and human cancer. *Adv. Genet., 36*,45-135.

[10] Barrett, M.T., Galipeau, P.C., Sanchez, C.A., Emond, M.J., and Reid, B.J. (1996). Determination of the frequency of loss of heterozygosity in esophageal adenocarcinoma by cell sorting, whole genome amplification and microsatellite polymorphisms. *Oncogene, 12*,1873-8.

[11] Boige, V., Laurent-Puig, P, Fouchet, P, Flejou, JF, Monges, G, Bedossa, P, Bioulac-Sage, P, Capron, F, Schmitz, A, Olschwang, S, and Thomas, G. (1997). Concerted nonsyntenic allelic losses in hyperploid hepatocellular carcinoma as determined by a high-resolution allelotype. *Cancer Res., 57*,1986-90.

[12] Paulson, T.G., Galipeau, P.C., and Reid, B.J. (1999). Loss of heterozygosity analysis using whole genome amplification, cell sorting, and fluorescence-based PCR. *Genome Res., 9*,482-91.

[13] Lu, N., Hu, N., Li, W.J., Roth, M.J., Wang, C., Su, H., Wang, Q.H., Taylor, P.R., and Dawsey, S.M. (2003). Microsatellite alterations in esophageal dysplasia and squamous cell carcinoma from laser capture microdissected endoscopic biopsies. *Cancer Lett., 189*,137-45.

[14] Vogelstein, B., Fearon, E.R., Kern, S.E., Hamilton, S.R., Preisinger, A.C., Nakamura, Y., and White, R. (1989). Allelotype of colorectal carcinomas. *Science, 244*,207-10.

[15] Fults, D., Pedone, C.A., Thomas, G.A., and White, R. (1990). Allelotype of human malignant astrocytoma. *Cancer Res., 50*,5784-9.

[16] Sato, T., Tanigami, A., Yamakawa, K., Akiyama, F., Kasumi, F., Sakamoto, G., and Nakamura, Y. (1990). Allelotype of breast cancer: Cumulative allele losses promote tumor progression in primary breast cancer. *Cancer Res., 50*,7184-9.

[17] Stanbridge, E.J. (1990). Human tumor suppressor genes. *Annu. Rev. Genet., 24*,615-57.

[18] Tsuchiya, E., Nakamura, Y., Weng, S.Y., Nakagawa, K., Tsuchiya, S., Sugano, H., and Kitagawa, T. (1992). Allelotype of non-small cell lung carcinoma-comparison between loss of heterozygosity in squamous cell carcinoma and adenocarcinoma. *Cancer Res., 52*,2478-81.

[19] Yamaguchi, T., Toguchida, J., Yamamuro, T., Kotoura, Y., Takada, N., Kawaguchi, N., Kaneko, Y., Nakamura, Y., Sasaki, M.S., and Ishizaki, K. (1992). Allelotype analysis in osteosarcomas: Frequent allele loss on 3q, 13q, 17p, and 18q. *Cancer Res. 52*,2419-23.

[20] Thrash-Bingham, C.A., Greenberg, R.E., Howard, S., Bruzel, A., Bremer, M., Goll, A., Salazar, H., Freed, J.J., and Tartof, K.D. (1995). Comprehensive allelotyping of human renal cell carcinomas using microsatellite DNA probes. *Proc. Natl. Acad. Sci.* U. S. A., *92*,2854-8.

[21] Kwon, M.S., Hong, S.J., Cho, H.A., Ahn, G.H., Lee, S.S., Lee, K.Y., and Rhyu, M.G. (2003). Extensive and divergent chromosomal losses in squamous and spindle-cell components of esophageal sarcomatoid carcinoma. *Virchows Arch., 443*,635-42.

[22] Attiyeh, E.F., London, W.B., Mosse, Y.P., Wang, Q., Winter, C., Khazi, D., McGrady, P.W., Seeger, R.C., Look, A.T., Shimada, H., Brodeur, G.M., Cohn, S.L., Matthay, K.K., Maris, J.M.; Children's Oncology Group. (2005). Chromosome 1p and 11q deletions and outcome in neuroblastoma. *N Engl J Med., 353*,2243-53.

[23] Chang, S.C., Lin, J.K., Yang, S.H., Wang, H.S., Li, A.F., and Chi, C.W. (2006). Relationship between genetic alterations and prognosis in sporadic colorectal cancer. *Int. J. Cancer, 118*,1721-7.

[24] Li, J., Liu, Z., Wang, Y., Yu, Z., Wang, M., Zhan, Q., and Liu, Z. (2005). Allelic imbalance of chromosome 1q in esophageal squamous cell carcinomas from China: a novel region of allelic loss and significant association with differentiation. *Cancer Lett., 220*,221-30.

[25] Ryan, B.M., McManus, R., Daly, J.S., Carton, E., Keeling, P.W., Reynolds, J.V., and Kelleher, D. (2001). A common p73 polymorphism is associated with a reduced incidence of oesophageal carcinoma. *Br. J. Cancer, 85*,1499-503.

[26] Fujiki, T., Haraoka, S., Yoshioka, S., Ohshima, K., Iwashita, A., and Kikuchi, M. (2002). p53 Gene mutation and genetic instability in superficial multifocal esophageal squamous cell carcinoma. *Int. J. Oncol., 20*,669-79.

[27] Hu, N., Roth, M.J., Polymeropolous, M., Tang, Z.Z., Emmert-Buck, M.R., Wang, Q.H., Goldstein, A.M., Feng, S.S., Dawsey, S.M., Ding, T., Zhuang, Z.P., Han, X.Y., Ried, T., Giffen, C., and Taylor, P.R. (2000). Identification of novel regions of allelic loss from a genomewide scan of esophageal squamous-cell carcinoma in a high-risk Chinese population. Gene. Chromosom. *Cancer, 27*,217-28.

[28] Huang, X.P., Zhao, C.X., Li, Q.J., Cai, Y., Liu, F.X., Hu, H., Xu, X., Han, Y.L., Wu, M., Zhan, Q.M., and Wang, M.R. (2005). Alteration of RPL14 in squamous cell carcinomas and preneoplastic lesions of the esophagus. Gene, in press.

[29] Kuroki, T., Trapasso, F., Yendamuri, S., Matsuyama, A., Alder, H., Mori, M., and Croce, C.M. (2003). Allele loss and promoter hypermethylation of VHL, RAR-beta, RASSF1A, and FHIT tumor suppressor genes on chromosome 3p in esophageal squamous cell carcinoma. *Cancer Res., 63*,3724-8.

[30] Menin, C., Santacatterina, M., Zambon, A., Montagna, M., Parenti, A., Ruol, A., and D'Andrea, E. (2000). Anomalous transcripts and allelic deletions of the FHIT gene in human esophageal cancer. *Cancer Genet. Cytogenet., 119*,56-61.

[31] Ogasawara, S., Maesawa, C., Tamura, G., and Satodate, R. (1995). Frequent microsatellite alterations on chromosome 3p in esophageal squamous cell carcinoma. *Cancer Res., 55*,891-4.

[32] Mori, T., Yanagisawa, A., Kato, Y., Miura, K., Nishihira, T., Mori, S., and Nakamura, Y. (1994). Accumulation of genetic alterations during esophageal carcinogenesis. *Hum Mol Genet., 3*,1969-71.

[33] Li, X.D., Huang, X.P., Zhao, C.X., Li, Q.J., Xu, X., Cai, Y., Han, Y.L., Rong, T.H., and Wang, M.R. (2004). Identification of a minimal deletion region on chromosome 5q in Chinese esophageal squamous cell carcinomas. *Cancer Lett., 215*,221-8.

[34] Si, H.X., Tsao, S.W., Poon, C.S., and Cheung, A.L. (2004). Human papillomavirus infection and loss of heterozygosity in esophageal squamous cell carcinoma. *Cancer Lett., 213*,231-9.

[35] Peralta, R.C., Casson, A.G., Wang, R.N., Keshavjee, S., Redston, M., and Bapat, B. (1998). Distinct regions of frequent loss of heterozygosity of chromosome 5p and 5q in human esophageal cancer. *Int. J. Cancer., 78*,600-5.

[36] Shibagaki, I., Shimada, Y., Wagata, T., Ikenaga, M., Imamura, M., and Ishizaki, K. (1994). Alleotype analysis of esophageal squamous cell carcinoma. *Cancer Res., 54*,2996-3000.

[37] Aoki, T., Mori, T., Du, X., Nisihira, T., Matsubara, T., and Nakamura, Y. (1994). Allelotype study of esophageal carcinoma. *Gene. Chromosom. Cancer, 10*,177-82.

[38] Nie, Y., Yang, G., Song, Y., Zhao, X., So, C., Liao, J., Wang, L.D., and Yang, C.S. (2001). DNA hypermethylation is a mechanism for loss of expression of the HLA class I genes in human esophageal squamous cell carcinomas. *Carcinogenesis, 22*,1615-23.

[39] Wang, S., Mori, Y., Sato, F., Yin, J., Xu, Y., Zou, T.T., Olaru, A., Kimos, M.C., Perry, K., Selaru, F.M., Deacu, E., Sun, M., Shi, Y.C., Shibata, D., Abraham, J.M., Greenwald, B.D., and Meltzer, S.J. (2003). An LOH and mutational investigation of the ST7 gene locus in human esophageal carcinoma. *Oncogene., 22*,467-70.

[40] Ishii, H., Baffa, R., Numata, S., Murakumo, Y., Rattan, S., Inoue, H., Mori, M., Fidanza, V., Alder, H., and Croce, C.M. (1999). The FEZ1 gene at chromosome 8p22 encodes a leucine-zipper protein, and its expression is altered in multiple human tumors. *Proc. Natl. Acad. Sci. U.S.A., 96*,3928-33.

[41] An, J.Y., Fan, Z.M., Gao, S.S., Zhuang, Z.H., Qin, Y.R., Li, J.L., He, X., Tsao, G.S., and Wang, L.D. (2005). Loss of heterozygosity in multistage carcinogenesis of esophageal carcinoma at high-incidence area in Henan Province, China. *World J. Gastroenterol., 11*,2055-60.

[42] Yang, L., Leung, A.C., Ko, J.M., Lo, P.H., Tang, J.C., Srivastava, G., Oshimura, M., Stanbridge, E.J., Daigo, Y., Nakamura, Y., Tang, C.M., Lau, K.W., Law, S., and Lung, M.L. (2005). Tumor suppressive role of a 2.4 Mb 9q33-q34 critical region and DEC1 in esophageal squamous cell carcinoma. *Oncogene, 24*,697-705.

[43] Hu, N., Wang, C., Su, H., Li, W.J., Emmert-Buck, M.R., Li, G., Roth, M.J., Tang, Z.Z., Lu, N., Giffen, C., Albert, P.S., Taylor, P.R., and Goldstein, A.M. (2004). High frequency of CDKN2A alterations in esophageal squamous cell carcinoma from a high-risk Chinese population. *Gene. Chromosom. Cancer, 39*,205-16.

[44] Lichun, Y., Ching Tang, C.M., Wai Lau, K., and Lung, M.L. (2004). Frequent loss of heterozygosity on chromosome 9 in Chinese esophageal squamous cell carcinomas. *Cancer Lett., 203*,71-7.

[45] Tokugawa, T., Sugihara, H., Tani, T., and Hattori, T. (2002). Modes of silencing of p16 in development of esophageal squamous cell carcinoma. *Cancer Res., 62*,4938-44.

[46] Giroux, M.A., Audrezet, M.P., Metges, J.P., Lozac'h, P., Volant, A., Nousbaum, J.B., Labat, J.P., Gouerou, H., Ferec, C., and Robaszkiewicz, M. (2002). Infrequent p16/CDKN2 alterations in squamous cell carcinoma of the oesophagus. *Eur. J. Gastroenterol. Hepatol., 14*,15-8.

[47] Mathew, R., Arora, S., Mathur, M., Chattopadhyay, T.K., and Ralhan, R. (2001). Esophageal squamous cell carcinomas with DNA replication errors (RER+) are associated with p16/pRb loss and wild-type p53. J. Cancer Res. Clin. Oncol., *127*,603-12.

[48] Nishiwaki, T., Daigo, Y., Kawasoe, T., and Nakamura, Y. (2000). Isolation and mutational analysis of a novel human cDNA, DEC1 (deleted in esophageal cancer 1), derived from the tumor suppressor locus in 9q32. *Gene. Chromosom. Cancer, 27*,169-76.

[49] Xing, E.P., Nie, Y., Wang, L.D., Yang, G.Y., and Yang, C.S. (1999). Aberrant methylation of p16INK4a and deletion of p15INK4b are frequent events in human esophageal cancer in Linxian, China. *Carcinogenesis, 20*,77-84.

[50] Tarmin, L., Yin, J., Zhou, X., Suzuki, H., Jiang, H.Y., Rhyu, M.G., Abraham, J.M., Krasna, M.J., Cottrell, J., and Meltzer, S.J. (1994). Frequent loss of heterozygosity on chromosome 9 in adenocarcinoma and squamous cell carcinoma of the esophagus. *Cancer Res., 54*,6094-6.

[51] Shima, H., Hiyama, T., Tanaka, S., Ito, M., Kitadai, Y., Yoshihara, M., Arihiro, K., Haruma, K., and Chayama, K. (2004). Frequent loss of heterozygosity on chromosome 10p14-p15 in esophageal dysplasia and squamous cell carcinoma. *Oncol. Rep., 12*,333-7.

[52] Lam, C.T., Tang, C.M., Lau, K.W., and Lung, M.L. (2002). Loss of heterozygosity on chromosome 11 in esophageal squamous cell carcinomas. Cancer Lett., *178*,75-81.

[53] Guo, G.M., Huang, X.P., Dong, J.T., Caia, Y., Xu, X., Han, Y.L., Wu, M., Zhan, Q.M., and Wang, M.R. (2005). A novel region of deletion on 13q33-q34 in esophageal squamous cell carcinoma. *Oncol. Rep., 14*,1639-46.

[54] Hu, N., Su, H., Li, W.J., Giffen, C., Goldstein, A.M., Hu, Y., Wang, C., Roth, M.J., Li, G., Dawsey, S.M., Xu, Y., Taylor, P.R., and Emmert-Buck, M.R. (2005). Allelotyping of esophageal squamous-cell carcinoma on chromosome 13 defines deletions related to family history. Gene. Chromosom. *Cancer, 44*,271-8.

[55] Hu, N., Goldstein, A.M., Albert, P.S., Giffen, C., Tang, Z.Z., Ding, T., Taylor, P.R., and Emmert-Buck, M.R. (2003). Evidence for a familial esophageal cancer susceptibility gene on chromosome 13. Cancer Epidemiol. *Biomarker. Prev., 12*,1112-5.

[56] Li, W.J., Hu, N., Su, H., Wang, C., Goldstein, A.M., Wang, Y., Emmert-Buck, M.R., Roth, M.J., Guo, W.J., and Taylor, P.R. (2003). Allelic loss on chromosome 13q14 and mutation in deleted in cancer 1 gene in esophageal squamous cell carcinoma. *Oncogene, 22*,314-8.

[57] Huang, X.P., Wei, F., Liu, X.Y., Xu, X., Hu, H., Chen, B.S., Xia, S.H., Han, Y.S., Han, Y.L., Cai, Y., Wu, M., and Wang, M.R. (2002). Allelic loss on 13q in esophageal squamous cell carcinomas from northern China. *Cancer Lett., 185*,87-94.

[58] Hu, N., Roth, M.J., Emmert-Buck, M.R., Tang, Z.Z., Polymeropolous, M., Wang, Q.H., Goldstein, A.M., Han, X.Y., Dawsey, S.M., Ding, T., Giffen, C., and Taylor, P.R. (1999). Allelic loss in esophageal squamous cell carcinoma patients with and without family history of upper gastrointestinal tract cancer. *Clin. Cancer Res., 5*,3476-82.

[59] Harada, H., Tanaka, H., Shimada, Y., Shinoda, M., Imamura, M., and Ishizaki, K. (1999). Lymph node metastasis is associated with allelic loss on chromosome 13q12-13 in esophageal squamous cell carcinoma. *Cancer Res., 59*,3724-9.

[60] Xing, E.P., Yang, G.Y., Wang, L.D., Shi, S.T., and Yang, C.S. (1999). Loss of heterozygosity of the Rb gene correlates with pRb protein expression and associates with p53 alteration in human esophageal cancer. *Clin. Cancer Res., 5*,1231-40.

[61] So, C.K., Nie, Y., Song, Y., Yang, G.Y., Chen, S., Wei, C., Wang, L.D., Doggett, N.A., and Yang, C.S. (2004). Loss of heterozygosity and internal tandem duplication mutations of the CBP gene are frequent events in human esophageal squamous cell carcinoma. *Clin. Cancer Res., 10*(1 Pt 1),19-27.

[62] Kuroki, T., Trapasso, F., Shiraishi, T., Alder, H., Mimori, K., Mori, M., and Croce, C.M. (2002). Genetic alterations of the tumor suppressor gene WWOX in esophageal squamous cell carcinoma. *Cancer Res., 62*,2258-60.

[63] Iwaya, T., Maesawa, C., Kimura, T., Ogasawara, S., Ikeda, K., Kimura, Y., Noda, Y., Ishida, K., Sato, N., Saito, K., and Masuda, T. (2005). Infrequent mutation of the human envoplakin gene is closely linked to the tylosis oesophageal cancer locus in sporadic oesophageal squamous cell carcinomas. *Oncol. Rep., 13*,703-7.

[64] von Brevern, M., Hollstein, M.C., Risk, J.M., Garde, J., Bennett, W.P., Harris, C.C., Muehlbauer, K.R., and Field, J.K. (1998). Loss of heterozygosity in sporadic oesophageal tumors in the tylosis oesophageal cancer (TOC) gene region of chromosome 17q. *Oncogene, 17*,2101-5.

[65] Iwaya, T., Maesawa, C., Ogasawara, S., and Tamura, G. (1998). Tylosis esophageal cancer locus on chromosome 17q25.1 is commonly deleted in sporadic human esophageal cancer. *Gastroenterology, 114*,1206-10.

[66] Karkera, J.D., Ayache, S., Ransome, R.J. Jr., Jackson, M.A., Elsayem, A.F., Sridhar, R., Detera-Wadleigh, S.D., and Wadleigh, R.G. (2000). Refinement of regions with allelic loss on chromosome 18p11.2 and 18q12.2 in esophageal squamous cell carcinoma. *Clin. Cancer Res., 6*,3565-9.

[67] Karkera, J.D., Balan, K.V., Yoshikawa, T., Lipman, T.O., Korman, L., Sharma, A., Patterson, R.H., Sani, N., Detera-Wadleigh, S.D., and Wadleigh, R.G. (1999). Systematic screening of chromosome 18 for loss of heterozygosity in esophageal squamous cell carcinoma. *Cancer Genet. Cytogenet., 111*,81-6.

[68] Mayama, T., Fukushige, S., Shineha, R., Nishihira, T., Satomi, S., and Horii, A. (2000). Frequent loss of copy number on the long arm of chromosome 21 in human esophageal squamous cell carcinoma. *Int. J. Oncol., 17*,245-52.

[69] Dolan, K., Garde, J., Walker, S.J., Sutton, R., Gosney, J., and Field, J.K. (2000). Histological and molecular mapping of adenocarcinoma of the oesophagus and

gastroesophageal junction: loss of heterozygosity occurs in histologically normal epithelium in the oesophagus and stomach. *Oncol Rep., 7*,521-8.

[70] Wijnhoven, B.P., Lindstedt, E.W., Abbou, M., Ijzendoorn, Y., de Krijger, R.R., Tilanus, H.W., Dinjens, W.N.; Rotterdam Esophageal Tumor Study Group. (2001). Molecular genetic analysis of the von Hippel-Lindau and human peroxisome proliferator-activated receptor gamma tumor-suppressor genes in adenocarcinomas of the gastroesophageal junction. *Int. J. Cancer, 94*,891-5.

[71] Dolan, K., Garde, J., Gosney, J., Sissons, M., Wright, T., Kingsnorth, A.N., Walker, S.J., Sutton, R., Meltzer, S.J., and Field, J.K. (1998). Allelotype analysis of oesophageal adenocarcinoma: loss of heterozygosity occurs at multiple sites. *Br. J. Cancer, 78*,950-7.

[72] Michael, D., Beer, D.G., Wilke, C.W., Miller, D.E., and Glover, T.W. (1997). Frequent deletions of FHIT and FRA3B in Barrett's mataplasia and esophageal adenocarcinomas. *Oncogene, 15*,1653-9.

[73] Hammoud, Z.T., Kaleem, Z., Cooper, J.D., Sundaresan, R.S., Patterson, G.A., and Goodfellow, P.J. (1996). Allelotype analysis of esophageal adenocarcinomas: evidence for the involvement of sequences on the long arm of chromosome 4. *Cancer Res., 56*,4499-502.

[74] Vieth, M., Schneider-Stock, R., Rohrich, K., May, A., Ell, C., Markwarth, A., Roessner, A., Stolte, M., and Tannapfel, A. (2004). INK4a-ARF alterations in Barrett's epithelium, intraepithelial neoplasia and Barrett's adenocarcinoma. *Virchow. Arch., 445*,135-41.

[75] Wong, D.J., Barrett, M.T., Stoger, R., Emond, M.J., and Reid, B.J. (1997). p16INK4a promoter is hypermethylated at a high frequency in esophageal adenocarcinomas. *Cancer Res., 57*,2619–22.

[76] Gonzalez, M.V., Artimez, M.L., Rodrigo, L., Lopez-Larrea, C., Menendez, M.J., Alvarez, V, Perez, R., Fresno, M.F., Perez, M.J., Sampedro, A., and Coto, E. (1997). Mutation analysis of the p53, APC, and p16 genes in the Barrett's oesophagus, dysplasia, and adenocarcinoma. *J. Clin. Pathol., 50*,212–7.

[77] Moskaluk, C.A., and Rumpel, C.A. (1998). Allelic deletion in 11p15 is a common occurrence in esophageal and gastric adenocarcinoma. *Cancer, 83*,232-9.

[78] Petty, E.M., Kalikin, L.M., Orringer, M.B., and Beer, D.G. (1998). Distal chromosome 17q loss in Barrett's esophageal and gastric cardia adenocarcinomas: implications for tumorigenesis. *Mol. Carcinog., 22*,222-8.

[79] Barrett, M.T., Sanchez, C.A., Prevo, L.J., Wong, D.J., Galipeau, P.C., Paulson, T.G., Rabinovitch, P.S., and Reid, B.J. (1999). Evolution of neoplastic cell lineages in Barrett oesophagus. *Nat. Genet., 22*,106-9.

[80] Sanz-Ortega, J., Hernandez, S., Saez, M.C., Sierra, E., Sanz-Ortega, G., Torres, A., Balibrea, J.L., Sanz-Esponera, J., and Merino, M.J. (2003). 3p21, 5q21, 9p21 and 17p13.1 allelic deletions are potential markers of individuals with a high risk of developing adenocarcinoma in Barrett's epithelium without dysplasia. *Hepatogastroenterology, 50*,404-7.

[81] Hage, M., Siersema, P.D., Vissers, K.J., Dinjens, W.N., Steyerberg, E.W., Haringsma, J., Kuipers, E.J., and van Dekken, H. (2006). Genomic analysis of Barrett's esophagus

after ablative therapy: persistence of genetic alterations at tumor suppressor loci. *Int. J. Cancer, 118*,155-60.

[82] Dolan, K., Garde, J., Walker, S.J., Sutton, R., Gosney, J., and Field, J.K. (1999). LOH at the sites of the DCC, APC, and TP53 tumor suppressor genes occurs in Barrett's metaplasia and dysplasia adjacent to adenocarcinoma of the esophagus. *Hum. Pathol., 30*,1508-14.

[83] Suspiro, A., Pereira, A.D., Afonso, A., Albuquerque, C., Chaves, P., Soares, J., and Leitao, C.N. (2003). Losses of heterozygosity on chromosomes 9p and 17p are frequent events in Barrett's metaplasia not associated with dysplasia or adenocarcinoma. *Am. J. Gastroenterol., 98*,728-34.

[84] Bian, Y.S., Osterheld, M.C., Fontolliet, C., Bosman, F.T., and Benhattar, J. (2002). p16 inactivation by methylation of the CDKN2A promoter occurs early during neoplastic progression in Barrett's esophagus. *Gastroenterology, 122*,1113–21.

[85] Galipeau, P.C., Prevo, L.J., Sanchez, C.A., Longton, G.M., and Reid, B.J. (1999). Clonal expansion and loss of heterozygosity at chromosomes 9p and 17p in premalignant esophageal (Barrett's) tissue. *J. Natl. Cancer Inst., 91*,2087-95.

[86] Reid, B.J., Blount, P.L., and Rabinovitch, P.S. (2003). Biomarkers in Barrett's esophagus. *Gastrointest. Endosc. Clin. N. Am., 13*,369-97.

[87] Reid, B.J., Prevo, L.J., Galipeau, P.C., Sanchez, C.A., Longton, G., Levine, D.S., Blount, P.L., and Rabinovitch, P.S. (2001). Predictors of progression in Barrett's esophagus II: baseline 17p (p53) loss of heterozygosity identifies a patient subset at increased risk for neoplastic progression. *Am. J. Gastroenterol., 96*,2839-48.

[88] Dunn, J., Garde, J., Dolan, K., Gosney, J.R., Sutton, R., Meltzer, S.J., and Field, J.K. (1999). Multiple target sites of allelic imbalance on chromosome 17 in Barrett's oesophageal cancer. *Oncogene, 18*,987-93.

[89] Campomenosi, P., Conio, M., Bogliolo, M., Urbini, S., Assereto, P., Aprile, A., Monti, P., Aste, H., Lapertosa, G., Inga, A., Abbondandolo, A., and Fronza, G. (1996). p53 is frequently mutated in Barrett's metaplasia of the intestinal type. *Cancer Epidemiol. Biomarker. Prev., 5*,559-65.

[90] Swift, A., Risk, J.M., Kingsnorth, A.N., Wright, T.A., Myskow, M., and Field, J.K. (1995). Frequent loss of heterozygosity on chromosome 17 at 17q11.2-q12 in Barrett's adenocarcinoma. *Br. J. Cancer, 71*,995-8.

[91] Roth, M.J., Hu, N., Emmert-Buck, M.R., Wang, Q.H., Dawsey, S.M., Li, G., Guo, W.J., Zhang, Y.Z., and Taylor, P.R. (2001). Genetic progression and heterogeneity associated with the development of esophageal squamous cell carcinoma. *Cancer Res., 61*,4098-104.

[92] Sunpaweravong, P., Sunpaweravong, S., Puttawibul, P., Mitarnun, W., Zeng, C., Baron, A.E., Franklin, W., Said, S., and Varella-Garcia, M. (2005). Epidermal growth factor receptor and cyclin D1 are independently amplified and overexpressed in esophageal squamous cell carcinoma. *J. Cancer Res. Clin. Oncol., 131*,111-9.

[93] Ishizuka, T., Tanabe, C., Sakamoto, H., Aoyagi, K., Maekawa, M., Matsukura, N., Tokunaga, A., Tajiri, T., Yoshida, T., Terada, M., and Sasaki, H. (2002). Gene amplification profiling of esophageal squamous cell carcinomas by DNA array CGH. *Biochem. Biophys. Res. Commun., 296*,152-5.

[94] Kitagawa, Y., Ueda, M., Ando, N., Ozawa, S., Shimizu, N., and Kitajima, M. (1996). Further evidence for prognostic significance of epidermal growth factor receptor gene amplification in patients with esophageal squamous cell carcinoma. *Clin. Cancer Res., 2*,909-14.

[95] Hollstein, M.C., Smits, A.M., Galiana, C., Yamasaki, H., Bos, J.L., Mandard, A., Partensky, C., and Montesano, R. (1988). Amplification of epidermal growth factor receptor gene but no evidence of ras mutations in primary human esophageal cancers. *Cancer Res., 48*,5119-23.

[96] Jin, Y., Jin, C., Law, S., Chu, K.M., Zhang, H., Strombeck, B., Yuen, A.P., and Kwong, Y.L. (2004). Cytogenetic and fluorescence in situ hybridization characterization of clonal chromosomal aberrations and CCND1 amplification in esophageal carcinomas. *Cancer Genet Cytogenet., 148*,21-8.

[97] Arai, H., Ueno, T., Tangoku, A., Yoshino, S., Abe, T., Kawauchi, S., Oga, A., Furuya, T., Oka, M., and Sasaki, K. (2003). Detection of amplified oncogenes by genome DNA microarrays in human primary esophageal squamous cell carcinoma: comparison with conventional comparative genomic hybridization analysis. *Cancer Genet. Cytogenet., 146*,16-21.

[98] Janssen, J.W., Imoto, I., Inoue, J., Shimada, Y., Ueda, M., Imamura, M., Bartram, C.R., and Inazawa, J. (2002). MYEOV, a gene at 11q13, is coamplified with CCND1, but epigenetically inactivated in a subset of esophageal squamous cell carcinomas. *J. Hum. Genet., 47*,460-4.

[99] Shinozaki, H., Ozawa, S., Ando, N., Tsuruta, H., Terada, M., Ueda, M., and Kitajima, M. (1996). Cyclin D1 amplification as a new predictive classification for squamous cell carcinoma of the esophagus, adding gene information. *Clin. Cancer Res., 2*,1155-61.

[100] Sheyn, I., Noffsinger, A.E., Heffelfinger, S., Davis, B., Miller, M.A., and Fenoglio-Preiser, C.M. (1997). Amplification and expression of the cyclin D1 gene in anal and esophageal squamous cell carcinomas. *Hum. Pathol., 28*,270-6.

[101] Adelaide, J., Monges, G., Derderian, C., Seitz, J.F., and Birnbaum, D. (1995). Oesophageal cancer and amplification of the human cyclin D gene CCND1/PRAD1. *Br. J. Cancer, 71*,64-8.

[102] Gramlich, T.L., Fritsch, C.R., Maurer, D., Eberle, M., and Gansler, T.S. (1994). Differential polymerase chain reaction assay of cyclin D1 gene amplification in esophageal carcinoma. *Diagn. Mol. Pathol., 3*,255-9.

[103] Tsuruta, H., Sakamoto, H., Onda, M., and Terada, M. (1993). Amplification and overexpression of EXP1 and EXP2/Cyclin D1 genes in human esophageal carcinomas. Biochem. Biophys. *Res. Commun., 196*,1529-36.

[104] Jiang, W., Kahn, S.M., Tomita, N., Zhang, Y.J., Lu, S.H., and Weinstein, I.B. (1992). Amplification and expression of the human cyclin D gene in esophageal cancer. *Cancer Res., 52*,2980-3.

[105] Imoto, I., Yuki, Y., Sonoda, I., Ito, T., Shimada, Y., Imamura, M., and Inazawa, J. (2003). Identification of ZASC1 encoding a Kruppel-like zinc finger protein as a novel target for 3q26 amplification in esophageal squamous cell carcinomas. *Cancer Res., 63*.5691-6.

[106] Bitzer, M., Stahl, M., Arjumand, J., Rees, M., Klump, B., Heep, H., Gabbert, H.E., and Sarbia, M. (2003). C-myc gene amplification in different stages of oesophageal squamous cell carcinoma: prognostic value in relation to treatment modality. *Anticancer Res., 23*(2B),1489-93.

[107] Yen, C.C., Chen, Y.J., Pan, C.C., Lu, K.H., Chen, P.C., Hsia, J.Y., Chen, J.T., Wu, Y.C., Hsu, W.H., Wang, L.S., Huang, M.H., Huang, B.S., Hu, C.P., Chen, P.M., and Lin, C.H. (2005). Copy number changes of target genes in chromosome 3q25.3-qter of esophageal squamous cell carcinoma: TP63 is amplified in early carcinogenesis but down-regulated as disease progressed. *World J. Gastroenterol., 11*,1267-72.

[108] Taniere, P., Martel-Planche, G., Puttawibul, P., Casson, A., Montesano, R., Chanvitan, A., and Hainaut, P. (2000). TP53 mutations and MDM2 gene amplification in squamous-cell carcinomas of the esophagus in south Thailand. *Int. J. Cancer, 88*,223-7.

[109] Shibagaki, I., Tanaka, H., Shimada, Y., Wagata, T., Ikenaga, M., Imamura, M., and Ishizaki, K. (1995). p53 mutation, murine double minute 2 amplification, and human papillomavirus infection are frequently involved but not associated with each other in esophageal squamous cell carcinoma. *Clin. Cancer Res., 1*,769-73.

[110] Imoto, I., Yang, Z.Q., Pimkhaokham, A., Tsuda, H., Shimada, Y., Imamura, M., Ohki, M., and Inazawa, J. (2001). Identification of cIAP1 as a candidate target gene within an amplicon at 11q22 in esophageal squamous cell carcinomas. Cancer Res., *61*,6629-34.

[111] Imoto, I., Pimkhaokham, A., Fukuda, Y., Yang, Z.Q., Shimada, Y., Nomura, N., Hirai, H., Imamura, M., and Inazawa, J. (2001). SNO is a probable target for gene amplification at 3q26 in squamous-cell carcinomas of the esophagus. *Biochem. Biophys. Res. Commun., 286*,559-65.

[112] Pimkhaokham, A., Shimada, Y., Fukuda, Y., Kurihara, N., Imoto, I., Yang, Z.Q., Imamura, M., Nakamura, Y., Amagasa, T., and Inazawa, J. (2000). Nonrandom chromosomal imbalances in esophageal squamous cell carcinoma cell lines: possible involvement of the ATF3 and CENPF genes in the 1q32 amplicon. *Jpn. J. Cancer Res., 91*,1126-33.

[113] Yang, Z.Q., Imoto, I., Fukuda, Y., Pimkhaokham, A., Shimada, Y., Imamura, M., Sugano, S., Nakamura, Y., and Inazawa, J. (2000). Identification of a novel gene, GASC1, within an amplicon at 9p23-24 frequently detected in esophageal cancer cell lines. *Cancer Res., 60*,4735-9.

[114] Shinomiya, T., Mori, T., Ariyama, Y., Sakabe, T., Fukuda, Y., Murakami, Y., Nakamura, Y., and Inazawa, J. (1999). Comparative genomic hybridization of squamous cell carcinoma of the esophagus: the possible involvement of the DPI gene in the 13q34 amplicon. *Gene. Chromosom. Cancer, 24*,337-44.

[115] Miller, C.T., Moy, J.R., Lin, L., Schipper, M., Normolle, D., Brenner, D.E., Iannettoni, M.D., Orringer, M.B., and Beer, D.G. (2003). Gene amplification in esophageal adenocarcinomas and Barrett's with high-grade dysplasia. *Clin. Cancer Res., 9*,4819-25.

[116] Dahlberg, P.S., Jacobson, B.A., Dahal, G., Fink, J.M., Kratzke, R.A., Maddaus, M.A., and Ferrin, L.J. (2004). ERBB2 amplifications in esophageal adenocarcinoma. *Ann. Thorac. Surg., 7,*1790-800.

[117] Owonikoko, T., Rees, M., Gabbert, H.E., and Sarbia, M. (2002). Intratumoral genetic heterogeneity in Barrett adenocarcinoma. *Am. J. Clin. Pathol., 117*,558-66.

[118] Walch, A., Specht, K., Braselmann, H., Stein, H., Siewert, J.R., Hopt, U., Hofler, H., and Werner, M. (2004). Coamplification and coexpression of GRB7 and ERBB2 is found in high grade intraepithelial neoplasia and in invasive Barrett's carcinoma. *Int. J. Cancer, 112*,747-53.

[119] Geddert, H., Zeriouh, M., Wolter, M., Heise, J.W., Gabbert, H.E., and Sarbia, M. (2002). Gene amplification and protein overexpression of c-erb-b2 in Barrett carcinoma and its precursor lesions. *Am. J. Clin. Pathol., 118*,60-6.

[120] al-Kasspooles, M., Moore, J.H., Orringer, M.B., and Beer, D.G. (1993). Amplification and over-expression of the EGFR and erbB-2 genes in human esophageal adenocarcinomas. *Int. J. Cancer, 54*,213-9.

[121] Brien, T.P., Odze, R.D., Sheehan, C.E., McKenna, B.J., and Ross, J.S. (2000). HER-2/neu gene amplification by FISH predicts poor survival in Barrett's esophagus-associated adenocarcinoma. *Hum. Pathol., 31*,35-9.

[122] Lin, L., Prescott, M.S., Zhu, Z., Singh, P., Chun, S.Y., Kuick, R.D., Hanash, S.M., Orringer, M.B., Glover, T.W., and Beer, D.G. (2000). Identification and characterization of a 19q12 amplicon in esophageal adenocarcinomas reveals cyclin E as the best candidate gene for this amplicon. *Cancer Res., 60*,7021-7.

[123] Lin, L., Miller, C.T., Contreras, J.I., Prescott, M.S., Dagenais, S.L., Wu, R., Yee, J., Orringer, M.B., Misek, D.E., Hanash, S.M., Glover, T.W., and Beer, D.G. (2002). The hepatocyte nuclear factor 3 alpha gene, HNF3alpha (FOXA1), on chromosome band 14q13 is amplified and overexpressed in esophageal and lung adenocarcinomas. *Cancer Res., 62*,5273-9.

[124] Miller, C.T., Aggarwal, S., Lin, T.K., Dagenais, S.L., Contreras, J.I., Orringer, M.B., Glover, T.W., Beer, D.G., and Lin, L. (2003). Amplification and overexpression of the dual-specificity tyrosine-(Y)-phosphorylation regulated kinase 2 (DYRK2) gene in esophageal and lung adenocarcinomas. *Cancer Res., 63*,4136-43.

[125] Hughes, S.J., Glover, T.W., Zhu, X.X., Kuick, R., Thoraval, D., Orringer, M.B., Beer, D.G., and Hanash, S. (1998). A novel amplicon at 8p22-23 results in overexpression of cathepsin B in esophageal adenocarcinoma. *Proc. Natl. Acad. Sci. U. S. A., 95*,12410-5.

[126] Wong, D.J., Paulson, T.G., Prevo, L.J., Galipeau, P.C., Longton, G., Blount, P.L. and Reid, B.J. (2001). p16(INK4a) lesions are common, early abnormalities that undergo clonal expansion in Barrett's metaplastic epithelium. *Cancer Res., 61*,8284–9.

[127] Klump, B., Hsieh, C.J., Holzmann, K., Gregor, M., and Porschen, R. (1998). Hypermethylation of the CDKN2/p16 promoter during neoplastic progression in Barrett's esophagus. *Gastroenterology, 115*,1381–6.

[128] Maley, C.C., Galipeau, P.C., Li, X., Sanchez, C.A., Paulson, T.G., and Reid, B.J. (2004). Selectively advantageous mutations and hitchhikers in neoplasms: p16 lesions are selected in Barrett's esophagus. *Cancer Res., 64*,3414-27.

[129] Kopelovich, L., Crowell, J.A., and Fay, J.R. (2003). The Epigenome as a Target for Cancer Chemoprevention. *J. Natl. Cancer Inst., 95*,1747-57.

[130] Boynton, R.F., Blount, P.L., Yin, J., Brown, V.L., Huang, Y., Tong, Y., McDaniel, T., Newkirk, C., Resau, J.H., Raskind, W.H., Haggitt, R.C., Reid, B.J. and Meltzer, S.J.

(1992). Loss of Heterozygosity Involving the APC and MCC Genetic Loci Occurs in the Majority of Human Esophageal Cancers. *Proc. Natl. Acad. Sci. U. S. A., 89*,3385-8.

[131] Huang, Y., Boynton, R.F., Blount, P.L., Silverstein, R.J., Yin, J., Tong, Y., McDaniel, T.K., Newkirk, C., Resau, J.H. and Sridhara, R. (1992). Loss of heterozygosity involves multiple tumor suppressor genes in human esophageal cancers. *Cancer Res., 52*,6525-30.

[132] Maesawa, C., Tamura, G., Suzuki, Y., Ogasawara, S., Ishida, K., Saito, K., and Satodate, R. (1994). Aberrations of tumor-suppressor genes (p53, apc, mcc and Rb) in esophageal squamous-cell carcinoma. *Int. J. Cancer, 57*,21-5.

[133] Zhuang, Z., Vortmeyer, A.O., Mark, E.J., Odze, R., Emmert-Buck, M.R., Merino, M.J., Moon, H., Liotta. L.A. and Duray P.H. (1996). Barrett's esophagus: metaplastic cells with loss of heterozygosity at the APC gene locus are clonal precursors to invasive adenocarcinoma. *Cancer Res., 56*,1961-4.

[134] Powell, S.M., Papadopoulos, N., Kinzler, K.W., Smolinski, K.N., and Meltzer, S.J. (1994). APC gene mutations in the mutation cluster region are rare in esophageal cancers. *Gastroenterology, 107*,1759–63.

[135] Ogasawara, S., Maesawa, C., Tamura, G., and Satodate, R. (1994). Lack of mutations of the adenomatous polyposis coli gene in oesophageal and gastric carcinomas. *Virchow. Arch., 424*,607–11.

[136] Tsuchiya, T., Tamura, G., Sato, K., Endoh, Y., Sakata, K., Jin, Z., Motoyama, T., Usuba, O., Kimura, W., Nishizuka, S., Wilson, K.T., James, S.P., Yin, J., Fleisher, A.S., Zou, T., Silverberg, S.G., Kong, D., and Meltzer, S.J. (2000). Distinct methylation patterns of two APC gene promoters in normal and cancerous gastric epithelia. *Oncogene, 19*,3642-6.

[137] Kawakami, K., Brabender, J., Lord, R.V., Groshen, S., Greenwald, B.D., Krasna, M.J., Yin, J., Fleisher, A.S., Abraham, J.M., Beer, D.G., Sidransky, D., Huss, H.T., Demeester, T.R., Eads, C., Laird, P.W., Ilson, D.H., Kelsen, D.P., Harpole, D., Moore, M.B., Danenberg, K.D., Danenberg, P.V., and Meltzer, S.J. (2000). Hypermethylated APC DNA in plasma and prognosis of patients with esophageal adenocarcinoma. *J. Natl. Cancer Inst., 92*,1805-11.

[138] Sarbia, M., Geddert, H., Klump, B., Kiel, S., Iskender, E., and Gabbert, H.E. (2004). Hypermethylation of tumor suppressor genes (p16INK4A, p14ARF and APC) in adenocarcinomas of the upper gastrointestinal tract. *Int. J. Cancer, 111*,224-8.

[139] Clement, G., Bosman, F.T., Fontolliet, C., and Benhattar, J. (2004). Monoallelic methylation of the APC promoter is altered in normal gastric mucosa associated with neoplastic lesions. *Cancer Res., 64*,6867-73.

[140] Clement, G., Braunschweig, R., Pasquier, N., Bosman, F.T., and Benhattar, J. (2006). Methylation of APC, TIMP3, and TERT: a new predictive marker to distinguish Barrett's oesophagus patients at risk for malignant transformation. *J. Pathol., 208*,100-7.

[141] Eads, C.A., Lord, R.V., Wickramasinghe, K., Long, T.I., Kurumboor, S.K., Bernstein, L., Peters, J.H., DeMeester, S.R., DeMeester, T.R., Skinner, K.A. and Laird, P.W. (2001). Epigenetic Patterns in the Progression of Esophageal Adenocarcinoma. *Cancer Res., 61*,3410-8.

[142] Sozzi, G., Pastorino, U., Moiraghi, L., Tagliabue, E., Pezzella, F., Ghirelli, C., Tornielli, S., Sard, L., Huebner, K., Pierotti, M.A., Croce, C.M., and Pilotti, S. (1998). Loss of FHIT function in lung cancer and preinvasive bronchial lesions. *Cancer Res., 58*,5032-7.

[143] Birrer, M.J., Hendricks, D., Farley, J., Sundborg, M.J., Bonome, T., Walts, M.J., and Geradts, J. (1999). Abnormal Fhit expression in malignant and premalignant lesions of the cervix. *Cancer Res., 59*,5270-4.

[144] Mori, M., Mimori, K., Shiraishi, T., Alder, H., Inoue, H., Tanaka, Y., Sugimachi, K., Huebner, K., and Croce, C.M. (2000). Altered expression of Fhit in carcinoma and precarcinomatous lesions of the esophagus. *Cancer Res., 60*,1177-82.

[145] Gatalica, Z., Lele, S.M., Rampy, B.A., and Norris, B.A. (2000). The expression of Fhit protein is related inversely to disease progression in patients with breast carcinoma. *Cancer, 88*,1378-83.

[146] Tanimoto, K., Hayashi, S., Tsuchiya, E., Tokuchi, Y., Kobayashi, Y., Yoshiga, K., Okui, T., Kobayashi, M., and Ichikawa, T. (2000). Abnormalities of the FHIT gene in human oral carcinogenesis. *Br. J. Cancer, 82*,838-43.

[147] Ahmadian, M., Wistuba, II., Fong, K.M., Behrens, C., Kodagoda, D.R., Saboorian, M.H., Shay, J., Tomlinson, G.E., Blum, J., Minna, J.D., and Gazdar, A.F. (1997). Analysis of the FHIT gene and FRA3B region in sporadic breast cancer, preneoplastic lesions, and familial breast cancer probands. *Cancer Res., 57*,3664-8.

[148] Gemma, A., Hagiwara, K., Ke, Y., Burke, L.M., Khan, M.A., Nagashima, M., Bennett, W.P., and Harris, C.C. (1997). FHIT mutations in human primary gastric cancer. *Cancer Res., 57*,1435-7.

[149] Yoshino, K., Enomoto, T., Nakamura, T., Nakashima, R., Wada, H., Saitoh, J., Noda, K., and Murata, Y. (1998). Aberrant FHIT transcripts in squamous cell carcinoma of the uterine cervix. *Int. J. Cancer, 76*,176-81.

[150] Muller, C.Y., O'Boyle, J.D., Fong, K.M., Wistuba, II., Biesterveld, E., Ahmadian, M., Miller, D.S., Gazdar, A.F., and Minna, J.D. (1998). Abnormalities of fragile histidine triad genomic and complementary DNAs in cervical cancer: association with human papillomavirus type. *J. Natl. Cancer Inst., 90*,433-439.

[151] Yoshino, K., Enomoto, T., Nakamura, T., Sun, H., Ozaki, K., Nakashima, R., Wada, H., Saitoh, J., Watanabe, Y., Noda, K., and Murata, Y. (2000). FHIT alterations in cancerous and non-cancerous cervical epithelium. *Int. J. Cancer, 85*,6-13.

[152] Huebner, K., Garrison, P.N., Barnes, L.D., and Croce, C.M. (1998). The role of the FHIT/FRA3B locus in cancer. *Annu. Rev. Genet., 32*,7-31.

[153] Sozzi, G., Huebner, K., and Croce, C.M. (1998). FHIT in human cancer. *Adv. Cancer Res., 74*,141-66.

[154] Tanaka, H., Shimada, Y., Harada, H., Shinoda, M., Hatooka, S., Imamura, M., and Ishizaki, K. (1998). Methylation of the 5' CpG island of the FHIT gene is closely associated with transcriptional inactivation in esophageal squamous cell carcinomas. *Cancer Res., 58*,3429-34.

[155] Nyberg, K.A., Michelson, R.J., Putnam, C.W., and Weinert, T.A. (2002). Toward maintaining the genome: DNA damage and replication checkpoints. *Annu. Rev. Genet., 36*,617-56.

[156] Sancar, A., Lindsey-Boltz, L.A., Unsal-Kacmaz, K., and Linn, S. (2004). Molecular mechanisms of mammalian DNA repair and the DNA damage checkpoints. *Annu. Rev. Biochem., 73*,39-85.

[157] Kastan, M.B., and Bartek, J. (2004). Cell-cycle checkpoints and cancer. *Nature, 432*,316-23.

[158] Bartek, J., and Lukas, J. (2003). Chk1 and Chk2 kinases in checkpoint control and cancer. *Cancer Cell, 3*,421-9.

[159] Wahl, G.M., and Carr, A.M. (2001). The evolution of diverse biological responses to DNA damage: insights from yeast and p53. *Nat. Cell. Biol., 3*,E277-86.

[160] Bartek, J., Bartkova, J., and Lukas, J. (1997). The retinoblastoma protein pathway in cell cycle control and cancer. *Exp. Cell. Res., 237*,1-6.

[161] Bartek, J., Lukas, C., and Lukas, J. (2004). Checking on DNA damage in S phase. *Nat Rev. Mol. Cell. Biol.,. 5*,792-804.

[162] Phillips, A.C., Bates, S., Ryan, K.M., Helin, K., and Vousden, K.H. (1997). Induction of DNA synthesis and apoptosis are separable functions of E2F-1. *Gene. Dev., 11*,1853-63.

[163] Johnson, D.G., Schwarz, J.K., Cress, W.D., and Nevins, J.R. (1993). Expression of transcription factor E2F1 induces quiescent cells to enter S phase. Nature, *365*,349-52.

[164] Johnson, D.G., Cress, W.D., Jakoi, L., and Nevins, J.R. (1994). Oncogenic capacity of the E2F1 gene. *Proc. Natl. Acad. Sci. U. S. A., 91*,12823-7.

[165] Kowalik, T.F., DeGregori, J., Schwarz, J.K., and Nevins, J.R. (1995). E2F1 overexpression in quiescent fibroblasts leads to induction of cellular DNA synthesis and apoptosis. *J. Virol., 69*,2491-500.

[166] Field, S.J., Tsai, F.Y., Kuo, F., Zubiaga, A.M., Kaelin, W.G. Jr, Livingston, D.M., Orkin, S.H., and Greenberg, M.E. (1996). E2F-1 functions in mice to promote apoptosis and suppress proliferation. *Cell, 85*,549-61.

[167] Yamasaki, L., Jacks, T., Bronson, R., Goillot, E., Harlow, E., and Dyson, N.J. (1996). Tumor induction and tissue atrophy in mice lacking E2F-1. *Cell, 85*,537-48.

[168] Pan, H., Yin, C., Dyson, N.J., Harlow, E., Yamasaki, L., and Van Dyke, T. (1998). Key roles for E2F1 in signaling p53-dependent apoptosis and in cell division within developing tumors. *Mol. Cell, 2*,283-92.

[169] Dalton, S. (1992). Cell cycle regulation of the human cdc2 gene. *EMBO J., 11*,1797-804.

[170] Ohtani, K., DeGregori, J., and Nevins, J.R. (1995). Regulation of the cyclin E gene by transcription factor E2F1. *Proc. Natl. Acad. Sci. U. S. A., 92*,12146-50.

[171] Geng, Y., Yu, Q., Whoriskey, W., Dick, F., Tsai, K.Y., Ford, H.L., Biswas, D.K., Pardee, A.B., Amati, B., Jacks, T., Richardson, A., Dyson, N., and Sicinski, P. (2001). Expression of cyclins E1 and E2 during mouse development and in neoplasia. Proc. *Natl. Acad. Sci. U. S. A., 98*,13138-43.

[172] Schulze, A., Zerfass, K., Spitkovsky, D., Middendorp, S., Berges, J., Helin, K., Jansen-Durr, P., and Henglein, B. (1995). Cell cycle regulation of the cyclin A gene promoter is mediated by a variant E2F site. *Proc. Natl. Acad. Sci. U. S. A., 92*,11264-8.

[173] Johnson, D.G., Ohtani, K., and Nevins, J.R. (1994). Autoregulatory control of E2F1 expression in response to positive and negative regulators of cell cycle progression. *Gene. Dev., 8*,1514-25.

[174] Furukawa, Y., Nishimura, N., Furukawa, Y., Satoh, M., Endo, H., Iwase, S., Yamada, H., Matsuda, M., Kano, Y., Nakamura, M. (2002). Apaf-1 is a mediator of E2F-1-induced apoptosis. *J. Biol. Chem., 277*,39760-8.

[175] Nahle, Z., Polakoff, J., Davuluri, R.V., McCurrach, M.E., Jacobson, M.D., Narita, M., Zhang, M.Q., Lazebnik, Y., Bar-Sagi, D., and Lowe, S.W. (2002). Direct coupling of the cell cycle and cell death machinery by E2F. *Nat. Cell Biol., 4*,859-64.

[176] Ishii, H., Mimori, K., Vecchione, A., Sutheesophon, K., Fujiwara. T., Mori, M., and Furukawa, Y. (2004). Effect of exogenous E2F-1 on the expression of common chromosome fragile site genes, FHIT and WWOX. *Biochem. Biophys. Res. Commun., 316*,1088-93.

[177] Yamazaki, K., Hasegawa, M., Ohoka, I., Hanami, K., Asoh, A., Nagao, T., Sugano, I., and Ishida, Y. (2005). Increased E2F-1 expression via tumour cell proliferation and decreased apoptosis are correlated with adverse prognosis in patients with squamous cell carcinoma of the oesophagus. *J. Clin. Pathol., 58*,904-10.

[178] Rigberg, D.A., Blinman, T.A., Kim, F.S., Cole, M.A., and McFadden, D.W. (1999). Antisense blockade of p21/WAF1 decreases radiation-induced G2 arrest in esophageal squamous cell carcinoma. *J. Surg. Res., 81*,6-10.

[179] Rigberg, D.A., Kim, F.S., Sebastian, J.L., Kazanjian, K.K., and McFadden, D.W. (1999). Hypophosphorylated retinoblastoma protein is associated with G2 arrest in esophageal squamous cell carcinoma. *J. Surg. Res., 84*,101-5.

[180] Miyata, H., Doki, Y., Shiozaki, H., Inoue, M., Yano, M., Fujiwara, Y., Yamamoto, H., Nishioka, K., Kishi, K., and Monden, M. (2000). CDC25B and p53 are independently implicated in radiation sensitivity for human esophageal cancers. *Clin. Cancer Res., 6*,4859-65.

[181] Nishioka, K., Doki, Y., Shiozaki, H., Yamamoto, H., Tamura, S., Yasuda, T., Fujiwara, Y., Yano, M., Miyata, H., Kishi, K., Nakagawa, H., Shamma, A., and Monden, M. (2001). Clinical significance of CDC25A and CDC25B expression in squamous cell carcinomas of the oesophagus. *Br. J. Cancer, 85*,412-21.

[182] Takeno, S., Noguchi, T., Kikuchi, R., Uchida, Y., Yokoyama, S., and Muller, W. (2002). Prognostic value of cyclin B1 in patients with esophageal squamous cell carcinoma. *Cancer, 94*,2874-81.

[183] Doak, S.H., Jenkins, G.J., Parry, E.M., Griffiths, A.P., Baxter, J.N., and Parry, J.M. (2004). Differential expression of the MAD2, BUB1 and HSP27 genes in Barrett's oesophagus-their association with aneuploidy and neoplastic progression. *Mutat Res., 547*,133-44.

[184] Flejou, J.F. (2005). Barrett's oesophagus: from metaplasia to dysplasia and cancer. Gut, *54*(Suppl 1),i6-12.

[185] Jankowski, J.A., Wright, N.A., Meltzer, S.J., Triadafilopoulos, G., Geboes, K., Casson, A.G., Kerr, D. and Young, L.S. (1999). Molecular Evolution of the Metaplasia-Dysplasia-Adenocarcinoma Sequence in the Esophagus. *Am. J. Pathol., 154*,965-73.

[186] Werner, M., Mueller, J., Walch, A., and Hofler, H. (1999). The molecular pathology of Barrett's esophagus. *Histol. Histopathol., 14*,553-9.

[187] Morales, C.P., Souza, R.F., and Spechler, S.J. (2002). Hallmarks of cancer progression in BarrettÅfs oesophagus. *Lancet Oncol., 360*,1587–9.

[188] Wijnhoven, B.P., Tilanus, H.W., and Dinjens, W.N. (2001). Molecular biology of BarrettÅfs adenocarcinoma. *Ann. Surg., 233*,322–37.

[189] Teodori, L., Gohde, W., Persiani, M., Ferrario, F., Tirindelli Danesi D., Scarpignato, C., Di Tondo, U., Alo, P., and Capurso, L. (1998). DNA/protein flow cytometry as a predictive marker of malignancy in dysplasia-free Barrett's esophagus: thirteen-year follow-up study on a cohort of patients. *Cytometry, 34*,257-63.

[190] Fennerty, M.B., Sampliner, R.E., Way, D., Riddell, R., Steinbronn, K., and Garewal, H.S. (1989). Discordance between flow cytometric abnormalities and dysplasia in Barrett's esophagus. *Gastroenterology, 97*,815-20.

[191] Chatelain, D., and Flejou, J. (2003). High-grade dysplasia and superficial adenocarcinoma in BarrettÅfs esophagus: histological mapping and expression of p53, p21 and Bcl-2 oncoproteins. *Virchow. Arch., 442*,18–24.

[192] Sherr, C.J. (1996). *Cancer cell cycles. Science, 274*,1672-7.

[193] Kim, R., Clarke, M.R., Melhem, M.F., Young, M.A., Vanbibber, M.M., Safatle-Ribeiro, A.V., Ribeiro, U. Jr, and Reynolds, J.C. (1997). Expression of p53, PCNA, and C-erbB-2 in Barrett's metaplasia and adenocarcinoma. *Dig. Dis. Sci., 42*,2453-62.

[194] Younes, M., Ertan, A., Lechago, L.V., Somoano, J.R., and Lechago, J. (1997). p53 Protein accumulation is a specific marker of malignant potential in Barrett's metaplasia. *Dig. Dis. Sci., 42*,697-701.

[195] Weston, A.P., Banerjee, S.K., Sharma, P., Tran, T.M., Richards, R., and Cherian, R. (2001). p53 protein overexpression in low grade dysplasia (LGD) in Barrett's esophagus: immunohistochemical marker predictive of progression. *Am. J. Gastroenterol., 96*,1355-62.

[196] Syrjanen, K.J. (1987). Human papillomavirus (HPV) infections and their associations with squamous cell neoplasia. *Arch. Geschwulstforsch., 57*,417-44.

[197] Chang, F., Shen, Q., Zhou, J., Wang, C., Wang, D., Syrjanen, S., and Syrjanen, K. (1990). Detection of human papillomavirus DNA in cytologic specimens derived from esophageal precancer lesions and cancer. Scand. *J. Gastroenterol., 25*,383-8.

[198] Chang, F., Syrjanen, S., Shen, Q., Ji, H.X., and Syrjanen, K. (1990). Human papillomavirus (HPV) DNA in esophageal precancer lesions and squamous cell carcinomas from China. *Int. J. Cancer, 45*,21-5.

[199] Ishii, H., Dumon, K.R., Vecchione, A., Fong, L.Y., Baffa, R., Huebner, K., and Croce, C.M. (2001). Potential cancer therapy with the fragile histidine triad gene: review of the preclinical studies. *JAMA, 286*,2441-9.

[200] Ozols, R. (1994). Esophageal cancer. *Curr. Problems Cancer, XVIII*(4),191-246.

[201] Shimada, Y., Sato, F., Watanabe, G., Yamasaki, S., Kato, M., Maeda M., and Imamura, M. (2000). Loss of fragile histidine triad gene expression is associated with progression of esophageal squamous cell carcinoma, but not with the patient's prognosis and smoking history. *Cancer, 89*,5-11.

[202] Zuo, L.F., Lin, P.Z., Qi, F.Y., Guo, J.W., and Liu, J.H. (2003). Flow cytometric analysis of DNA, telomerase content and multi-gene expression in esophageal epithelial dysplasia. *World J. Gastroenterol., 9*,2409-12.

[203] Roth, J.A., and Cristiano, R.J. (1997). Gene therapy for cancer: what have we done and where are we going?. *J. Natl. Cancer Inst., 89*,21-39.

[204] Anklesaria, P. (2000). Gene therapy: a molecular approach to cancer treatment. *Curr. Opin. Mol. Therapeu.*, *2*,426-32.

[205] Levine, A.J. (1997). p53, the cellular gatekeeper for growth and division. *Cell, 88*,323-31.

[206] Rocco, J.W., and Sidransky, D. (2001). p16(MTS-1/CDKN2/INK4a) in cancer progression. *Exp. Cell Res., 264*,42-55.

[207] Ji, L., Fang, B., Yen, N., Fong, K., Minna, J.D., and Roth, J.A. (1999). Induction of apoptosis and inhibition of tumorigenicity and tumor growth by adenovirus vector-mediated fragile histidine triad (FHIT) gene overexpression. *Cancer Res., 59*,3333-9.

[208] Huebner, K., and Croce, C.M. (2001). FRA3B and other common fragile sites: the weakest link. Nat. Rev. *Cancer, 1*,214-221.

[209] Shimada, H., Shimizu, T., Ochiai, T., Liu, T.L., Sashiyama, H., Nakamura, A., Matsubara, H., Gunji, Y., Kobayashi, S., Tagawa, M., Sakiyama, S., and Hiwasa, T. (2001). Preclinical study of adenoviral p53 gene therapy for esophageal cancer. *Surg. Today, 31*,597-604.

[210] Schrump, D.S., Chen, G.A., Consuli, U., Jin, X., and Roth, J.A. (1996). Inhibition of esophageal cancer proliferation by adenovirally mediated delivery of $p16^{INK4}$. *Cancer Gene Ther., 3*,357-64.

[211] Wu, C.M., Huang, T.H., Xie, Q.D., Wu, D.S., and Xu, X.H. (2004). Construction of pETNF-P16 plasmid and its expression properties in EC9706 cell line induced by X-ray irradiation. *World J. Gastroenterol., 10*,2927-30.

[212] Tsunoo, H., Komura, S., Ohishi, N., Yajima, H., Akiyama, S., Kasai, Y., Ito, K., Nakao, A., Yagi, K. (2002). Effect of transfection with human interferon-beta gene entrapped in cationic multilamellar liposomes in combination with 5-fluorouracil on the growth of human esophageal cancer cells in vitro. *Anticancer Res., 22*,1537-43.

[213] Stickle, R.L., Epperly, M.W., Klein, E., Bray, J.A., Greenberger, J.S. (1999). Prevention of irradiation-induced esophagitis by plasmid/liposome delivery of the human manganese superoxide dismutase transgene. *Radiat. Oncol. Investig., 7*,204-17.

[214] Epperly, M.W., Sikora, C., Defilippi, S., Bray, J., Koe, G., Liggitt, D., Luketich, J.D., and Greenberger, J.S. (2000). Plasmid/liposome transfer of the human manganese superoxide dismutase transgene prevents ionizing irradiation-induced apoptosis in human esophagus organ explant culture. *Int. J. Cancer, 90*,128-37.

[215] Epperly, M.W., Gretton, J.A., DeFilippi, S.J., Greenberger, J.S., Sikora, C.A., Liggitt, D., and Koe, G. (2001). Modulation of radiation-induced cytokine elevation associated with esophagitis and esophageal stricture by manganese superoxide dismutase-plasmid/liposome (SOD2-PL) gene therapy. *Radiat. Res., 155*(1 Pt 1),2-14.

[216] Epperly, M.W., Defilippi, S., Sikora, C., Gretton, J., and Greenberger, J.S. (2002). Radioprotection of lung and esophagus by overexpression of the human manganese superoxide dismutase transgene. *Mil. Med., 167*(2 Suppl),71-3.

[217] Epperly, M.W., Tyurina, Y.Y., Nie, S., Niu, Y.Y., Zhang, X., Kagan, V., and Greenberger, J.S. (2005). MnSOD-plasmid liposome gene therapy decreases ionizing irradiation-induced lipid peroxidation of the esophagus. *In Vivo, 19*,997-1004.

[218] Dumon, K.R., Ishii, H., Fong, L.Y.Y., Zanesi, N., Fidanza, V., Mancini, R., Vecchione, A., Baffa, R., Trapasso, F., During, M.J., Huebner, K., and Croce, C.M. (2001). FHIT gene therapy prevents tumor development in Fhit-deficient mice. *Proc Natl. Acad. Sci. U. S. A.,. 98*,3346-51.

[219] Marsman, W.A., Buskens, C.J., Wesseling, J.G., Van Lanschot, J.J., and Bosma, P.J. (2005). Gene therapy for barrett's esophagus: adenoviral gene transfer in different intestinal models. *Cancer Gene Ther., 12*,778-86.

[220] Ning, S.B., Zhang, Z.B., Xie, W.F., and Xin, S.L. (2004). Construction of a recombinant adenovirus carrying endothelial nitric oxide synthase gene and its expression and control in esophageal smooth muscle cells. *Chin. J. Dig. Dis., 5*,17-21.

[221] Schmid, R.M., Weidenbach, H., Draenert, G.F., Liptay, S., Luhrs, H., and Adler, G. (1997). Liposome mediated gene transfer into the rat oesophagus. *Gut, 41*,549-56.

[222] Heideman, D.A., Snijders, P.J., Craanen, M.E., Bloemena, E., Meijer, C.J., Meuwissen, S.G., van Beusechem, V.W., Pinedo, H.M., Curiel, D.T., Haisma, H.J., and Gerritsen, W.R. (2001). Selective gene delivery toward gastric and esophageal adenocarcinoma cells via EpCAM-targeted adenoviral vectors. *Cancer Gene Ther., 8*,342-51.

[223] Tadokoro, K., Koizumi, Y., Miyagi, Y., Kojima, Y., Kawamoto, S., Hamajima, K., Okuda, K., Tanaka, S., Onari, K., Wahren, B., Aoki, I., and Okuda, K. (2001). Rapid and wide-reaching delivery of HIV-1 env DNA vaccine by intranasal administration. Viral *Immunol., 14*,159-67.

[224] Byrnes, C.K., Nass, P.H., Shim, J., Duncan, M.D., Lacy, B., and Harmon, J.W. (2002). Novel nuclear shuttle peptide to increase transfection efficiency in esophageal mucosal cells. *J. Gastrointest. Surg., 6*,37-42.

[225] Marsman, W.A., Buskens, C.J., Wesseling, J.G., Offerhaus, G.J., Bergman, J.J., Tytgat, G.N., van Lanschot, J.J., and Bosma, P.J. (2004). Gene therapy for esophageal carcinoma: the use of an explant model to test adenoviral vectors ex vivo. *Cancer Gene Ther., 11*,289-96.

[226] Niu, Y., Shen, H., Epperly, M., Zhang, X., Nie, S., Cao, S., and Greenberger, J.S. (2005). Protection of esophageal multi-lineage progenitors of squamous epithelium (stem cells) from ionizing irradiation by manganese superoxide dismutase-plasmid/liposome (MnSOD-PL) gene therapy. *In Vivo, 19*,965-74.

[227] Buskens, C.J., Marsman, W.A., Bosma, P.J., and van Lanschot, J.J. (2005). The current state of cancer gene therapy and its application in esophageal carcinoma. *Dig. Surg., 22*,222-33.

[228] Buter, J., and Giaccone, G. (2005). EGFR inhibitors in lung cancer. Oncology (Williston Park). *19*,1707-11; *discussion* 1711-2, 1720-3.

[229] Tang, P.A., Tsao, M.S., and Moore, M.J. (2006). A review of erlotinib and its clinical use. *Expert Opin. Pharmacother., 7*,177-93.

[230] Sutter, A.P., Hopfner, M., Huether, A., Maaser, K., and Scherubl, H. (2006). Targeting the epidermal growth factor receptor by erlotinib (Tarcevatrade mark) for the treatment of esophageal cancer. *Int. J. Cancer, 118*,1814-22.

[231] Ciardiello, F., and De Vita, F. (2005). Epidermal growth factor receptor (EGFR) inhibitors in cancer therapy. *Prog. Drug Res., 63*,93-114.

[232] Teraishi, F., Kagawa, S., Watanabe, T., Tango, Y., Kawashima, T., Umeoka, T., Nisizaki, M., Tanaka, N., and Fujiwara, T. (2005). ZD1839 (Gefitinib, 'Iressa'), an epidermal growth factor receptor-tyrosine kinase inhibitor, enhances the anti-cancer effects of TRAIL in human esophageal squamous cell carcinoma. *FEBS Lett., 579*,4069-75.

In: Esophageal Cancer Research Developments
Editor: Henry K. Kinner, pp. 207-225
ISBN: 1-60021-309-X

Chapter VII

p53 Gene Therapy for Esophageal Cancer

Hideaki Shimada*[*], *Takenori Ochiai
Department of Frontier Surgery, Chiba University Graduate School of Medicine, Chiba, Japan

Abstract

Despite improvement of surgical treatment and application of multi-modality therapies to advanced esophageal cancer, the prognosis is extremely poor in patients with un-resectable tumors. Based on the genetic background of esophageal cancer, we have developed various gene therapy strategies against human esophageal cancer. In this article, we reviewed molecular events of esophageal cancer and gene therapy approaches for its treatment. First, we analyzed p53 genetic alterations and angiogenesis in esophageal cancer. Second, we tested a p53 recombinant adenoviral vector (Ad5CMV-p53). Significant growth suppression was observed following infection with Ad5CMV-p53 in human esophageal squamous cell carcinoma cell lines. This observation suggests that Ad5CMV-p53 may be a potentially effective therapeutic agent for locally advanced esophageal cancer. Promising avenues for investigation include double gene therapy and adjuvant use of gene therapy with radiation therapy. Third, based on recent reports of clinical trials of p53 gene therapy for lung cancer and head and neck cancer, we developed a clinical protocol for p53 gene therapy for un-resectable advanced esophageal cancer. This clinical trial was planned to evaluate vector tolerability and efficacy. A total of 10 patients were enrolled into this phase I/II trial. The overall responses were stable in six and progressive in four cases. Using PCR analyses, gene transfer and *p53* specific

[*] Correspondence to Hideaki Shimada M.D.; Department of Frontier Surgery, Chiba, University Graduate School of Medicine, Chiba 260-8670, Japan. FAX: 81-43-226-2113; TEL: 81-43-226-2110; E-mail: hshimada@faculty.chiba-u.jp

transgene expression were detected in tumor biopsy tissue from all patients. mRNA levels of *p53*, *p21* and *MDM2* increased in all but one case. Three patients showed absence of disease upon repeat biopsies. No serious adverse events related to Ad5CMV-p53 have occurred in these patients, and the trial has been safely conducted. Substantial improvement in swallowing was observed in one patient with stenotic lesions. Intra-tumoral injection of Ad5CMV-*p53* is safe, feasible, and biologically active when administered in multiple doses to patients with esophageal cancer. Observations from this study indicate that this treatment results in local antitumor effects in chemo-radiation-resistant esophageal squamous cell carcinoma.

Key words: p53, gene therapy, esophageal cancer, clinical study

Introduction

Esophageal cancer is the sixth most common cancer in the world, and it is one of the most lethal tumors [1]. In more than 90% of all patients with esophageal cancer, the tumor is detected in an advanced stage [2,3]. Conventional treatments are not adequate for the majority of esophageal cancer patients. Despite improvements in surgical techniques, rapid fatal recurrence is common in patients with advanced esophageal cancer [4]. Because surgical resection alone rarely results in long-term survival for advanced cancer, efforts are now focused on combined multi-modality treatments in an attempt to improve local control and eliminate micro-metastasis present at the time of surgery [5-8]. Recently, the use of neoadjuvant chemoradiotherapy followed by esophagectomy has become a widespread treatment owing to several favorable reports regarding this treatment. Although combined treatment produced a lower rate of cancer-related deaths, it produced a higher rate of treatment-associated mortality. The prognosis is extremely poor in unresectable patients. The five-year survival rate in T4-stage (where cancer tissue infiltration affects other organs according to the esophageal cancer TNM classification system) patients treated with radiotherapy alone is 3.1% [9]. The five-year survival rate after radical radiotherapy was 0% in Stage IV patients [10] and T4 patients [11]. Thus, no therapy that shows sufficient clinical efficacy has been established at present. In this context, it will be necessary to enhance the anti-tumor efficacy of chemotherapy and radiotherapy and reduce their systemic adverse effects in order to improve the outcome of multidisciplinary therapy.

Molecular Events in Esophageal Cancer

Abundant evidence exists that the process of malignant transformation is mediated by genetic alterations. The major lesions detected in cancer cells occur in dominant oncogenes and tumor suppressor genes [12]. Oncogenes are the result of genetic alterations in a class of genes called proto-oncogenes, which participate in the critical functions of normal cells, including signal transduction and transcription. Oncogenes are typically dominant in their mode of action and, therefore, genetic alteration of both alleles is not necessary for cellular transformation. Tumor suppressor genes appear to require homozygous loss of function by mutation, deletion, or a combination of both, for transformation to occur. It is also possible that modification of the expression levels of dominant oncogenes and tumor suppressor genes may influence certain characteristics of cells that contribute to the malignant phenotype.

p53 Alterations in Esophageal Cancer

The p53 gene is the most frequently mutated gene yet identified in human esophageal cancer. p53 abnormalities are observed in 40-60% of patients with esophageal cancer [13-15]. Indeed, p53 alteration is a good predictor for treatment response and survival in esophageal cancer [16,17]. The five-year survival of our patient cohort demonstrated that the survival rate of the p53 mutation-negative group (n=28) is significantly higher than that of the p53 mutation-positive group (n=14) (51% vs 36%, respectively, $P<0.05$) [18]. The p53 gene abnormalities usually consist of partial or total deletion of the short arm of chromosome 17, associated with the simultaneous occurrence of minimal mutations like point mutation in p53 alleles, leading to the inactivation of its tumor suppressor function.

Angiogenic Factor in Esophageal Cancer

Angiogenic factors and growth factors also play an essential role in the process of growth and metastasis of solid tumors [19,20]. Among several angiogenic factors, vascular endothelial growth factor (VEGF) and thymidine phosphorylase (dThdPase) have been shown to be vital for pathological angiogenesis. Immunohistochemical analyses of esophageal carcinomas have revealed that VEGF expression, dThdPase expression and angiogenesis are prognostic factors [21-24]. Serum VEGF concentration as well as serum dThdPase concentration are significantly elevated in patients with primary esophageal cancer. High serum VEGF concentration as well as serum dThdPase concentration were associated with poor treatment response and poor survival [25,26]. Over-expression of VEGF protein is, therefore, partially responsible for the malignant potential of esophageal cancer and represents a useful prognostic marker. Based on the reports that indicate a significant association between VEGF and p53 expression [27], wild type p53 gene transfer into cancer cells is supposed to down regulate angiogenesis [28]. Because a significant association between dThdPase expression and p53 mutation was observed, wild type p53 gene transfer would also down regulate dThdPase expression [29].

p53 Gene Therapy for Human Esophageal Cancer Cells

Based on the concept of genetic alteration in carcinogenesis and angiogenesis, cancer gene therapy has rapidly developed as an alternative to conventional cancer therapy. We previously reported a growth inhibitory effect and an enhancement of the sensitivity to radiation and anti-cancer drugs after transduction of esophageal squamous carcinoma cell lines with retroviral vectors expressing wild-type p53 [30]. Clinically, the recombinant adenoviral vector is expected to be a good agent for introducing genes of interest into cancer cells, due to its high transduction efficiency. Since few studies have been reported on esophageal cancer gene therapy, preclinical studies of adenoviral-mediated p53 gene delivery are required to confirm the anti-cancer effect on esophageal cancer [31].

In vitro Cytotoxic Effect of Ad.CMV-p53 Infection on Cell Growth

The growth properties of human esophageal cancer cells (T.Tn), which revealed a p53 mutation at codon 258 (G to T, resulting in translational termination) in exon 7 of p53 gene, were examined at various MOIs of Ad5CMV-p53, ranging from 10 to 1000. AV1.0CMV-βgal infected cells revealed similar growth rates to those of non-infected cells at MOIs 0, 10, 30 and 100. However, Ad5CMV-p53 infected cells revealed partial growth suppression at an MOI of 30, and complete growth suppression at an MOI of 100.

In vivo Anti-tumor Effect of Ad5CMV-p53 Infection

The *in vivo* inhibition of tumorigenicity of T.Tn cells by injection of Ad5CMV-p53 or AV1.0CMV-βgal was examined. T.Tn cells inoculated with AV1.0CMV-β-gal at an MOI of 300 developed subcutaneous tumors in nude mice within 14 days post-inoculation (5-8 mm in diameter) in 5 of 5 mice (100%). T.Tn cells inoculated with Ad5CMV-p53 at an MOI of 30 developed tumors in 4 of 5 mice (80%) within 28 days post-inoculation. However, significant growth retardation was observed with this lower MOI. T.Tn cells inoculated with Ad5CMV-p53 at an MOI of 100 produced tumor in only one mouse (20%). T.Tn cells inoculated with Ad5CMV-p53 at an MOI of 300 developed no tumors. Tumor growth was thus significantly suppressed in the group treated with Ad5CMV-p53 inoculation.

On days 14 and 16, either an adenovirus vector or Dulbeco's Modified Eagle Medium was injected into two peritumoral sites at an MOI of 100. In this experiment, the tumors injected with medium, AV1.0CMV/293 or AV1.0CMV-βgal showed no growth retardation. The intra-tumoral injections of Ad5CMV-p53 significantly inhibited tumor growth and the tumor volumes remained stable for 10 days after treatment. Two injections of Ad5CMV-p53 suppressed tumor growth by 60-70% compared with the growth observed in tumors injected with AV1.0CMV/293, medium or AV1.0CMV-βgal.

In the above set of experiments, we demonstrated that adenovirus-mediated wild-type p53 gene transduction efficiently suppressed tumor growth of human esophageal cancer cell lines both in vitro and in vivo. The in vitro cytotoxic effects on normal esophageal epithelial cells were assessed and slight growth suppression was observed at an MOI of 300 but not at an MOI of 100 when using Ad5CMV-p53. The mechanism by which overexpression of wild-type p53 protein induces the above growth suppressive effect appears to be the induction of apoptosis. Precise determination of the mechanism underlying the growth suppressive effect of p53 will require further investigation. In vitro study showed that the therapeutic level of MOI was greater than 30 MOI and the transduction rate was approximately 60% at 100 MOI. Tumor cells inoculated with Ad5CMV-p53 demonstrated a significant suppression of tumorigenicity at an MOI of 100 and growth retardation at an MOI of 30. This strategy could be useful in a situation where relatively curative resection for locally advanced esophageal cancer has been performed and possible micro residual cancer cells are present.

Clinical Study of p53 Gene Therapy

Recently, clinical studies of p53 gene therapy for non-small cell lung cancer showed promising results and no severe side effects [32,33]. In vitro data regarding the efficacy of p53 gene therapy in the setting of esophageal cancer are similar to those obtained with lung and head and neck cancers. In clinical trials of lung cancer patients in the US, the protocol for the present study calls for administration twice or three times per cycle.

A phase I clinical trial of p53 gene transduction via retroviral vector for patients with lung cancer conducted at M.D. Anderson Cancer Center (Texas University, U.S.A) showed good tolerability and tumor size reduction in a subset of patients [32,33]. Basic studies of lung cancers showed that transduction with Ad5CMV-p53 reduced the proliferative capacity of cancer cells that had normal p53 gene status as well [34]. This finding was corroborated by findings obtained in a Phase I clinical trial in which Ad5CMV-p53 showed good antitumor effects with only slight adverse reactions, independently of the p53 status of the patients' tumors. A phase I study of adenoviral p53 gene therapy, conducted by Swisher SG et al, demonstrated a partial response in two patients (8%) and disease stabilization in 16 patients (64%) [32]. The therapeutic objective of the present trial was to inhibit the tumor cell growth and induce apoptosis by infecting the cancer cells with the p53-encoding adenovirus. Phase I clinical trials in the US conducted to date for head and neck cancer and non-small cell lung cancer confirmed good tolerability of Ad5CMV-p53 and showed no severe side effects.

Clinical Protocol of p53 Esophageal Cancer Gene Therapy

Patient Eligibility

Study patients included men and women between the ages of 20 and 80 with pathologically documented SCC of the esophagus. Subjects were eligible for the study if they were not candidates for radical surgery. Patients with non-resectable advanced esophageal SCC that was resistant to definitive chemoradiotherapy, more than 60Gy, were eligible. The tumors, revealed stable disease (SD) or progressive disease (PD) after completion of chemoradiation, were defined as chemoradiation-resistant tumor. The tumors, started to grow after partial response (PR) or complete response (CR), were also defined as chemoradiation-resistant tumor. Patients were judged to be able to survive for 12 weeks or more with performance status (PS) of 0-2 defined as follows: PS0, Normal activity, asymptomatic; PS1, Symptomatic, fully ambulatory; PS2, Symptomatic, in bed $< 50\%$ of time; PS3, Symptomatic, in bed $>50\%$ of time, not bedridden. Patients who refused previous treatment were not enrolled. Patients provided written informed consent. Eligibility criteria included ability to administer Ad5CMV-*p53* into tumors under endoscopy or ultrasonography. Tumors ($\leq$ 10 cm in its major axis) were detected and evaluated by physical examinations, endoscopy, ultrasonography, X-rays or computed tomography scanning. Normal bone marrow counts, liver enzymes, and renal function were maintained by adherence to the following criteria: neutrophil count $>1{,}500m^3$; platelet count $\geq 100{,}000/mm^3$; hemoglobin ≥ 8 g/dl; total bilirubin ≤ 1.5 times the upper limit of normal; ALT and AST ≤ 2 times the upper limit of normal; alkaline phosphatase ≤ 5 times the upper limit of normal; creatinine ≤ 1.5 mg/dL; and PT and PTT within the normal range. The presence of a *p53* mutation in the tumor was not a requirement for study entry. Clinical classification of the primary tumor, the degree of lymph node involvement, and organ metastasis were examined according to the TNM/UICC classification.

Study Design and Treatment Plan

This was a single-center, phase I/II study of Ad5CMV-*p53* intra-tumoral injection. The primary objective was to determine the feasibility and safety of this therapy. The secondary objective was to observe biological responses and antitumor effects. The duration of efficacy, time to tumor progression, and survival period were also determined. The study vector was supplied by Introgen Therapeutics, Inc (Houston, TX). The protocol was approved by Chiba University Graduate School of Medicine Institutional Review Board and the Recombinant DNA Advisory Committee of the National Institutes of Health Office of Biotechnology Activities. Written informed consent was obtained from all participants.

The replication-defective adenoviral vector Ad5CMV-*p53* contains the cytomegalovirus promotor, wild-type human *p53* cDNA, and a SV40 polyadenylation signal in a minigene cassette inserted into the E1-deleted region of modified adenovirus-5. Individual doses were 10 to 25×10^{11} viral particles (VP) dependent on tumor size. The vector was suspended in phosphate-buffered saline, which was kept on ice until the time of administration. On a 28-day cycle, intra-tumoral injections of Ad5CMV-*p53* were administered on days 1 and 3 and treated for up to five cycles. The dose level for each patient was determined based on the estimated tumor size (sum of major axis of all measurable lesions). According to the results of clinical trials on head and neck cancer and lung cancer, dose levels were fixed as follows: ≤2cm, $5x10^{11}$VP; 2 to 4cm, $10x10^{11}$VP; 4 to 6cm, $15x10^{11}$VP; 6 to 8cm, $20x10^{11}$VP; and 8 to 10cm, $25x10^{11}$VP. The dosing volume was adjusted using a diluent to be 0.1-0.2 ml/cm^3. This dosing volume was approximately 20% (10-30%) of the estimated tumor volume. An endoscope was inserted under pharyngeal anesthesia, and a prepared Ad5CMV-*p53* solution was injected using a fine needle. In the case of lesions that allowed passage of the endoscope, injections were made at 1-cm intervals so that the entire lesion was covered. Biopsies were taken before installation from tumor and the subsequent intra-tumoral injections were performed through an endoscope when the treatment coincided with a follow-up endoscopic examination on day 3. All treatments were performed in a negative-pressure environment with biosafety precautions. Patients who were free of progression at day 56 were offered additional treatment up to a maximum of five cycles. In the case of lymph nodes, injections were made under ultrasonographic guidance.

Objective Responses

A patient was regarded as a responder if complete response or partial response was obtained during the study, treatment period, or during the 12-month follow-up period and if responses lasted for at least 4 weeks. Evaluation was made to determine if the efficacy was attributable to Ad5CMV-*p53* and not due to adjuvant therapy prohibited by the protocol (surgery, chemotherapy, radiotherapy, etc.). "Guidelines for clinical and pathological studies on carcinoma of the esophagus, ninth edition" was applied to measurable tumors and inmeasurable tumors for tumor efficacy assessment.

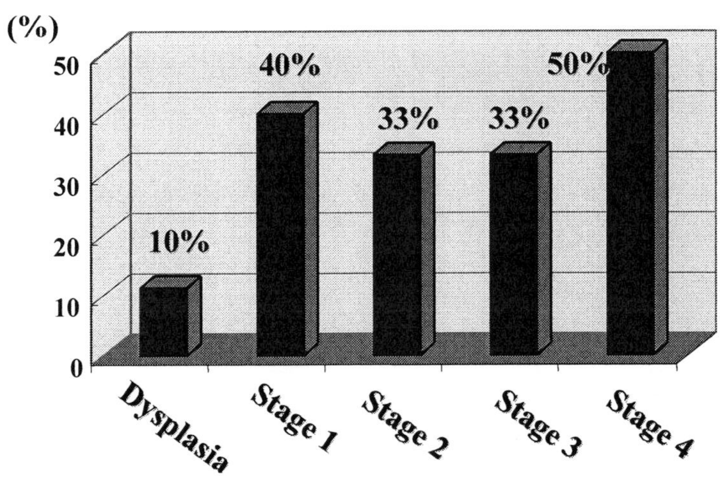

Figure 1. Frequency of p53 mutation according to TNM stages in esophageal squamous cell carcinoma.

Reverse Transcription-PCR and real-Time PCR

Tissue samples were either formalin-fixed or fresh-frozen at the time of resection. For RT-PCR, RNA was extracted from thawed, homogenized, and DNase-digested tissue. After reverse transcription, PCR was performed using primers specific to the Ad5CMV-*p53* vector or to glyceraldehyde phosphate dehydrogenase (GAPDH) as described previously. The esophageal cancer cell line T.Tn was transfected *in vitro* and used as a positive control. To determine the copy number of Ad5CMV-*p53* virus, viral DNA extracted from Ad5CMV-*p53* was used as an absolute standard, and the GAPDH gene was used a reference gene to count cell numbers. The vector-*p53* transcript contains viral sequences from the CMV promoter, which serve to distinguish it from endogenous *p53* mRNA. For RT-PCR detection of vector-*p53* gene expression, primers were constructed to bridge the viral and human transcribed sequences as follows: forward primer, 5'-TGGAGGAGCCGCAGTCAGAT-3'; reverse primer, 5'-ATATCGTCCGGGGACAGC-3'; and probe, 5'-TGCCGTCCCAAGCAATGG ATGA-3'. For detection of *p21* gene expression, primers were constructed as follows: forward primer, 5'-CACTGGAGGGTGACTTCG-3'; reverse primer, 5'-CGTTTGGAGTGGTAGAAATC-3'; and probe, 5'-CCTTGGCCTGCCCAAGCTCT-3'. For detection of *MDM2* gene expression, primers were constructed as follows: forward primer, 5'-CTCACAGATTCCAGCTTCGG-3'; reverse primer, 5'-ACAGAGAAGCTT GGCACGC-3'; and probe, 5'-GGTTAGACCAAAGCCATTGCTTTTGAAG-3'. Expression of mRNA was analyzed by quantitative real time PCR using the Light-Cycler system (Roche, Basel, Switzerland). The PCR products were detected by measuring the fluorescence of SYBR Green I, which selectively bound to double-stranded DNA and emitted greatly enhanced fluorescence. PCR products were resolved on 1% agarose gels and visualized by ethidium bromide staining.

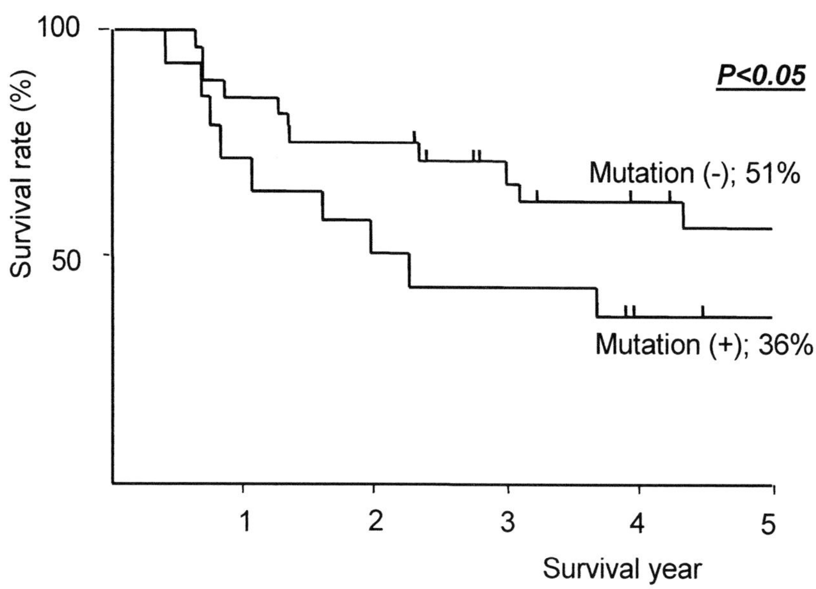

Figure 2. Survival of patients with esophageal squamous cell carcinoma based on p53 gene mutation status.

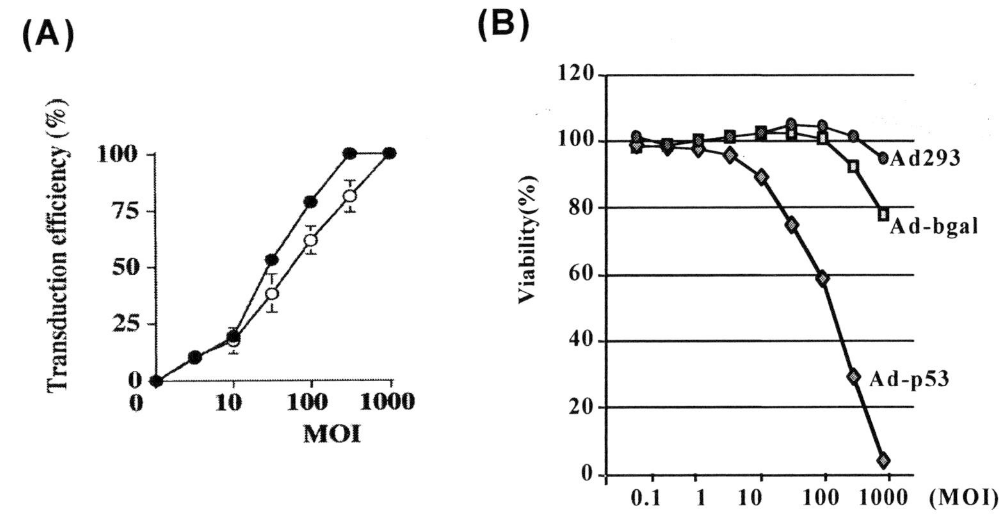

Figure 3. (A) Transduction efficiency of recombinant adenovirus on esophageal carcinoma cell lines ECGI-10 (○) and T.Tn (•). AV1.0CMVβ-gal was used to infect the cells at different MOIs raging from 1 to 1000. The percentage of β-gal-positive cells was obtained from scoring 500 cells each on replicate dishes. (B) In vitro growth suppression of the Ad5CMV-p53-infected human esophageal carcinoma cell line (T.Tn). The cells were inoculated at densities of 5x105 in 60 mm culture dishes 24 h before infection. The cells were infected with Ad5CMV-p53 (□), Ad1.0CMV-βgal (□) or Ad293 (•) at different MOIs raging from 1 to 1000. Triplets of each cell line for each treatment were counted on days 3.

Clinical Monitoring

Patients were monitored for adverse events for a minimum of 56 days. A follow-up endoscopy was performed on days 3, 29, and 57 for all patients. Patients were seen for tumor assessment every month, which included endoscopic examination until there was evidence of progression. Hematology, serum chemistry (including electrolytes, ALT, AST, lactate dehydrogenase, total bilirubin, urea nitrogen, and creatinine), and urinalysis were performed before treatment and during follow-up. Follow-up information was collected from the time of study entry until patient death or through May 1, 2006. Toxicity was graded according to NCI common toxicity criteria (version 1.0). Toxicity not included in the toxicity scale was scored as grade 3 if hospitalization was required and grade 4 if toxicity was regarded as life threatening. Overall survival was defined as the interval between the first treatment and death or last follow-up visit. Time to progression was defined as the interval between first treatment and appearance of a new metastasis, increased size of an established metastasis, or increased size of an established esophageal tumor (25% or greater increase in the product or sum of axes of measurable lesions). Any changes in tumor dimension, radiographic appearance, or endoscopic appearance of an index lesion were noted.

The objective of this study was to evaluate the safety and therapeutic efficacy in patients with non-resectable esophageal cancer after injection of a normal *p53* gene expressing adenovirus vector into tumors. Safety was monitored for 28 days after completion of the final cycle of Ad5CMV-*p53* treatment. Long-term follow-up was conducted every two months until patient death, other treatments become necessary, or expiration of a one-year period after starting treatment.

A total of 10 patients were enrolled into this trial. After providing informed consent, the first patient received injections of Ad5CMV-p53 on December 19, 2000 (Table 1). All patients had histologically confirmed SCC of the esophagus. Patients had been previously treated with definitive chemo-radiation therapy, which included more than 60Gy of radiation and concurrent 5-fluorouracil plus cisplatin. Initial response to chemoradiation revealed complete response in three, partial response in two and stable disease in four patients. The other one patient (#10), received chemoradiation because of positive margin in endoscopic resected specimen. At the time of starting gene therapy, five patients had tumor invasion into adjacent organs (Patient 3, 4, 7, 9, 10) and four had high risk factors for surgery (Patients 1, 2, 5, 8). The other one had T3 tumor with multiple lymph node metastases (Patient 6). Pretreatment tissue samples were obtained from all patients. Eight patients had tumors with *p53* mutations among exons 5 to 8.

Nine of 10 patients completed at least two cycles or four injections of treatment according to the protocol. Treatment cycles ranged from one to five cycles with the median number of cycles as two. Patient 6 did not undergo a second cycle of treatment because he showed rapid progression of distant metastases after the first cycle. Injected lesions were large and ranged from 25-100 mm (median=43 mm in its longest axis).

(a)

Day	7	14	21	28	35	42
300MOI (β gal)	4	5	5	5	5	5
30MOI (p53)	0	0	0	4	4	4
100MOI (p53)	0	0	0	0	0	1
300MOI (p53)	0	0	0	0	0	0

(b)

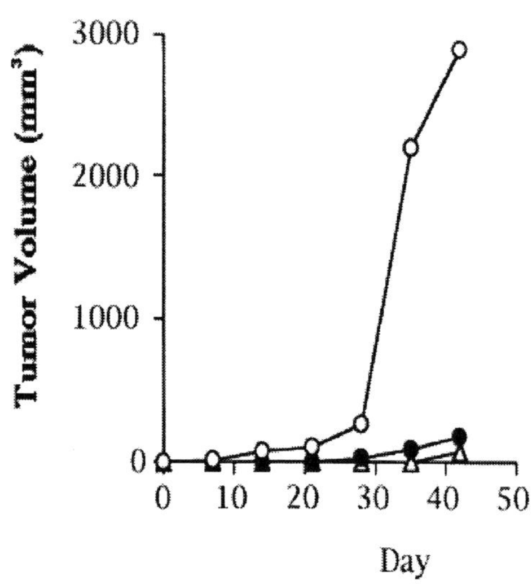

Figure 4. Inhibition of in vivo tumorigenicity of T.Tn cells by Ad5CMV-p53 infection. On day 0, Ad5CMV-p53 was coinjected at the same site of mice with T.Tn xenografts at various MOI (30, 100 and 300). AV1.0CMV-βgal was coinjected at 300 MOI as a control. (a) The number of mice per five mice with tumor growth. (b) The mean tumor volumes per five mice following injection were plotted against the number of days since subcutaneous transplantation; bars, standard deviation.

Table 1. Characteristics of the patients with esophageal squamous cell carcinoma treated with AD.CMV-*p-53*

Patient No.	Age & Gender	Time to Tumor Progression[a]	Site of p53 Mutation	Viral Particles	Treatment Cycles	Tumor Size	Local & Overall Response[d]	Prognosis after GT[e]
1	64M	7M	Exon 5	15×10^{11}	2	43mm	SD & SD	19M dead
2	71M	7M	Exon 7	10×10^{11}	5	25mm	SD & SD	15M dead
3	62M	5M	Negative[b]	15×10^{11}	3	50mm	SD & PD	3M dead
4	78M	2M	Exon 8	10×10^{11}	2	30mm	PD & PD	6M dead
5	66M	2M	Exon 7	10×10^{11}	3	40mm	SD & SD	35M alive
6	60M	1M	Exon 7	20×10^{11}	1	70mm	SD & PD	2M dead
7	67F	4F	Exon 5	10×10^{11}	2	38mm[c]	SD & SD	13M dead
8	58M	8M	Exon 6	10×10^{11}	2	40mm	SD & SD	15M dead
9	48M	2M	Exon 7	25×10^{11}	4	100mm	SD & SD	12M dead
10	77M	12M	Negative[b]	20×10^{11}	2	68mm	SD & PD	2M dead

a. Time to tumor progression after completion of chemo-radiation therapy.
b. No mutation among exon 5, 6, 7 and 8.
c. Tumor size was sum of esophageal tumor and lymph node.
d. Treatment response was determined 4 weeks after completion of therapy by external review board. SD, stable disease; PD, progressive disease.
e. Survival months after first injection of Ad.CMV-p53.

Table 2. Highest grade adverse event observed in 10 patients during AD.CMV-p53

Adverse events	Grade 1	Grade 2	Grade 3	Grade 4	Total
Fever	1 (10%)	9(90%)	0	0	10(100%)
Pain	1 (10%)	2(20%)	0	0	3(30%)
Hyperglycemia	3(30%)	0	0	0	3(30%)
Hypocalcemia	2(20%)	0	0	0	2(20%)
PTT elongation	1(10%)	0	0	0	1(10%)
Amylase	1(10%)	0	0	0	1(10%)
Creatinine	1(10%)	0	0	0	1(10%)

Adverse Events

Adverse events attributed to treatment with Ad.5CMV-*p53* generally were mild to moderate. The most common adverse events were fever and local pain (NCI grade 1 or 2). Fever was observed in all patients and pain in 30% (Table 2). Overall, drug administration was feasible and well tolerated. In one patient, the study had to be discontinued prematurely because of disease progression. This patient died 61 days after starting treatment (59 days after the second and last dose), but death was not related to treatment with Ad5CMV-*p53*. Three patients showed hyperglycemia, which was likely attributable to total parental nutrition. Two patients revealed hypocalcemia. One patient each experienced partial thromboplastin time elongation, an increase in serum amylase or creatinine. There were no other significant laboratory abnormalities detected on follow-up evaluations.

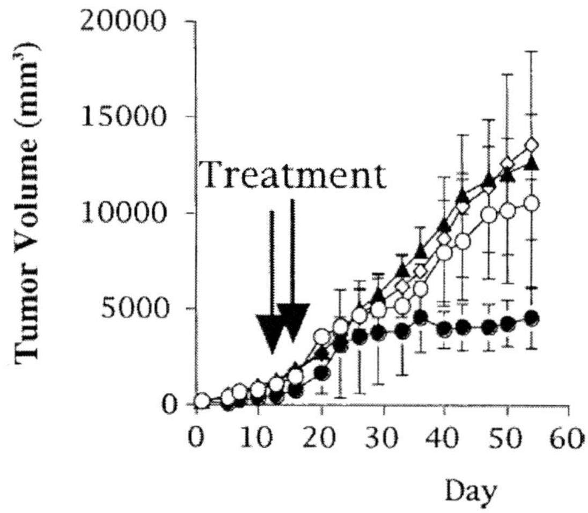

Figure 5. The effect of Ad5CMV-p53 infection on in vivo tumor growth of T.Tn cells. Established tumors were treated after reaching a size of 200mm3 with two injections delivered on days 14 and 16 of Dulbecco's Modified Eagle Medium/F-12 medium or the virus (total dose, 2x1010 PFU). The mean tumor volumes per five mice following injection were plotted against the number of days since subcutaneous transplantation; bars, standard deviation. DMEM;Dulbecco's Modified Eagle Medium/F-12 medium.

Detection of *p53* Transgene, *p21* Induction and *MDM2* Induction

Tissue samples were obtained and fresh-frozen in liquid nitrogen before and after treatment. All patients demonstrated successful *p53* gene transfer by DNA-PCR on day 3 (Table 3), whereas all pre-treatment samples were negative. DNA levels were quantitiated in biopsies from patients 1-3 and demonstrated an average of 8.2×10^4 copies/10^5 gene copies. Post-treatment biopsies were obtained on day 3, day 29 and day 57. We found vector-specific *p53* expression in post-treatment esophageal biopsies from all patients. Expression levels increased in seven patients, but expression levels did not increase on day 3 in the other three patients (Patient 3, 7, 8) (Table 3). The expression level of *p53* on day 29 was higher in all but two patients. *P21* and *MDM2* expression levels were increased in six patients on day 3 (Patients 1, 4, 5, 6, 7, 10). Although the peaks of average mRNA levels of each gene were slightly different, all three levels increased after treatment (Figure 1A). The ratio of expression between *p21* and *MDM2* increased on day 29 (Figure 1B). Biopsies taken on day 29 showed no evidence of disease in three patients (2, 3 and 5). Patients 2 and 3 also showed the highest levels of gene transfer.

Long-term Follow up

Nine of the 10 eligible patients have died of disease. One patient had local tumor progression and four patients had systemic progression during the initial 56 days. Six of 10 patients were alive for more than one year (Table 1). Three patients showed no evidence of tumor on multiple biopsies after treatment (Patients 2, 3, and 5). Patient 2 had a tumor with deep ulcer before treatment. After three or four cycles, ulcer was changed to mild shape. No viable cancer cells were observed in biopsy specimens. Stable local tumor and no residual SCC were found on esophageal biopsies during the five cycles of treatment. Patient 3 had a large 50 mm mass and was unable to swallow at the initiation of gene therapy (Figure 7A). Analysis of biopsies after the first cycles of treatment showed mild dysplasia and necrotic tissue (Table 3). After two injections, he was able to swallow liquid and meals (Figure 7B). Although the endoscope could not pass through the stenotic portion of the lesion before treatment, it passed through after treatment (Figure 7C and D). Although the local tumor was controlled during treatment, distant metastases occurred and the patient died at 3 months. Patient 5 did not show tumor progression for 24 months after initial gene therapy. Although slight viable cancer cells were observed in biopsy specimens at 24 months, the tumor was well controlled by argon-plasma coagulation.[17] He remained progression-free and alive 50 months after completion of treatment as of today. He received a total of three courses with 10^{11} viral particles on days 1 and 3 of each course. With an intent to treat analysis, overall survival rate was 60% at one year (Figure 8). Four patients have developed clinically diagnosed distant metastases (two bone metastasis, one liver and lymph node, and one lung). Three of these four patients, with development of distant metastases, died of disease within three months after initial viral treatment. Median time to local as well as distant progression was 6 months.

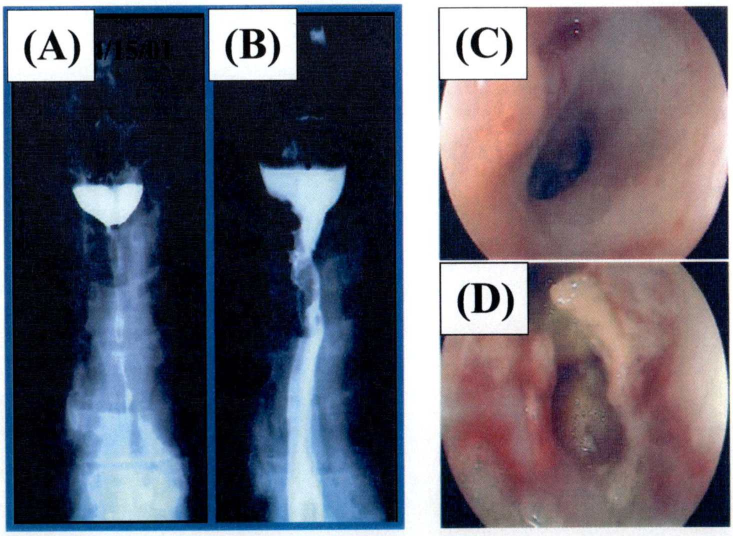

Figure 7. Patient No. 3: Upper thoracic esophageal cancer unable to undergo surgery because of tracheal invasion. (A) Esophagography shows complete obstruction before p53 gene transfer. (B) Esophageal obstruction was reduced after treatment. Esophagoscopic view before treatment (C) and after treatment (D).

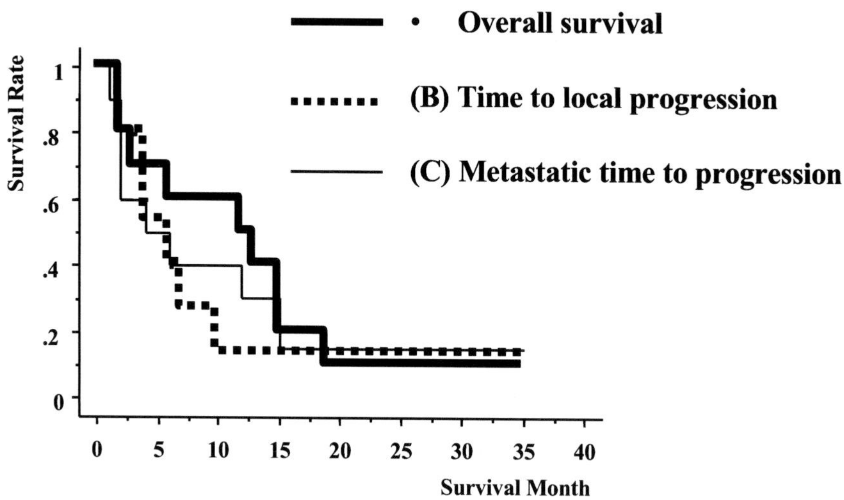

Figure 8. Overall survival (A), time to local progression (B) and time to metastatic progression of all patients (C).

Future Clinical Trial of p53 Gene Therapy Combined with Chemoradiotherapy

According to recent cancer gene therapy reports, adenovirus-mediated p53 gene transfer is frequently used, together with cis-dichloro-diammineplatinum(II) administration [35, 36] or ionizing radiation [37]. We herein report the findings of a preclinical study that reveal the efficacy of Ad5CMV-p53 treatment on human esophageal squamous cell carcinomas both in vitro and *in vivo*. Although the transduction efficiency was relatively low on esophageal carcinoma, an MOI of 100 was sufficient to show efficacy in clinical trials. Since an MOI of 100 may represent a relatively high viral concentration for clinical use, combination therapy using CDDP or other apoptotic agents should also be investigated. Further experimentation is required to develop a new combination therapy with a low dose CDDP and/or 5-FU and Ad5CMV-p53 at a low MOI and also a new combination therapy with concurrent chemoradiotherapy. A phase II trial of adenoviral-mediated p53 gene transfer in conjunction with radiation therapy for lung cancer patients was safe and effectiveness [37]. We also confirmed a synergistic anti-tumor effect of Ad5CMV-p53 and radiation therapy for human esophageal cancer [38].

Double Gene Transfer to Enhance p53 Gene Therapy

Several reports have shown that double gene transfer enhanced anti-tumor activity of single gene transfer [39,40,41]. Some reports suggested that p14, p16 and p33 genes were able to enhance apoptotic tumor cell death of p53 gene therapy [40,41,42]. We suggested a role of Mdm2 in the synergistic effect of p33 gene transfer in combination with p53 gene transfer [42]. Although stable expression of both genes was claimed for double gene therapy, this mode of gene therapy should undergo clinical study in the near future.

We herein reviewed the findings of a preclinical study of esophageal cancer gene therapy and reviewed a protocol of clinical p53 gene therapy for unresectable advanced esophageal squamous cell carcinomas.

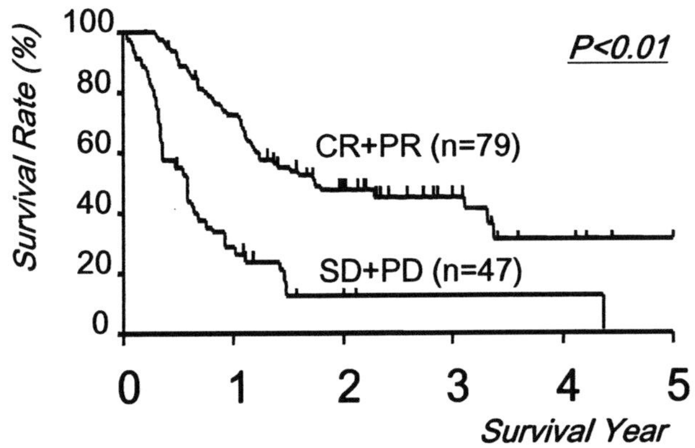

Figure 9. Treatment response and survival of patients with thoracic esophageal carcinoma. A total of 126 patients were treated with chemoradiation between 1997 and 2001.

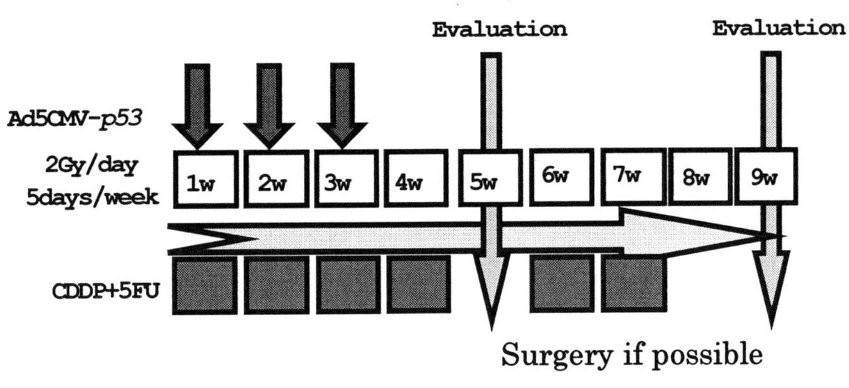

Figure 10. Protocol of p53 gene therapy combined with chemoradiation for T4 esophageal carcinoma.

Acknowledgements

This work was supported, in part, by the 21st Century COE (Center of Excellence) Programs Grant from the Ministry of Education, Culture, Sports, Science and Technology of Japan.

References

[1] Parkin DM, Laara E, Muir CS. Estimates of the worldwide frequency of sixteen major cancers in 1980. *Int J Cancer* 1988;41:184-97.

[2] Sugimachi K, Ohno S, Matsuda H, Mori M, Matsuoka H, Kuwano H. Clinicopathologic study of early stage esophageal carcinoma. Surgery 1989;105:706-10.

[3] Kato H. Diagnosis and treatment of esophageal neoplasms. *Jpn J Cancer Res.* 1995;86:993-1009.

[4] Isono K, Sato H, Nkaayma K. Results of a nationwide study on the three-field lymph node dissection of esophageal cancer. *Oncol* 1991;48: 411-20.

[5] Ancona E, Ruol A, Castoro C, Chiarion-Sileni V, Merigliano S, Santi S, Bonavina L, Peracchia A. First-line chemotherapy improves the resection rate and long-term survival of locally advanced (T4, any N, M0) squamous cell carcinoma of the thoracic esophagus: final report on 163 consecutive patients with 5-year follow-up.*Ann Surg.* 1997;226:714-24.

[6] Ide H, Nakamura T, Hayashi K, Eguchi R, Tanigawa K, Ota M. Neoadjuvant chemotherapy with cisplatinum/5-fluorouracil/low-dose leucovorin for advanced squamous cell carcinoma of the esophagus. *Semin Surg Oncol.* 1997; 13:263-9.

[7] Poplin EA, Jacobson J, Herskovic A, Panella TJ, Valdivieso M, Hutchins LF, Macdonald JS. Evaluation of multimodality treatment of locoregional esophageal carcinoma by Southwest Oncology Group 9060. *Cancer.* 1996; 78:1851-6.

[8] Toh U, Yamana H, Sueyoshi S, Tanaka T, Niiya F, Katagiri K, Fujita H, Shirozou K, Itoh K. Locoregional cellular immunotherapy for patients with advanced esophageal cancer. *Clin Cancer Res.* 2000; 6:4663-73.

[9] Okawa T, Dokiya T, Nishio M, Hishikawa Y, Morita K.Multi-institutional randomized trial of external radiotherapy with and without intraluminal brachytherapy for esophageal cancer in Japan. Japanese Society of Therapeutic Radiology and Oncology (JASTRO) Study Group. *Int J Radiat Oncol Biol Phys.* 1999; 45:623-8.

[10] Yamada S, Takai Y, Nemoto K, Kakuto Y, Ogawa Y,Hoshi A, Sakamoto S. *J. Jpn. Soc. Cancer Ther.* 1994;29: 1819-25.

[11] Zenone T, Romestaing P, Lambert R, Gerard JP. Curative non-surgical combined treatment of squamous cell carcinoma of the oesophagus. *Eur J Cancer.* 1992;28A(8-9):1380-6.

[12] Soussi T, Legros Y, Lubin R, Ory K, Schlichtholz B. Multifactorial analysis of p53 alteration in human cancer: a review. *Int J Cancer* 1994;57:1-9.

[13] Hollstein MC, Metcalf RA, Welsh JA, Montesano R, Harris CC. Frequent mutation of the p53 gene in human esophageal cancer. *Proc Natl Acad Sci U S A.* 1990; 87:9958-61.

[14] Bennett WP, Hollstein MC, Metcalf RA, Welsh JA, He A, Zhu SM, Kusters I, Resau JH, Trump BF, Lane DP, et al. p53 mutation and protein accumulation during multistage human esophageal carcinogenesis. *Cancer Res.* 1992;52:6092-7.

[15] Huang Y, Meltzer SJ, Yin J, Tong Y, Chang EH, Srivastava S, McDaniel T, Boynton RF, Zou ZQ. Altered messenger RNA and unique mutational profiles of p53 and Rb in human esophageal carcinomas. *Cancer Res.* 1993;53:1889-94.

[16] Sarbia M, Porschen R, Borchard F, Horstmann O. and Willers R. p53 protein expression and prognosis in squamous cell carcinoma of the esophagus. *Cancer* 1994; 74:2218-23.

[17] Ribeiro U Jr, Finkelstein SD, Safatle-Ribeiro AV, Landreneau RJ, Clarke MR, Bakker A, Swalsky PA, Gooding WE, Posner MC. p53 sequence analysis predicts treatment response and outcome of patients with esophageal carcinoma. *Cancer.* 1998;83:7-18.

[18] Kobayashi S, Koide Y, Endo M, Isono K, Ochiai T. The p53 gene mutation is of prognostic value in esophageal squamous cell carcinoma patients in unified stages of curability. *Am J Surgery* 1999;177: 497-502.

[19] Weidner N. Angiogenesis as a predictor of clinical outcome in cancer patients. *Hum Pathol.* 2000;31:403-5. Review.

[20] Hanahan D, Folkman J. Patterns and emerging mechanisms of the angiogenic switch during tumorigenesis. *Cell* 1996;86:353-64.

[21] Inoue K, Ozeki Y, Suganuma T, Sugiura Y, Tanaka S. Vascular endothelial growth factor expression in primary esophageal squamous cell carcinoma. Association with angiogenesis and tumor progression. *Cancer* 1997;79:206-13.

[22] Shih CH, Ozawa S, Ando N, Ueda M, Kitajima M. Vascular endothelial growth factor expression predicts outcome and lymph node metastasis in squamous cell carcinoma of the esophagus. *Clin Cancer Res.* 2000;6:1161-8.

[23] Takebayashi Y, Natsugoe S, Baba M, Akiba S, Fukumoto T, Miyadera K et. al. Thymidine phosphorylase in human esophageal squamous cell carcinoma. *Cancer* 1999; 85:282-9.

[24] Igarashi M, Dhar DK, Kubota H, Yamamoto A, El-Assel O, Nagasue N. The prognostic significance of micro vessel density and thymidine phosphorylase expression in squamous cell carcinoma of the esophagus. *Cancer* 1998; 82:1225-32.

[25] Shimada H, Takeda A, Nabeya Y, Okazumi S, Matsubara H, Funami Y et al. Clinical significance of serum vascular endothelial growth factor in esophageal squamous cell carcinoma. *Cancer* 2001;92:663-9.

[26] Shimada H, Takeda A, Nabeya Y, Okazumi S, Matsubara H, Funami Y et al. Prognostic significance of serum thymidine phosphorylase concentration in esophageal squamous cell carcinoma. *Cancer* 2002;94:1947-54.

[27] Linderholm BK, Lindahl T, Holmberg L, Klaar S, Lennerstrand J, Henriksson R, Bergh J. The expression of vascular endothelial growth factor correlates with mutant p53 and poor prognosis in human breast cancer. *Cancer Res* 2001;61:2256-60.

[28] Bouvet M, Ellis LM, Nishizaki M, Fujiwara T, Liu W, Bucana CDet. al. Adenovirus-mediated wild-type p53 gene transfer down-regulates vascular endothelial growth factor expression and inhibits angiogenesis in human colon cancer. *Cancer Res.*1998;58:2288-92.

[29] Oota M, Shimada H, Takeda A, Nabeya Y, Okazumi S, Matsubara H et al. Prognostic significance of thymidine phosphorylase and p53 co-expression in esophageal squamous cell carcinoma. *Oncology Report* 2002;9:23-28.

[30] Matsubara H, Kimura M, Sugaya M, Asano T, Ochiai T, Sakiyama S, Tagawa M. Expression of wild-type p53 gene confers increased sensitivity to radiation and chemotherapeutic agents in human esophageal carcinoma cells. *Int J Oncolgy* 1999;14: 1081-5.

[31] Shimada H, Shimizu T, Ochiai T, Liu T, Sashiyama H, Nakamura A et al. Preclinical study of adenoviral p53 gene therapy for esophageal cancer. *Surg Today* 2001;31:597-604.

[32] Swisher SG, Roth JA, Nemunaitis J, Lawrence DD, Kemp BL, Carrasco CH et. al. Adenovirus-mediated p53 gene transfer in advanced non-small-cell lung cancer. *J Natl Cancer Inst.* 1999;91:763-71.

[33] Schuler M, Herrmann R, De Greve JL, Stewart AK, Gatzemeier U, Stewart DJ et al. Adenovirus-mediated wild-type p53 gene transfer in patients receiving chemotherapy for advanced non-small-cell lung cancer: results of a multicenter phase II study. *J Clin Oncol.* 2000;19:1750-8.

[34] Fujiwara T, Grimm EA, Mukhopadhyay T, Zhang WW, Owen-Schaub LB, Roth JA. Induction of chemosensitivity in human lung cancer cells in vivo by adenovirus-mediated transfer of the wild-type p53 gene. *Cancer Res* 1994;54:2287-91.

[35] Nguyen DM, Spitz FR, Yen N, Cristiano RJ, Roth JA. Gene therapy for lung cancer: enhancement of tumor suppression by a combination of sequential systemic cisplatin and adenovirus-mediated p53 gene transfer. *J Thorac Cardiovasc Surg.* 1996; 112:1372-6; discussion 6-7.

[36] Ogawa N, Fujiwara T, Kagawa S, Nishizaki M, Morimoto Y, Tanida T et al. Sensitization of cis-platinum by a recombinant adenovirus vector expressing wild-type p53 gene in human ovarian carcinomas. *Int J Cancer* 1997; 7:367-70.

[37] Swisher SG, Roth JA, Komaki R, et al. A phase II trial od adnoviral mediated p53 gene transfer (RPR/ING/IN 201) in conjunction with radiation therapy in patients with localized non-small cell lung cancer (NSCLC). *Proc Am Soc Clin Oncol* 2000;19:461a.

[38] Oohira G, Yamada S, Ochiai T, Matsubara H, Okazumi S, Ando K, Tsujii H, Hiwasa T, Shimada H. Growth suppression of esophageal squamous cell carcinoma induced by heavy carbon-ion beams combined with p53 gene transfer. *Int J Oncol.* 2004;25:563-9.

[39] Shimizu T, Shimada H, Ochiai T, Hamada H. Adenovirus-mediated gene therapy for esophageal carcinoma cells with fusion gene (Uracil Phosphoribosyl-transferase and herpes simplex virus thymidine kinase). *Cancer Gene Therapy* 2001;8:512-21.

[40] Saegusa M, Machida B D, Okayasu I. Possible associations among expression of p14(ARF), p16(INK4a), p21(WAF1/CIP1), p27(KIP1), and p53 accumulation and the balance of apoptosis and cell proliferation in ovarian carcinomas. *Cancer.* 2001;92:1177-89.

[41] Sandig V, Brand K, Herwig S, Lukas J, Bartek J, Strauss M. Adenovirally transferred p16INK4/CDKN2 and p53 genes cooperate to induce apoptotic tumor cell death. *Nat Med.* 1997;3:313-9.

[42] Shimada H, Liu T, Ochiai T, Shimizu T, Haupt Y, Hiwasa T et al. Stimulation of adenoviral wild-type p53-induced apoptotic cell death by overexpression of p33ING1 in T.Tn human esophageal carcinoma cells. *Oncogene* 2002;21:552-7.

In: Esophageal Cancer Research Developments
Editor: Henry K. Kinner, pp. 227-246

ISBN: 1-60021-309-X

Chapter VIII

Oesophageal Cancer Treatment Using Intraluminal Ultrasound Applicator

D. Melodelima[1*], F. Prat[2] and D. Cathignol[1]

[1]INSERM, U556, Lyon, F-69003, France

[2]Hopital Cochin, Paris, F-75014, France

Abstract

Oesophageal tumours generally bear a poor prognosis. Radical surgery is generally the only curative method available but is not feasible in the majority of patients; palliative therapy with stent placement is generally performed. It has been demonstrated that High Intensity Ultrasound (HIU) can induce rapid, complete and well-defined coagulation necrosis. Thus, for the treatment of this cancer, we designed an ultrasound applicator that uses an intraluminal approach to fill up this therapeutic gap. Thermal ablation is performed with water-cooled ultrasound transducers operating at a frequency of 10 MHz. Elementary lesions extend from the transducer surface up to 10 mm in depth when applying an intensity of 16 W/cm2 for 10s. A lumen inside the therapy applicator provides path for an endoscopic ultrasound imaging probe. The rotation of the applicator around its axis enables treatment of sectorial or cylindrical volumes. This method is thus particularly suitable for oesophageal tumours that may develop only on a portion of the oesophageal circumference. Feasibility experiments were conducted from bench to *in vivo* studies on pig oesophagi. Four patients with oesophageal tumours included in a pilot study have been treated using this ultrasound applicator. The treatment of oesophageal tumours is done under fluoroscopic guidance and ultrasound imaging. Tumour response was observed in all cases. This clinical work demonstrated the efficacy of interstitial high intensity ultrasound therapy for local tumour destruction in the oesophagus.

* Corresponding author: David Melodelima, INSERM U556, 151 cours Albert Thomas, 69003 Lyon, France; Email: melodelima@lyon.inserm.fr; Tel: +33 (0) 4 72 68 19 30; Fax: +33 (0) 4 72 68 19 31

Keywords: Oesophagus, tumour, ultrasound, thermal ablation, coagulation necrosis, intraluminal, sector-based, transducer, therapy.

Introduction

Oesophageal tumours develop inside the circumference of the lumen and are very difficult to treat for two reasons. They are sometimes sector-based and the tumour thickness can vary significantly from one point to another (three millimetres to 12 mm in most cases). Furthermore, most occurrences of this type of cancer are not amenable to curative resection because of the extent of the tumour at diagnosis and the presence of a comorbid condition. Radiochemotherapy has not demonstrated any survival advantage so far. Photocoagulation by laser is sometimes used [1,2], but it is relatively expensive and demanding since the tissue destruction is superficial and several sessions are generally needed to achieve symptomatic improvement. Thus, for this form of cancer, a minimally invasive local treatment, leading to an immediate, deep and complete destruction of the targeted tissues, could be curative in the earliest-detected cases. If the tumour has reached a metastatic stage, the treatment could be envisaged as a palliative strategy to delay the use of stents and increase the median survival time. Among the interstitial heating modalities for therapeutic thermal ablation, tissue coagulation by high intensity ultrasound is a well-established method of cancer treatment. Treatments by ultrasound are well-delimited; damages to surrounding tissues are minimal. Excellent results have been obtained (both experimentally and clinically) in inducing homogeneous and reproducible tumour destruction by thermal coagulation necrosis [3-6]. World-wide clinical applications of focused ultrasound surgery (FUS) include the treatment of glaucoma [7], benign prostatic hyperplasia (BPH) [8,9], treatment of localised prostatic cancer of the prostate [10], ocular tumours [11], breast tumours [12] and brain tumours [13]. Recently, several authors have suggested using miniaturised applicators for therapeutic ultrasound [14-16], which makes it an approach that is particularly appropriate for carcinoma of the oesophageal ducts.

Some miniaturised ultrasound applicators use a tubular transducer inducing cylindrical coagulation necrosis [17]. Sector-based necroses are generated by remodelling irreversibly a tubular transducer and the radiated acoustic field. The method consists in cutting or paralysing part of the transducer [18,19]. In contrast, a plane transducer generates elementary parallelepiped-shaped lesions in front of the emitting face. Since the ultrasonic beam does not diverge in the near field, the pressure drop along the ultrasonic axis is only dependent on the energy attenuation in tissues [17,20]. Each elementary lesion can be induced in less time (10 s), to a greater depth, and with less perfusion-dependence [21] compared with lesions produced by tubular transducers. Furthermore, by rotating the applicator around its axis after each ultrasound exposure, a sector-based or cylindrical volume can be coagulated.

The aim of this study was first to conduct *ex vivo* experiments on pig liver in order to demonstrate the ability of the applicator to generate any angular sector-based shapes of coagulation necroses and to link the depth of the cylindrical lesions with the exposure conditions. In the case of sector-based tumours, both cancerous and thin healthy zones are irradiated. The deposited thermal dose should not induce perforation in the healthy zone, yet it should still coagulate the immediately adjacent cancerous tissues. Consequently, the second aim of this study was to determine the maximal thermal dose that can be delivered to a healthy oesophagus without causing perforation. The thermal dose determined by these in vivo trials was compared with recent studies [16] to verify whether or not the exposure conditions were high enough to kill malignant cells. Finally an open pilot trial which was conducted between Sept, 2002 and Nov, 2004 on four patients.

Intraluminal Ultrasound Applicator

The head of the applicator is made of brass and round-shaped to permit transoesophageal application without risk of injury (Figure 1). The ultrasound emitter is a $15\times8\ mm^2$ piezoceramic plane transducer (Quartz & Silice, P7-62, Nemours, France) which operates at 10.35 MHz. This air-backed transducer was sealed with epoxy resin (Emerson & Cuming, Stycast 2651, Westerlo, Belgium) to ensure the applicator was watertight. The rotation of the applicator is controlled remotely with negligible angular loss using an 80-cm long, 10-mm OD, flexible metallic shaft. A biocompatible envelope (1.3 mm thick) covers this metallic shaft. This envelope conforms to medical device safety requirements USP VI (U.S. Pharmacopoeia, class VI). Connections for the transducer supply are done with a PVC casing attached on the opposite end of the applicator. This casing is mounted on an UR 100 microcontrol unit (Microcontrol, Evry, France) to monitor the rotation of the applicator. Electrical connections were realised via a miniaturised 50-Ω coaxial cable that was 80-cm long with a 0.9-mm OD. A tube (1-mm o.d.) over the whole length of the applicator is for holding a guide wire previously introduced in the oesophagus. The active part of the applicator was covered with a 65-μm thick latex balloon attached using watertight seals. This envelope attenuated the ultrasound pressure by about 9% [22] and was inflated with degassed water to provide acoustic coupling between the transducer and the tissues. This envelope can be inflated via a syringe connected to the tank. During treatments, the transducer was cooled using a continuous flow of degassed water at room temperature (25°C) at a rate of 0.2 L/min. A peristaltic Masterflex pump (L/S model 7518-60, Cole-Parmer Instruments Co., Chicago, IL, USA) drove the water around a closed cooling circuit and through a 1-liter watertight tank of degassed water at room temperature. The treatment is monitored using an ultrasound imaging probe. A tube inside the applicator provides path for an UM2R ultrasound imaging probe (Olympus, Japan) which goes through the therapy applicator and perform a visualisation of the tissues in front of the therapy transducer.

The electrical power was delivered in continuous mode by a Kalmus model 150 CF power amplifier (Engineering International, Woodinville, WA, USA) driven by a Hameg HM8131 sinusoidal wave generator (Hameg, Frankfurt, Germany). Directional power meter (wattmeter/reflectometer (Rohde & Schwarz, Munich, Germany) fitted with a bidirectional coupling device probe (NAP Z7, Rohde & Schwarz, Munich, Germany) were inserted into the line between the amplifier output and the applicator to determine with the aid of a built-in directional coupler the incident and reflected electrical power. The generator was controlled by a timer, which enabled emissions to be triggered manually and cut off automatically at the end of the desired shot time. A radiation force balance, to the design of Hill [23], was used to measure the total acoustic power output of the ultrasound beam in degassed water at 21°C. An electro-acoustic efficiency of 61% at 10.35 MHz was measured.

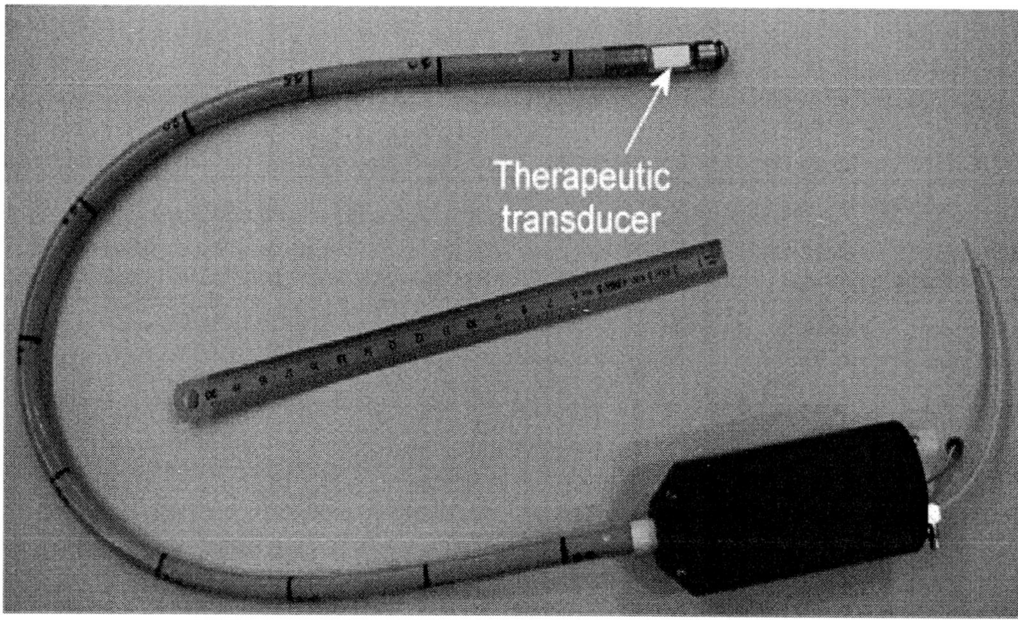

Figure 1. Intraluminal high intensity ultrasound applicator.

Ex Vivo Experiments

Method

Tests were performed *ex vivo* on 48 samples of pig liver. Before experiment, each sample of liver was degassed in a vacuum pump (0.7 bars for 30 min) at room temperature to remove bubbles (which can reflect the beam) initially present in the tissue and vessels. During the tests, each sample was immersed in a tank of degassed water maintained at 37°C. A thermocouple was introduced into each sample to ensure that temperature was close to 37°C before commencing the firings. A hole is produced into the centre of each sample with a PVC awl (12-mm OD). The ultrasound applicator was introduced into the centre of each parallelepiped shaped section of liver (100 x 100 x 30 mm^3) through the hole (Figure 2). 20 exposures were done with 18°-increments of angular rotation in order to induce cylindrical zone of coagulation necrosis. The time between the end of one exposure and the beginning of the next exposure after the rotation is 10 s. These exposure conditions come from our experience in the treatment of biliary cancers [24]. This method provides a homogeneous necrosis across the complete circumference. The duration of the elementary exposures was set to 5, 10, 15, and 20 s and the acoustic intensity to 12, 14, and 16 W/cm^2. A set of three experiments was carried out for each combination of exposure duration and acoustic intensity to ensure reproducibility of the results. A total of 36 circular lesions were created.

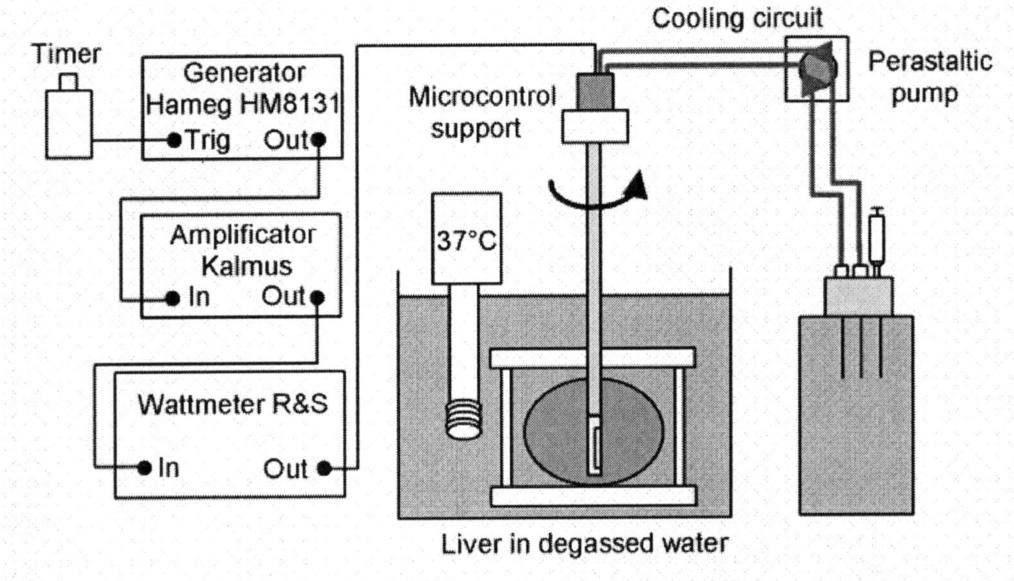

Figure 2. Experimental set up and electrical equipment for *in vitro* experiments.

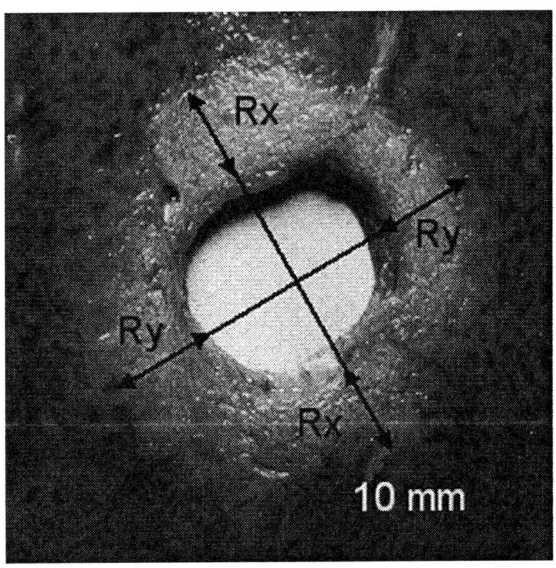

Figure 3. Liver tissue treated *ex vivo* (section perpendicular to the axis of the applicator). Twenty ultrasound exposures at 14 W/cm² with an angle of 18° between sonications. The exposure time was set at 10 seconds for all sonications.

Identical exposure conditions were chosen to induce sector-based lesions but the number of exposures depended of the treated angle. The acoustic intensity in this case was fixed at 14 W/cm². The exposure time was set at 20 seconds for 10 s. The time between the end of one exposure and the beginning of the next exposure after the rotation is 10 s. The sectors of necrosis chosen were 72, 108, 198, and 216°. A total of 12 sector-based lesions were produced. The coagulation necroses were inspected macroscopically.

The distinction between the surrounding non-treated tissues can clearly be seen in a medium like pig liver; the lesion has an off-white colour, sometimes dark at the most heated points, similar to that of cooked liver. Liver samples were frozen just after treatment to make the lesion easier to cut and examine. When the tissue temperature was close to –6°C, the tissues were sliced along a plane perpendicular to the applicator rotation axis situated at mid-height from the transducer and the depth of the lesion was measured. To account for the fact that the depth could be not constant around the whole circumference, the measurements taken along two orthogonal axes in the cutting plane were averaged (Figure 3). Two axes issuing from the centre of the applicator and touching the outer edges of the lesion allowed the value of the angular section of each sector-based lesion.

Results

The dimension of the lesions along the applicator axis corresponded to the height of the transducer i.e. 15 mm. Cuts perpendicular to the axis of the applicator (Figure 3) allowed measurement of the radius of the coagulation necrosis volumes, which represents the maximal lesion depth along the acoustic propagation axis. This cross-section demonstrates that the therapeutic depth is constant around the applicator. Figure 3 shows a typical lesion induced by a series of 20 ultrasound exposures at an acoustic intensity of 12 W/cm² for 10 s. The average value of the treatment depth, with standard deviation, is given in Figure 4 as a function of intensity and exposure duration (12, 14, and 16 W/cm² for 5, 10, 15, and 20 s). Cylindrical 13 mm-deep lesions can be induced applying 16 W/cm² for 20 s. Figure 5 shows sector based lesions. These results were obtained by rotating the transducer over angles of 72, 108, 198 and 216° respectively. The correspondence between the desired angle and the experimental result is given in table 1. The most significant deviation was measured at ±9°.

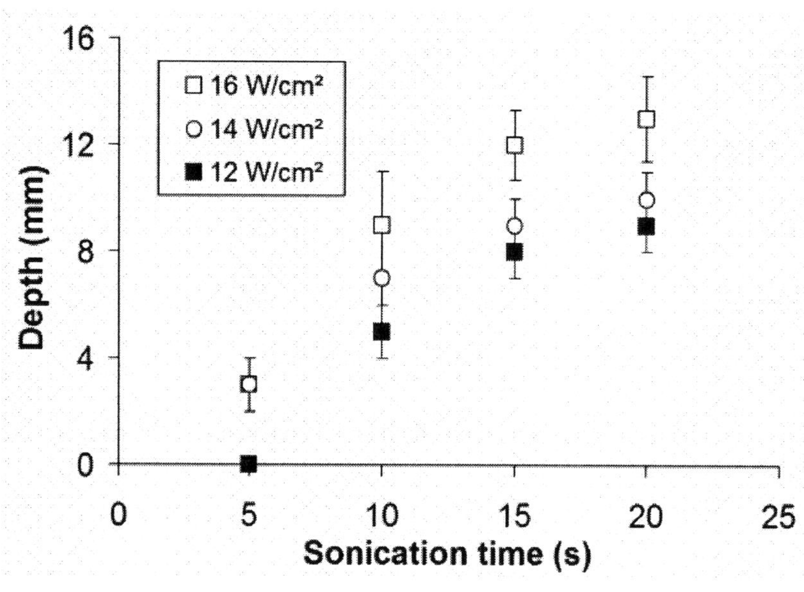

Figure 4. Depth of circular lesions as a function of the acoustic intensity.

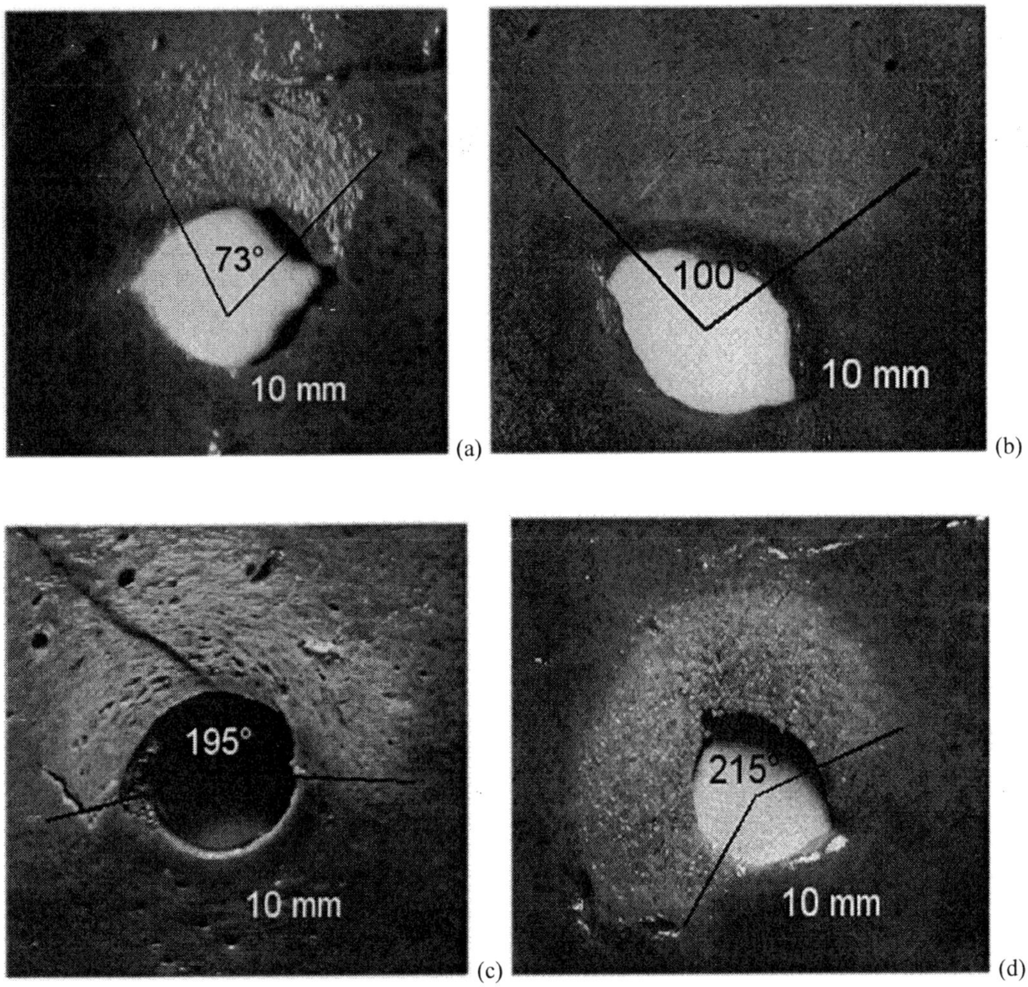

Figure 5. Shape of sector-based lesions generated by the plane transducer. The ultrasound exposures were performed at 14 W/cm² with an angle of 18° between sonications. (a) 5 exposures (72°), (b) 7 exposures (108°), (c) 12 exposures (198°), (d) 13 exposures (216°).

Table 1. Comparison between the experimental and the desired angles of sector based lesions

Preplanned treated angle (Degrees)	Experimental result (Degrees)
72°	(73±2)°
108°	(100±2)°
198°	(195±2)°
216°	(215±2)°

In Vivo Trials

Method

In vivo trials were carried out on seven healthy pig oesophagi. Male pigs with an average weight of 30 kg were used for the study. Anaesthesia was obtained with a mixture of 4 ml of ketamine (50 mg/ml) and 6 ml of 2% xylazine, administered intramuscularly behind the ear, followed by a slow intravenous infusion of nesdonal (0.4 g/100 ml). Oxygenation was supplied with an assisted ventilation system. The coupling balloon was inflated via a syringe after the applicator's introduction in the oesophagus. Ultrasound exposures started at 20 mm from the cardia. They were realised every 50–60 mm. The target zone was considered as a cylinder, whose treatment required twenty ultrasound exposures for 5, 10 or 15 s. In order to treat cylindrical or sector-based volumes, the applicator was rotated around its axis with an angle of 18° between sonications. Acoustical intensities tested were 12, 14 and 16 W/cm^2. During these firing sequences, proper functioning of the cooling circuit was verified by the reflected power reading on the wattmeter. A significant rise in reflected power is typical of the appearance of microbubbles across from the transducer, as well as all other effects of vaporisation by overheating. The pigs were sacrificed under general anaesthesia. The times between treatment and euthanasia were staggered at 24h (two pigs), 48h (one pig), 72h (one pig) and 96h (three pigs). In total four cylindrical lesions per animal were created. Each cylindrical lesion was obtained by overlapping twenty elementary lesions, thus eighty elementary lesions were created per animal.

Autopsy was performed to observe the development of the thermal damage administered. The oesophagus was then removed and sliced lengthwise. The effects of ultrasound were firstly visually inspected. The height of the lesions was measured and thermal damage (homogeneity of the lesion, perforations...) were observed macroscopically. Secondly, if there is no perforation, histological sections were carried out. Nucleus, cytoplasm and collagen of the cells were stained with Hemalun-Eosine and Saffron (HES) respectively to determine the effective treatment depth on the oesophagus, the homogeneity of the thermal damage, the presence or absence of perforation and the ability of the tissue to heal after treatment.

These animal experiments conform to the requirements of the INSERM Office of Animal Experimentation and are in accordance with the legal conditions of the National Commission on Animal Experimentation.

Results

Table 2 summarises the results obtained on the pig oesophagi and proposes a synthesis of the influence of exposure parameters. The effects of the different firing parameters on the healthy oesophagus were examined according to the homogeneity of the thermal damage, the treatment of a circular area, the presence or absence of perforation and the proper functioning of the cooling circuit. These experiments show (Figure 7 and 10) that it is possible to deliver thermal doses that do not prevent healing of healthy exposed tissues. On Figure 7, the scar does not appear perfectly cylindrical as a reject from the pig made the applicator move slightly along the oesophagus during treatment. The corresponding thermal dose consisted of a series of 20 shots, one shot every 20 s at an acoustic intensity of 14 W/cm² for an exposure duration of 10 s. An identical firing sequence with a longer exposure time (15 s) or a higher acoustical intensity (16 W/cm² for 10 s) results in oesophageal perforations (Figure 8). For all other firing sequences, oesophageal and adjacent tissues were preserved. The height of the thermal damage was measured as being the same as the height of the transducer (15 mm) as shown in figure 6. A constant reflected power reading on the wattmeter confirmed an effective cooling of the transducer and a good coupling with the targeted tissues. The highest thermal dose that can be safety delivered to the part of the oesophagus adjacent to the trachea was found to be 10 s at 12 W/cm² for circular-type lesions. For this shot sequence, the depths of *ex vivo* thermal damage were 5±1 mm and 4±1 mm in the oesophagus. This latter value is close to the thickness of the oesophagus and was determined by histological analysis. In the other parts of the oesophagus, the maximum safety thermal dose was found to be 10 s at 14 W/cm² for circular-type lesions. This dose is known to be sufficient to induce the necrosis of tumour tissue [16]. This shot sequence is associated with the recovery of healthy tissue. The safety of this HIU treatment was confirmed by histological analysis. It was proved (Figure 9) that the underlying muscle layers were intact. Figure 9 also shown that, as it was observed for all the firing sequences, mucosal layers are necrosed. Figure 10 shows the very well defined boundary between healthy and necrosed cells.

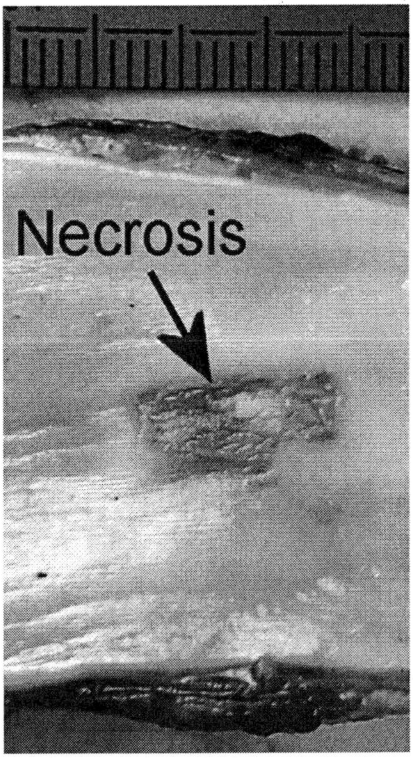

Figure 6. Elementary lesion in the oesophagus. The ultrasound exposures was performed at 14 W/cm² for 10 seconds. Height: 13 mm, width: 5 mm.

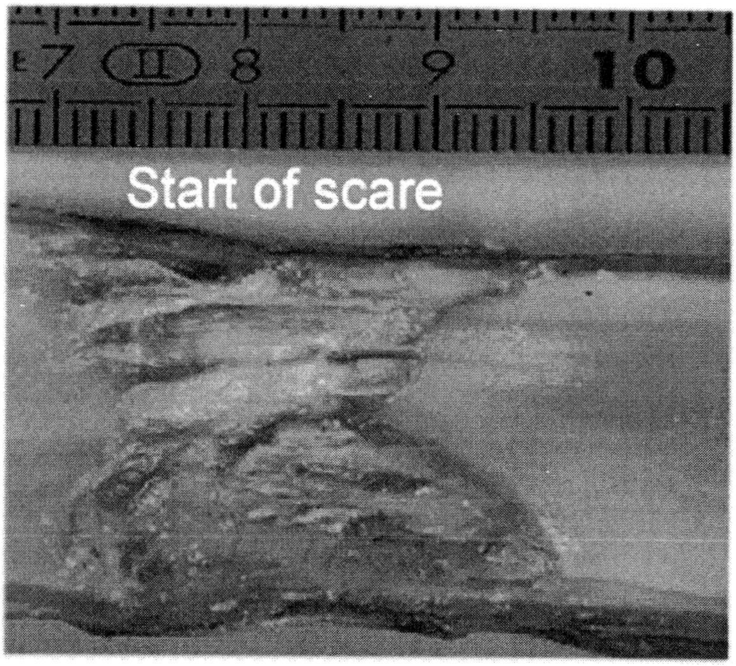

Figure 7. Pig oesophageal tissue treated *in vivo*. Twenty ultrasound exposures at 14 W/cm² with an angle of 18° between sonications. The exposure time was set at 10 seconds for all sonications. Autopsy was performed 72 hours after treatment. The treated tissues are healing

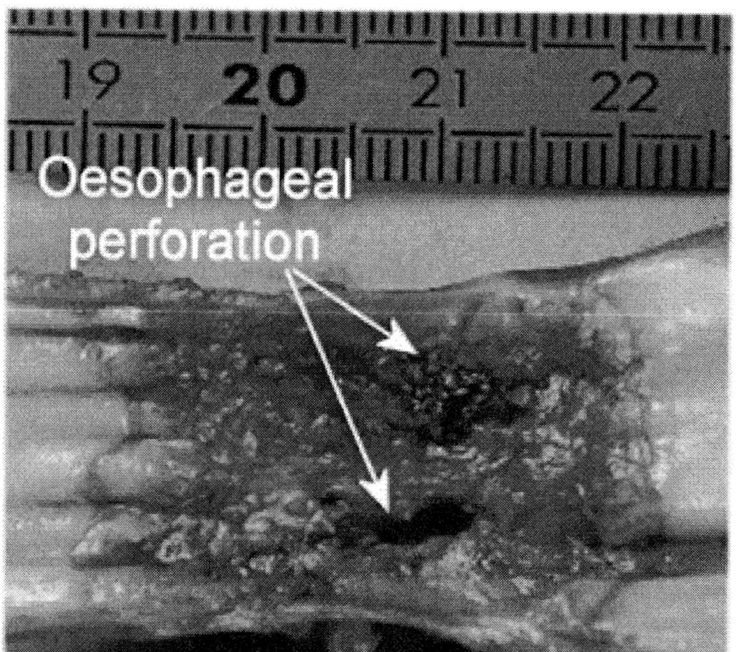

Figure 8. Oesophageal tissue treated *in vivo*. Twenty ultrasound exposures at 14 W/cm² with an angle of 18° between sonications. The exposure time was set at 15 seconds for all sonications. Autopsy was performed 72 hours after treatment. Two oesophageal fistulas appear.

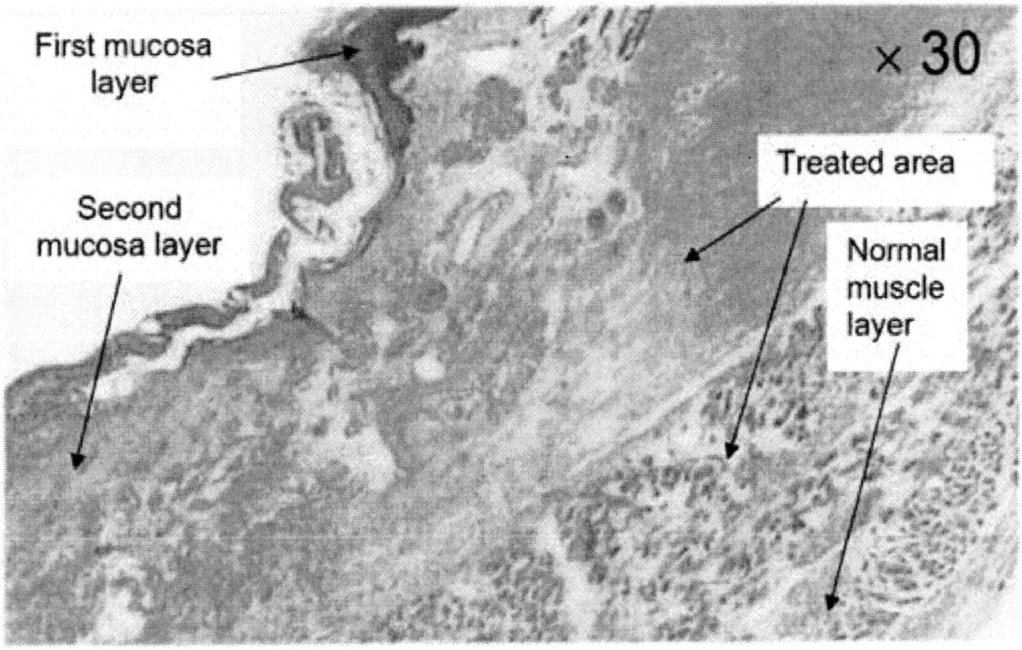

Figure 9. Histological view (×30) of an oesophagus sonicated with twenty ultrasound exposures at 14 W/cm^2 with an angle of 18° between sonications. The exposure time was set at 10 seconds for all sonications. Autopsy was performed 48 hours after treatment.

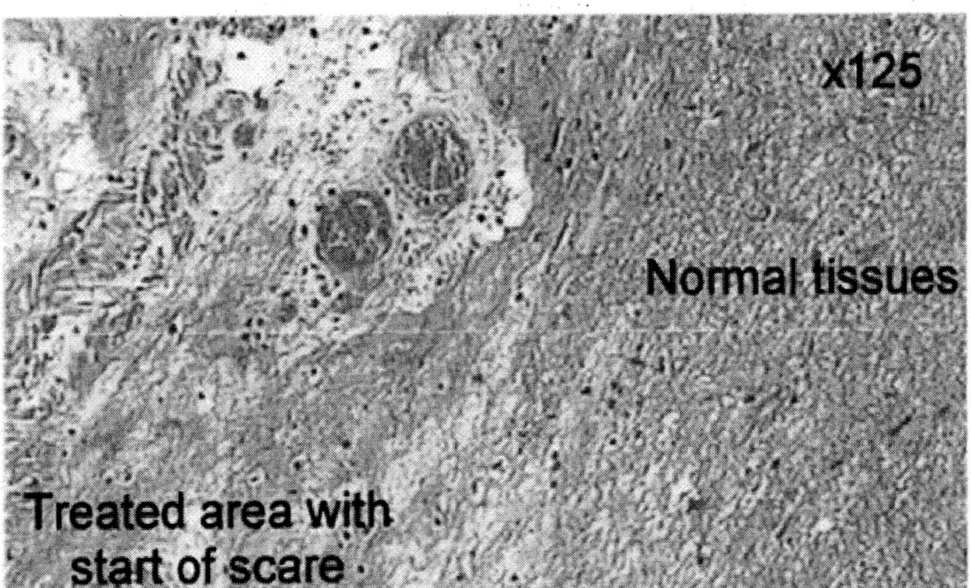

Figure 10. Histological view (×125) of an oesophagus sonicated twenty ultrasound exposures at 14 W/cm^2 with an angle of 18° between sonications. The exposure time was set at 10 seconds for all sonications. Autopsy was performed 72 hours after treatment.

Table 2. Summary of *in vivo* experiments

Time between treatment and autopsy	24 h			72 h		96 h	
Acoustical intensity (W/cm^2)	14		16	12	14	12	14
Exposure time (s)	10	15	10	10	10	10	5
Oesophageal perforation	No	Yes	Yes	No	No	No	No
Homogeneous thermal damages	Yes	Burned areas	Burned areas	Yes	Yes	Yes	Yes
Circular treatment	Yes	Yes	Yes	Yes	Yes	Yes	Yes
Efficient cooling	Yes	Yes	Yes	Yes	Yes	Yes	Yes
Tracheal perforation	Yes	Yes	Yes	No	Yes	No	Yes
Surrounding tissues	Intact	Lesion	Lesion	Intact	Intact	Intact	Intact

Clinical Pilot Study

These animal experiments led to an open pilot trial which was conducted between Sept, 2002 and Nov, 2004, after IRB approval from INSERM, Paris. Criteria for inclusion were the presence of a histologically proven cancer of the oesophagus, of any T, N and M stage, without invasion of the tracheo-bronchial tract on CT-scan or EUS, considered non operable as a consequence of local or distant extent or of associated comorbidities. Patients with tracheo-bronchial invasion, tracheo-oesophageal fistula, patients with ASA grade > 3 or Karnofsky index <50%, patients with major coagulation impairment or bearing an oesophageal stent were excluded. It was planned to include 5 patients in this preliminary trial. The study endpoints were the feasibility of the treatment (with emphasis on the tumour visualization and position control during treatment) and the presence of effective tumour destruction in accordance with the visible tumour localization. Improvement in comfort (dysphagia visual scale graded 0-5) and quality of food intake, as well as undesirable side effects were also noted as secondary endpoints. The treatment was planned as a one session therapy, with a possibility of repeated sessions in case of initial positive but partial result. Treatment was performed with the rotating transducer described above, mounted on 10mm large introducer and guided on a .035" guidewire, after bougienage when necessary (9 or 11mm bougie). Gross positioning of the transducer was achieved under fluoroscopy after lipiodol injection in the upper and lower ends of the tumour as assessed by endoscopy, and positioning of metallic clips on the patient's skin (Figure 11). Fine positioning and sectorial imaging, prior to and during treatment was achieved by use of a 12MHz miniprobe (UM2R, Olympus, Tokyo, Japan) introduced in the dedicated channel as described in animal experiments (Figure 12). One or several levels of ultrasound applications were performed depending on tumour length as estimated by EUS miniprobe and fluoroscopy. High intensity US application was circumferential or sectorial depending on EUS miniprobe and endoscopic appearance.

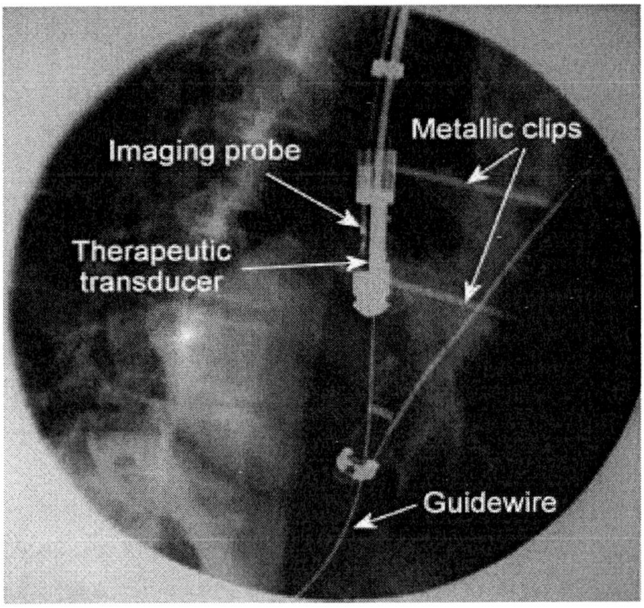

Figure 11. Gross positioning of the therapeutic ultrasound applicator achieved under fluoroscopy after positioning metallic clips on the patient's skin.

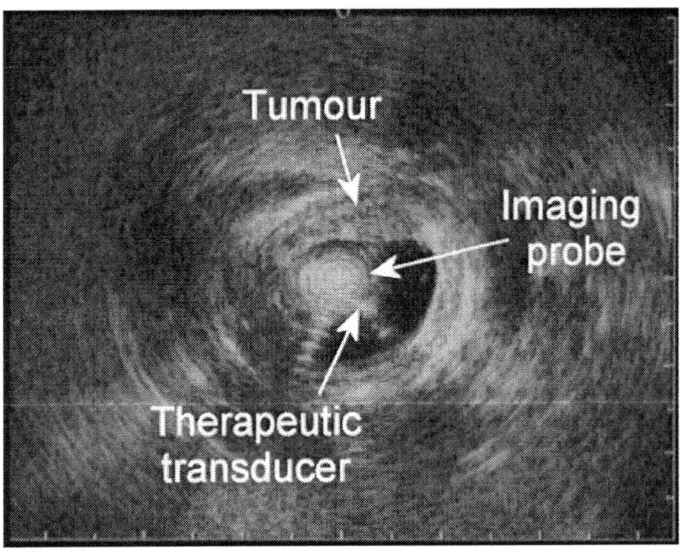

Figure 12. Sonogram of an oesophageal tumour obtained during treatment.

Follow-up was based on oesophagoscopy with biopsies 8 days, 1 month and 3 months after therapy. Patients included in radiochemotherapy protocols underwent otherwise usual follow-up.

Five patients have been included but only 4 have been treated with ultrasound. One patient included with a tumour of the oesophagus at the level of the oro-pharyngeal junction was found to have a tracheo-oesophageal fistula and was excluded. The other 4 patients are described below:

Patient 1

An 80-year old man presenting with complete dysphagia was found to have a squamous cell carcinoma (SCC) of the median 1/3 of the oesophagus extending over 45mm in length; the stricture was circumferential over the distal 30mm and hemi-circumferential over the proximal 15mm. The estimated tumour transversal diameter was 30mm on CT-scan and EUS, and the TNM stage was usT3N1M0. Upper endoscopy revealed an associated adenocarcinoma of the antrum, which was classified as T1N0. Ultrasound was applied on the 2 distal levels over 360° and on the proximal level over 180°, under EUS miniprobe control. However, tumour imaging was suboptimal by EUS miniprobe and the rotation of the applicator was not fully satisfactory, with fits and starts instead of a smooth rotation. Finally, a dislocation of the sheath was noted during treatment and required repositioning to complete the session under fluoroscopy. The patient recovered uneventfully. Dysphagia was unchanged during the first week and on upper endoscopy at day 8, tumour necrosis was observed on the whole length of the tumour, except at the proximal part where necrosis appeared to be incomplete. A second focus of incomplete necrosis was noted on 5mm in the centre of the tumour. The necrosis was pushed distally with the endoscope and the stricture was subsequently passed easily without need for further dilation. Dysphagia improved considerably during the following 2 weeks, allowing for solid food intake, but recurred at 4 weeks: control upper endoscopy at 1 month showed an obstructive nodule at the upper part of the tumour, while the more distal part remained persistently patent. Despite a partial efficacy of the US therapy, a second session was declined and radiochemotherapy was started, along with the insertion of a self-expandable covered stent. Control endoscopy at four months showed no tumour progression and the patient presented no dysphagia, with a patent stent.

Patient 2

A 55 year-old man had previously undergone radiochemotherapy (5FU-CDDP-based) for a double localization of a locally advanced SCC of the upper and median 1/3 of the oesophagus. A complete response had been established in 2002. One year later, follow-up endoscopy revealed recurrences on both initial sites, without significant symptoms. On EUS, the upper tumour was graded T1N0 and the lower tumour appeared to infiltrate the muscularis propria, without evidence of lymph node involvement. Both tumours did not exceed 4mm in thickness. It was decided to attempt ultrasound therapy. A first treatment session was undertaken with decreased power output in order to prevent transmural necrosis and possible fistula. At control endoscopy at 8 days, the tumours appeared to be only superficially ablated, and biopsies showed persistence of carcinoma cells. A second treatment session was subsequently proposed, with the same treatment parameters. The control endoscopy 10 days after this second treatment demonstrated non significant change and still positive biopsies. Three months later, the patient presented with mild dysphagia and a moderate weight loss. On the patient's request, a 3rd treatment session was decided; since the tumour thickness had increased in the meantime to reach a maximum of 7mm, usual treatment parameters were applied. Recovery was uneventful, a mild cervical pain requiring

2g paracetamol IBD during 3 days was noted, and dysphagia disappeared completely within 10 days, along with an improvement in the patient's performance status. Control endoscopy at 10 days showed a complete necrosis of the proximal lesion and an estimated 80% necrosis of the more distal one, although with still positive biopsies in non necrotic areas (Figure 13). At 1 month,. the patient maintained the same performance status (Karnofsky 90%), had a steady weight and no dysphagia. Endoscopy showed no stenosis, an infiltrating, non-stenotic aspect of the distal lesion and a nearly complete regression of the proximal lesion. Two months later (3 months after the third ultrasound session), the patient presented with recurrent dysphagia and deteriorating general condition. He was admitted in a primary referral centre, where no endoscopy was done, and a complementary radiotherapy was discussed. During this hospital stay, a sepsis developed with a septic shock and coma. Blood cultures and cephalo-rachidian fluid were positive for E.coli sp., leading to the diagnosis of meningo-encephalitis secondary to a septicaemia of unknown origin.

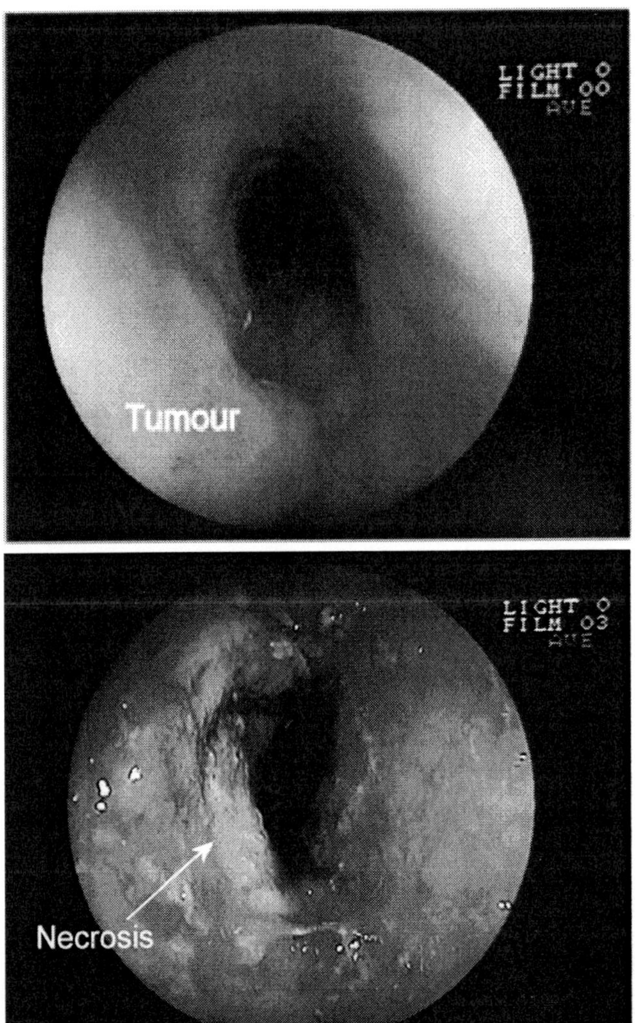

(a)

(b)

Figure 13. Control endoscopy of an oesophageal tumour. (a) before and (b) 10 days after the treatment.

Patient 3

An 83 year-old man presenting with dysphagia and a major weight loss (-15 kg) was diagnosed to have an adenocarcinoma of the distal oesophagus developed on Barrett's oesophagus. The tumour was staged usT3N1 on EUS and N2M1 on CT-scan (cervical lymph nodes and multiple liver metastases). The oesophageal stricture was measured at 30mm in length and 12mm in thickness. Additionally, the patient presented a stenosis of benign appearance of the upper 1/3 of the oesophagus. Although no specific cause was demonstrated to explain this second stenosis, a mediastinal tumour extent was suspected. In view of the poor prognosis, the strategy included an endoscopic stenting of the upper oesophageal stenosis, an oral chemotherapy based on oral 5FU and an ultrasound therapy of the lower oesophageal stricture with the aim of comfort improvement. The ultrasound treatment was performed on 2 levels of the ultrasound transducer with a satisfactory visualization of the tumour and an oesophageal stent was implanted. The patient recovered uneventfully and his dysphagia improved significantly within 10 days, allowing for resuming a solid diet. However, the performance status decreased continuously with the progressive development of jaundice due to the hepatic metastases. Subsequently, the control endoscopy was contra-indicated and the patient died 3 weeks after ultrasound therapy.

Patient 4

An 86 year-old woman presenting with severe dysphagia and a weight loss of 14kg was found to have an oesophageal SCC of the lower 1/3 of the oesophagus extending over 20mm in length and 7 to 12mm in thickness (25mm in overall diameter). The tumour was staged usT3Nx on EUS and M0 on CT-scan. Radiochemotherapy was precluded and ultrasound therapy was proposed. The treatment was achieved over 2 levels of the US transducer; the visualization of the tumour was poor, making the device positioning cumbersome. However, the treatment was completed and the patient recovered uneventfully. After 8 days, the dysphagia score regressed from 5 to 2 and the follow-up endoscopy demonstrated a large necrotic area on the whole length of the tumour, predominantly in the centre of the lesion, and a regression of the percentage of oesophageal obstruction from 90% to an estimated 50%, with a stricture length reduced from 2 to 1cm. At 1 month, the dysphagia score was identical, the patient had lost another 2 kg, and the endoscopic observations were unchanged as compared to the first follow-up endoscopy. After 3 months, the patient's general condition remained fair, but she had lost another 2 kg, her dysphagia had increased from a score of 2 to 3, allowing only a soft diet and the endoscopy showed an increase in the degree of the stricture estimated at 70% without any increase in length (10mm as previously). Biopsies were negative. Bougienage was done up to 13mm without further stenting.

Conclusion

Ex vivo experiments have shown that the 360° lesions induced are circular in shape when the applicator is rotated through an angle of 360°. Moreover, the height of the zone of necrosis is close to that of the transducer, as was reported previously [25]. The depths of lesions are reproducible with a standard deviation of less than 10 %. Lesional tissue was clearly distinguishable from untreated tissue with the border even perceptible at the macroscopic level. A relationship was established between the radius of the cylindrical lesion and the parameters of sonication (intensity and shot duration). The greatest depths which we were able to generate were 8, 10 and 13 mm for surface intensity levels of 12, 14, and 16 W/cm^2 respectively, all with a sonication time of 20 s. Beyond 20 s, the depth of the lesion does not increase very much because thereafter, the rate at which the heat is being delivered is close to the rate of thermal diffusion. For this reason, it is preferable to choose a sonication time in the shallow region of the curve, which describes the variation of lesion radius as a function of shot parameters. It is also important to bear in mind that increasing the sonication time tends to render the lesion more susceptible to the effect of blood flow.

Furthermore, these *ex vivo* experiments show that mechanical rotation of the applicator makes it possible to induce coagulation necrosis with a well-defined angle. The difference between the angle chosen and the angle experimentally obtained was no greater than ±9° meaning that sector-based tumours can be treated accurately. The round shape at the edge of the lesion and the difference between the planed treatment angle and the observed result can be explained, firstly by small angular losses during the mechanical rotation of the applicator, and secondly by thermal diffusion, i.e. the heat delivered beyond the treatment margin. The constancy of the reflected power shows that the flow rate of 0.21 litres per minute provides adequate transducer cooling, which is confirmed by the fact that the tissue near to the transducer was not vaporized. An even higher flow rate would cool the superficial layer of the tissue being treated and increase penetration.

A series of *in vivo* experiments were performed to assess the effects of different thermal doses delivered to healthy oesophageal tissue. The specific effects depended on the region of the oesophagus being treated. Tracheal tissue contains extensive cartilage and has a far higher attenuation coefficient compared with oesophageal tissue. In consequence, coagulation necrosis is induced by smaller doses in this area, a phenomenon which can lead to a life-threatening situation if the organ is perforated. A thermal dose of 12 W/cm^2 for 10 s for circular-type lesions does not cause perforation in either the oesophagus or the trachea and allows regeneration of healthy tissue yet it is known to induce necrosis in tumour tissue [16]. For this shot sequence, the depth of *ex vivo* and *in vivo* thermal damages are close together because the sonication times are so short that blood flow is only a minor factor when it comes to lesion formation. Other parts of the oesophagus can tolerate much higher thermal doses (10 s at 14 W/cm^2 for circular-type lesions) because other neighbouring tissues do not attenuate as efficiently as the trachea. All these experiments were conducted on healthy tissue. For the treatment of tumours, an acoustic intensity of 12 W/cm^2 will be used for oesophageal tissue near the trachea, and one of 14 W/cm^2 will be used for other parts of the oesophagus. Sonication times will be chosen on the basis of the dimensions of the tumour being treated. The *in vivo* short-term study has determined, by the way of histological studies, the maximal thermal dose, which never induce a perforation in healthy oesophagus. As the underlying muscle layers were intact, the probability that at long term a perforation will appear in the oesophagus is very low. Moreover, Chen [26] clearly shown in a long-term study (4 months) a regeneration of tissues if an underlying structure was intact after a high intensity focused ultrasound treatment.

The clinical pilot trial was planned to be continued if the first results had been enough satisfactory. We observed several limitations, related in particular to the positioning and the rotation of the applicator. Both of these 2 criteria were found unsatisfactory in 2 out of the 4 patients actually treated. This is explained by a reduced image quality of the US miniprobe when introduced into the applicator, which is worsened by the presence of the balloon. This is responsible for an impairment in tumour imaging and subsequent longitudinal positioning of the applicator, as well as transversal positioning in case of sectorial tumour infiltration. The uneven rotation of the probe was due to frictions over the applicator induced by the presence of tumour tissue, thus precluding a smooth and accurate transducer glide. None of these factors could have been foreseen from preclinical studies, since there was no animal model of oesophageal tumour available. However, this clinical study has shown that the treatment was feasible and induced tissue necrosis as well as a significant and relatively rapid symptomatic improvement in these severely ill patients (Table 3). We have decided to resume these clinical studies as soon as the above mentioned problems will be appropriately addressed. Our main goals are now to develop multitransducer applicators which can be operated without any rotation, once positioned longitudinally, and to prefer MRI to diagnostic US to achieve tumour imaging [25].

Table 3. Summary of clinical pilot study

Patient N°	Age	Type of tumour	Number of US sessions	Quality of probe positioning	Quality of probe rotation	Dysphagia improvement (maxi > mini score)	Tumour necrosis documented	Tumour destruction completed
1	80	SCC	1	Poor	Poor	5 > 1	Yes	Partial
2	55	SCC	3	Good	Good	3 > 0	Yes	Partial
3	83	AC	1	Good	Good	5 > 2	No	Not documented
4	86	SCC	1	Poor	Poor	5 > 2	Yes	Partial

AC: adenocarcinoma.

SCC: squamous cell carcinoma.

References

[1] De Palma G. D., di Matteo E., Romano G., Fimmano A., Rondinone G. and Catanzano C., "Plastic prosthesis versus expandable metal stents for palliation of inoperable oesophageal thoracic carcinoma: a controlled prospective study," *Gastrointest Endosc* 1996;*43*:478-482.

[2] Konigsrainer A., Riedmann B., De Vries A., Ofner D., Spechtenhauser B., Aigner F., Fritsch E. and Margreiter R., "Expandable metal stents versus laser combined with radiotherapy for palliation of unresectable oesophageal cancer: a prospective randomized trial," *Hepatogastroenterology* 2000;*47*:724-727.

[3] Gelet A., Chapelon J. Y., Poissonnier L., Bouvier R., Rouviere O., Curiel L., Janier M. and Vallancien G., "Local recurrence of prostate cancer after external beam radiotherapy: early experience of salvage therapy using high-intensity focused ultrasonography," *Urology* 2004;*63*:625-629.

[4] Illing R. O., Kennedy J. E., Wu F., ter Haar G. R., Protheroe A. S., Friend P. J., Gleeson F. V., Cranston D. W., Phillips R. R. and Middleton M. R., "The safety and feasibility of extracorporeal high-intensity focused ultrasound (HIFU) for the treatment of liver and kidney tumours in a Western population," *Br J Cancer* 2005;*93*:890-895.

[5] Jolesz F. A., Hynynen K., McDannold N. and Tempany C., "MR imaging-controlled focused ultrasound ablation: a noninvasive image-guided surgery," *Magn Reson Imaging Clin N Am* 2005;*13*:545-560.

[6] Ross A. B., Diederich C. J., Nau W. H., Rieke V., Butts R. K., Sommer G., Gill H. and Bouley D. M., "Curvilinear transurethral ultrasound applicator for selective prostate thermal therapy," *Med Phys* 2005;*32*:1555-1565.

[7] Burgess S. E., Silverman R. H., Coleman D. J., Yablonski M. E., Lizzi F. L., Driller J., Rosado A. and Dennis P. H., Jr., "Treatment of glaucoma with high-intensity focused ultrasound," *Ophthalmology* 1986;*93*:831-838.

[8] Madersbacher S. and Marberger M., "High-energy shockwaves and extracorporeal high-intensity focused ultrasound," *J Endourol* 2003;*17*:667-672.

[9] Diederich C. J., "Thermal ablation and high-temperature thermal therapy: overview of technology and clinical implementation," *Int J Hyperthermia* 2005;*21*:745-753.

[10] Thuroff S., Chaussy C., Vallancien G., Wieland W., Kiel H. J., Le Duc A., Desgrandchamps F., De La Rosette J. J. and Gelet A., "High-intensity focused ultrasound and localized prostate cancer: efficacy results from the European multicentric study," *J Endourol* 2003;*17*:673-677.

[11] Lizzi F. L., Deng C. X., Lee P., Rosado A., Silverman R. H. and Coleman D. J., "A comparison of ultrasonic beams for thermal treatment of ocular tumours," *Eur J Ultrasound* 1999;*9*:71-78.

[12] Gianfelice D., Khiat A., Amara M., Belblidia A. and Boulanger Y., "MR imaging-guided focused US ablation of breast cancer: histopathologic assessment of effectiveness-- initial experience," *Radiology* 2003;*227*:849-855.

[13] Jolesz F. A., "Future perspectives in intraoperative imaging," *Acta Neurochir Suppl* 2003;*85*:7-13.

[14] Melodelima D., Lafon C., Prat F., Theillere Y., Arefiev A. and Cathignol D., "Transoesophageal ultrasound applicator for sector-based thermal ablation: first in vivo experiments," *Ultrasound Med Biol* 2003;*29*:285-291.

[15] Nau W. H., Diederich C. J., Ross A. B., Butts K., Rieke V., Bouley D. M., Gill H., Daniel B. and Sommer G., "MRI-guided interstitial ultrasound thermal therapy of the prostate: a feasibility study in the canine model," *Med Phys* 2005;*32*:733-743.

[16] Prat F., Lafon C., Melodelima D., Theilliere Y., Fritsch J., Pelletier G., Buffet C. and Cathignol D., "Endoscopic treatment of cholangiocarcinoma and carcinoma of the duodenal papilla by intraductal high-intensity US: Results of a pilot study," *Gastrointest Endosc* 2002;*56*:909-915.

[17] Diederich C. J., Khalil I. S., Stauffer P. R., Sneed P. K. and Phillips T. L., "Direct-coupled interstitial ultrasound applicators for simultaneous thermobrachytherapy: a feasibility study," *Int J Hyperthermia* 1996;*12*:401-419.

[18] Diederich C. J., Nau W. H., Ross A. B., Tyreus P. D., Butts K., Rieke V. and Sommer G., "Catheter-based ultrasound applicators for selective thermal ablation: progress towards MRI-guided applications in prostate," *Int J Hyperthermia* 2004;*20*:739-756.

[19] Ross A. B., Diederich C. J., Nau W. H., Gill H., Bouley D. M., Daniel B., Rieke V., Butts R. K. and Sommer G., "Highly directional transurethral ultrasound applicators with rotational control for MRI-guided prostatic thermal therapy," *Phys Med Biol* 2004;*49*:189-204.

[20] Lafon C., Chavrier F., Prat F., Chapelon J. Y. and Cathignol D., "Theoretical comparison of two interstitial ultrasound applicators designed to induce cylindrical zones of tissue ablation," *Med Biol Eng Comput* 1999;*37*:298-303.

[21] Billard B. E., Hynynen K. and Roemer R. B., "Effects of physical parameters on high temperature ultrasound hyperthermia," *Ultrasound Med Biol* 1990;*16*:409-420.

[22] Melodelima D., Salomir R., Mougenot C., Prat F., Theillere Y., Moonen C. and Cathignol D., "Intraluminal ultrasound applicator compatible with magnetic resonance imaging "real-time" temperature mapping for the treatment of oesophageal tumours: an ex vivo study," *Med Phys* 2004;*31*:236-244.

[23] Hill C. R., "Calibration of ultrasonic beams for bio-medical applications," *Phys Med Biol* 1970;*15*:241-248.

[24] Prat F., Lafon C., Margonari J., Gorry F., Theillere Y., Chapelon J. Y. and Cathignol D., "A high-intensity US probe designed for intraductal tumour destruction: experimental results," *Gastrointest Endosc* 1999;*50*:388-392.

[25] Melodelima D., Salomir R., Chapelon J. Y., Theillere Y., Moonen C. and Cathignol D., "Intraluminal high intensity ultrasound treatment in the oesophagus under fast MR temperature mapping: in vivo studies," *Magn Reson Med* 2005;*54*:975-982.

[26] Chen L., ter Haar G., Robertson D., Bensted J. P. and Hill C. R., "Histological study of normal and tumour-bearing liver treated with focused ultrasound," *Ultrasound Med Biol* 1999;*25*:847-856.

Index

A

B

C

D

E

F

G

H

I

J

K

L

M

Q

R

S

T

U

V

W

X

Y

Z